Birkhäuser

Arthur Knoebel

Sheaves of Algebras over Boolean Spaces

Birkhäuser

Arthur Knoebel
8412 Cherry Hills Drive, N.E.
Albuquerque, NM 87111-1027
USA
a.knoebel@member.ams.org

ISBN 978-0-8176-4218-1 e-ISBN 978-0-8176-4642-4
DOI 10.1007/978-0-8176-4642-4
Springer New York Dordrecht Heidelberg London

Library of Congress Control Number: 2011940976

Mathematics Subject Classification (2010): 03C05, 06E15, 08-02, 08A05, 08A30, 08A40, 08B26, 18A22, 18F20

Printed on acid-free paper

Springer is part of Springer Science+Business Media
(www.birkhauser-science.com)

To the memory of
ALFRED L. FOSTER,
who set me to work
representing algebras.

Preface

My involvement in the line of research leading to this book began in 1963 when I was a graduate student working under the direction of Alfred Foster, and was first learning about representing algebras as subdirect products. In particular, for a starter I learned that Stone's representation theorem was valid not just for Boolean algebras but for any class of algebras satisfying the identities of a primal algebra. Foster perceived in these algebras a Boolean part whose representation theory could be levered into representing many other kinds of algebras.

The broad motivation was to break up a complicated algebra into simpler pieces; if the pieces could be understood, then hopefully so could the whole algebra. The obvious decomposition to try first is a direct product. The advantage of direct products is the simplicity of their construction. The overwhelming disadvantage is that most algebras are indecomposable in this sense, and even when decomposable there may be no ultimate refinement. Subdirect products overcome both of these liabilities, as first demonstrated by Garrett Birkhoff.

The main drawback to subdirect products is that, while factors may be commonplace and well understood, the transfer of an argument from the components to the whole algebra may fail because one may not know in sufficient detail how the components fit together to form the original algebra. Thus one grafts topological spaces onto subdirect products to form significantly superior sheaves. Elements of the subdirect product become continuous functions, and are easier to recognize. Boolean spaces are often used since they arise naturally in representing Boolean algebras and have been the key to many other representation theorems. However, the topological spaces of algebra are intuitively quite different from the more traditional topological spaces such as manifolds with a local Euclidean topology. They may be totally disconnected or not even Hausdorff.

The question we address then is, how far can one go in representing arbitrary algebras by sheaves over general topologies, and in particular Boolean spaces? The overall structure of a given algebra should come from a systematic synthesis of the components, that is, the stalks of the sheaf. Many questions about any algebra in such a class should be answerable by analyzing locally what is happening in the components, rather than working globally with formulas over the whole algebra.

My first exposure to sheaves over Boolean spaces was in a seminar run by Joseph Kist in the spring of 1972, in which he presented the seminal paper of Stephen Comer. Here I learned of the rich and productive world of ring spaces as expounded by Richard Pierce in his memoir.

It was in this seminar that I discovered factor elements, which generalize central idempotents in rings, and how they correspond to factor congruences. Later, factor bands, ideals, and sesquimorphisms were added. The goal was to extend the classical representation of regular commutative rings as subdirect products of fields.

Although general tools are developed, applicable to all algebras, the best efforts come from settling on those that I dub 'shells', which assert the existence of a zero and a one for a multiplicative operation and perhaps an addition that otherwise need not satisfy any of the usual identities such as commutativity and associativity. In this context, one can generalize well beyond ring theory a number of classical results on biregularity, strong regularity, and lack of nilpotents.

This monograph adapts the intuitive idea of a metric space to universal algebra, leading to the useful device of a complex. Then a sheaf is constructed directly from a complex.

The core of this book does not look at all congruences of an algebra, but at only some of them comprising a Boolean subsemilattice of congruences, and more typically, at others splitting the algebra into a product of complementary factors. Thus there are no restrictions on the whole lattice of congruences, but only on parts of it. This is one of the themes of this monograph.

Over the course of time, terms and notations tend to grow like Topsy. In synthesizing disparate fields and even extending them, inconsistencies across them pose a dilemma for an author. Should he completely streamline the terminology, thereby shutting out the casual reader who is merely browsing but already knows something of the traditional notation? Or, should he leave every term as it has originally arisen, thereby making it difficult for the serious reader to correlate similar ideas? I have taken a middle course, respecting most terms and notations already in the literature, but occasionally changing some to better reflect the overall picture. For example, congruences that *permute* elsewhere *commute* here since other internal factor objects, such as idempotent endomorphisms, always commute when creating a product. But I left unchanged directly *indecomposable* and subdirectly *irreducible*, although one ought to have a common root word for the many kinds of algebraic atoms. The definitions of the rather general algebras, *shells* and *half-shells*, have broadened over time as weaker and weaker conditions were observed to create sheaves that would accomplish most of the same ends. *Nullity* is used for an element annihilating a binary operation as a zero does in ring theory. And *unity* is the term used where others might use 'unit element' or 'identity;' it even means 'object' in categories. Likewise, the adjective *unital* adds a unity to a ring or shell.

Many exercises and problems have been included. The distinction between them is as follows. On the one hand, the exercises come from notes I wrote to myself while trying to understand the relationships between new concepts. There was no attempt to create other exercises that might fill out the book; thus the density of exercises varies from section to section. The reader may enjoy more healthy exercise by filling in wherever a proof trails off with a phrase such as 'straightforward to prove', 'trivial', or 'left to the reader'. This is especially so in the categorical sections establishing adjointness and equivalence.

On the other hand, the problems are open questions that I have not resolved because I did not take the time. Thus, such problems may range from the trivial to the significant, perhaps to promising research to pursue. I have not attempted to distinguish these possibilities.

As for prerequisites, a reader should have a nodding acquaintance with universal algebra, logic, categories, topology, and Boolean algebra. By recalling useful facts about these topics, prerequisites have been kept to a minimum. All concepts beyond these are defined. However, as the goal is new theorems, and the ideas already in the literature are lightly illustrated here, the prospective reader will be well motivated if he is familiar with some of the classical results that are being generalized.

I am thankful to the participants who asked penetrating questions in algebra seminars at New Mexico State University, Tennessee Technological University, the University of Tennessee and Vanderbilt University; some of these led to additional insights and examples. Fruitful conversations with Joseph Kist have cleared up a number of murky points. Mai Gehrke pointed out non sequiturs, and shortened several long-winded proofs. Isadore Fleischer corrected several of the early chapters. Paul Cohn offered suggestions on the history of the subject, and Ross Willard pointed out a significant extension of the concept of a shell. Diego Vaggione quickly dispatched several of the original open problems. All of these, including three anonymous reviewers, deserve warm handshakes for their many comments and thought-provoking suggestions. As for remaining faux pas that I should have caught, may the sympathetic reader forgive me for any difficulties they might cause.

Albuquerque, New Mexico Arthur Knoebel

Contents

I

Introduction

This chapter has two sections. The first is a history of the ideas and previous theorems upon which this monograph is based. The second is a survey of the principal results presented in this book.

1. History

To set the stage, we take a short historical jaunt. This will not be a literal, detailed history, but a genetic reconstruction of key events that have come to play a role in this book. There are three areas, as befits its title: sheaves, algebras, and Boolean algebras. We begin with the last.

The attempt to decompose an involved problem into workable parts is an old one – it is called the reductionist philosophy. A good starting point for examining attempts at symbolic decomposition is the work of Gottfried von Leibniz [**Leib66**] [**Mido65**]. While his efforts did not lead directly to the analysis of algebras, the motivations and flawed solutions shed light on our work in this book. Leibniz's dream of a universal calculus of logic for deriving facts mechanically by combining together basic concepts is, in a sense, a precursor of Boolean algebra.

This view that algebra could carry the burden of logical manipulation is already seen in George Peacock's definition of algebra as "the science of general reasoning by symbolical language". [**Peac30**, p. 1] A part of this dream was realized independently by Boole [**Bool47**] and Augustus

A. Knoebel, *Sheaves of Algebras over Boolean Spaces*,
DOI 10.1007/978-0-8176-4642-4_I,

De Morgan [**DeMo47**] [**MacH85**, pp. 68–71]. Boole's book developed an algebra of logic, which bears his name, although 'Boolean algebras' today are not what he described [**Burr00**].

We have mentioned Boolean algebras at the outset since they will subsequently provide a calculus for decomposing algebras by sheaves. At a higher and more powerful level of logic, the successful application of the first-order predicate calculus to mathematics came later; but unfortunately this does not solve problems in the generality envisioned by Leibniz. Kurt Gödel and Jacques Herbrand showed how limited automatic problem solving could be [**Mend64**].

Having discussed an algebra of logic, we now move on to the discoveries in linear algebra, such as quaternions, vectors and exterior algebras, which paved the way to modern algebra. Hermann Grassmann [**Gras44**], William Hamilton [**Hami44**], and later Benjamin Peirce [**Peir70**], J. Willard Gibbs, [**Gibb81**] and Oliver Heaviside [**Heav93**] invented and studied many different kinds of linear algebras, thereby opening a path to the study of noncommutative and nonassociative systems. Peirce introduced what is now called the right Peirce decomposition of a linear algebra: $A = iA + (1-i)A$, for an idempotent i, which need not be central. Also important as another example of a noncommutative system is Arthur Cayley's [**Cayl54**] attempted axiomatization of abstract groups, which arose from the study of the permutation of roots of a polynomial equation.

In another direction, Richard Dedekind [**Dede97**] first recognized the notion of a lattice in the context of number theory. Lattice theory is significant for our history in providing us with laws similar to but not identical with those of arithmetic, and in generalizing Boolean algebras to nonlogical examples. The history of algebra in the nineteenth century is rich and varied; there is much we could mention that would lead into our research, but, to keep this part of the book short, we refer the interested reader to the fine histories of Luboš Nový [**Nový73**] and B. L. van der Waerden [**vdWa85**].

The twentieth century saw the flowering of six fields that have influenced our work and provided examples: rings, lattices, universal algebra, algebraic geometry, sheaf theory, and functional analysis. The oldest of these fields is commutative ring theory.[1] We mention two papers as samples of the influence of David Hilbert and Emmy Noether on the development of modern algebra with its distinctive perspective and abstract axiomatics. Hilbert's work [**Hilb96**] on invariants cut through the Gordian knot of case-by-case construction of bases for polynomial invariants by indirect, qualitative methods – non-constructive, if you like.

[1] Rings are assumed, in this section, to have a unity, since historically they always had one; otherwise, throughout the rest of this monograph, they will not, unless designated 'unital'.

A most significant event in the history of modern algebra occurred with the publication of Noether's [1921] *Idealtheorie in Ringbereichen.* Noether's mathematical philosophy [**Noet21**] was to replace arguments that manipulated elements by structural proofs using ideals, thereby creating a powerful theory predicated only on the ascending chain condition on ideals. Here, for the first time, an ideal in an abstract ring is decomposed as a product of primary ideals. This notion was systematized by Wolfgang Krull into a principle: whether a ring $\boldsymbol{R}$ is indecomposable in a certain sense is equivalent to determining whether the intersection of a certain class of ideals in $\boldsymbol{R}$ is the null ideal [**Krull35**]. This idea was exploited by Garrett Birkhoff as the construction of subdirect products in the context of universal algebra [**Birk44**].

Another perspective on looking for representations is to specify the kind of rings we want to draw components from and the candidates in the way of ideals and congruences that are initially proposed to obtain these building blocks. A classic example is the class of semisimple rings. The factors must be quotients by maximal ideals whose intersection is the trivial ideal. But in general the intersection of all maximal ideals is not the zero ideal – witness local rings – so the representation is not faithful. Historically, to overcome this, one appropriately restricts the class of rings, for example, to those that are Artinian (that is, they satisfy a descending chain condition on ideals) and contain no nilpotent ideals other than the null ideal. The Wedderburn-Artin theorem then concludes that such a ring is a direct sum of a finite number of ideals each of which is isomorphic to the ring of all linear transformations of a finite-dimensional vector space over a division ring [**Jaco80**, vol. 2, p. 203].

In 1929, Krull was one of the first to attain theorems without chain conditions; these had the advantage of giving representations with an infinite number of factors [**Krull29**]. Gottfried Köthe defines the notion of a transcendent reducible ring and proves that each transcendent reducible commutative ring is a direct product of fields [**Köthe30**, p. 548]. Here, we see the first theorem in which the quotients have no divisors of zero, which will be a recurring theme later in this book. John von Neumann defined regularity of rings and discovered the isomorphism of the lattice of factors with the lattice of central idempotents [**vonN36**].

In 1936, Marshall Stone found his far-reaching representation theorem for Boolean algebras and rings: every such algebra is a subdirect power of a two-element algebra [**Stone36**]. Thus, the kernel into which all Boolean algebras decompose is the two-element Boolean algebra (or ring) of traditional truth values. Stone's paper is the spur and inspiration for much of the work that leads to the work explored in this monograph. Compare the length of Stone's original paper to Birkhoff's much shorter proof of the same result [**Birk44**]; this is a considerable distillation to take place within a decade. In Stone's paper of the next year, he explored the duality between

Boolean algebras and Boolean spaces [**Stone37**]. The power of this theory to tackle problems about Boolean algebras by going to their corresponding Boolean spaces was subsequently illustrated by Paul Halmos [**Halm63**, chap. 28], when he used it to demonstrate William Hanf's result that Boolean algebras need not have unique roots, in the sense that $\boldsymbol{A}^2 \cong \boldsymbol{B}^2$ need not imply $\boldsymbol{A} \cong \boldsymbol{B}$ [**Hanf57**].

Representing Boolean algebras both as rings of sets and topological spaces stimulated a number of mathematicians. Over a two-year period, 1937–1938, several papers appeared, apparently independently of each other, giving subdirect representations of special classes of commutative regular rings in terms of fields, without assuming either the ascending or descending chain conditions on ideals. The first result of this kind was the theorem of Neal McCoy and Deane Montgomery, who proved that any p-ring (commutative, $px = 0$ and $x^p = x$) is a subdirect product of prime fields $\mathbb{Z}_p$ [**McCMo37**]. A theorem of this type without regard to characteristic is due to McCoy for commutative rings: any commutative, von Neumann regular ring is a subdirect product of fields [**McCoy38**]. More generally, McCoy, using a lemma of [**Krull29**], showed that any commutative ring without nilpotent elements is a subdirect product of integral domains. Since each integral domain is embeddable in a field, it follows that any such ring is a subring of a direct product of fields. Birkhoff systematized the presentation of such results by proving a lemma suggested by McCoy: a subdirectly irreducible commutative ring without nilpotents is a field [**Birk44**]. Shortly thereafter, Alexandra Forsythe and McCoy extended this result to the noncommutative case: any regular ring without nilpotents is a subdirect product of division rings [**ForMc46**].

Since we will be talking about variants of regular rings shortly, we should mention the relationship between regularity and nilpotents in the case of a commutative ring $\boldsymbol{R}$. On the one hand, it is easy to show that if $\boldsymbol{R}$ is regular, then it has no nilpotent elements. On the other hand, by the result of McCoy above, if $\boldsymbol{R}$ has no nilpotent elements, then it is a subring of a product of fields, which are always regular. Hence, if $\boldsymbol{R}$ has no nilpotent elements, then $\boldsymbol{R}$ is embeddable in a regular ring, still having no nilpotents.

Richard Arens and Irving Kaplansky give examples showing that, in the noncommutative case, biregular rings and regular rings are independent notions [**AreKa48**]. They prove in their theorem 6.2 that, if A is an algebra over the field $\boldsymbol{GF}(p)$ in which every element a of A satisfies the equation $a^{p^n} = a$ for a fixed n, then there is a locally compact, zero-dimensional space $\boldsymbol{X}$ with a homeomorphism σ for which $\sigma^n = 1$, such that $\boldsymbol{A}$ is isomorphic to the algebra of all continuous functions f from $\boldsymbol{X}$ to $\boldsymbol{GF}(p^n)$ that vanish outside a compact set and respect σ: $f(\sigma x) = [f(x)]^p$. The authors go on to concoct a counterexample built around $\boldsymbol{GF}(4)$ showing that the equation, $a^4 = a$, is necessary.

Reinhold Baer studied the condition that each annihilator is generated by a projection, which will be significant in some of our applications [**Baer52**]. Somewhat earlier, the application of topological methods to non-Boolean rings began with Israel Gel'fand and George Šilov generalizing the Stone topology of prime ideals in Boolean rings to commutative normed rings [**Gel'Ši41**]. Shortly thereafter, Nathan Jacobson generalizes the Stone topology by adapting it to the set of primitive ideals in an arbitrary ring, not necessarily with a unity [**Jaco45**]. It is no longer Hausdorff; but if there is a unity, it will still be compact.

The writing of a history of representing general algebras is complicated by the history not being linear. Instead, the history may be thought of as more like a braided stream with many strands and rivulets, some of them running in parallel, some bifurcating, and others merging back together. Certainly, to judge from the references absent in published papers, there must have been considerable independent effort. How to identify these strands and how many to pay attention to are matters of opinion. In any case, a strictly chronological account would be confusing and misleading. So we must often follow one strand for a while, then back up in time to pursue another. We now go back to look at the origins of universal algebra.

Alfred North Whitehead, in writing his book on universal algebra, also had a lofty but less sweeping goal than Leibniz: he wished to create a theory of algebra capable of unifying and comparing the many linear algebras that had been proposed in the nineteenth century [**Whit98, Fear82**].

We follow the current view that the concept of universal algebra as it is recognized today, despite Whitehead's treatise by the same name, began with the two seminal papers of Garrett Birkhoff, who showed that there were significant theorems simultaneously covering groups, rings, fields and vector spaces as well as lattices and Boolean algebras. Birkhoff formulated the concept of a general algebraic system as we know it today [**Birk35**]. In his next paper on the subject [**Birk44**], Birkhoff presented the theorem that every algebra is a subdirect product of subdirectly irreducible algebras. This theorem is fundamental to our purposes and illustrates how the finitary nature of algebra makes for a good theory with many applications.

Another stream flowed into universal algebra from logic via Emil Post's generalization of classical two-valued logic [**Post21**]. Post algebras, as defined by Paul Rosenbloom, are to Boolean algebras as n-valued logic is to two-valued logic [**Rosen42**]. Following close upon [**Birk44**], L. I. Wade established that any Post algebra is a subdirect power of a primal Post algebra [**Wade45**], Rosenbloom having proven this first for finite Post algebras.

The work of Alfred Foster, influenced by that of Wade and McCoy, was a watershed in the way we view representations of algebras [**Fost53**].

different methods, shows that this last result is still true even when $\boldsymbol{R}$ is noncommutative [**Ledb77**]. Hofmann surveys much that is known about sheaves of rings [**Hofm72**]. To generalize such results beyond rings, new techniques are needed, which are discussed next.

Comer realized that the construction of the Pierce sheaf for rings could be extended to a rather broad class of algebras [**Comer71**]. These are algebras whose factor congruences form a Boolean algebra. Besides rings, there are lattices, semilattices, and the shells of this book. This came out of Comer's investigation into the question of the decidability of the theory of cylindric algebras [**Comer72**]. Stanley Burris and Ralph McKenzie make some unique comments on this portion of our history [**BurMc81**, pp. 15–20, 67–70].

Brian Davey pushed the work of Comer further by realizing that all we need in order to obtain a sheaf over a Boolean space is a Boolean sublattice of congruences [**Davey73**]. In this very general set-up, there is the Gel'fand morphism, named for his work [**Gel'f41**], that takes the original algebra to the algebra of all global sections of the sheaf; it is injective, but not necessarily surjective. Davey notes further, however, that if we start with a Boolean sublattice of *factor* congruences, then the Gel'fand morphism is indeed surjective. Since most of the remaining contributors to the unfolding of sheaves in general algebra in the 1970s will be discussed more fully in later chapters, we only list them here: Klaus Keimel [**Keim70**], Maddana Swamy [**Swam74**], Albrecht Wolf [**Wolf74**], William Cornish [**Corn77**], and Peter Krauss and David Clark [**KraCl79**]. For a survey see [**Keim74**].

Another area with many examples of interest is this problem: for which equational classes can one express each algebra as a Boolean product of a finite number of finite algebras, depending only on the original class? Unfortunately, as shown by the work of Burris and McKenzie, this traditional situation in both classical and universal algebra – a Hausdorff sheaf over a Boolean space with a finite number of finite stalks – is limited by the generators having to be simple Abelian or quasiprimal [**BurMc81**].

Sheaves have proved their worth in model theory by establishing theorems in decidability, elementary equivalence and embedding, preservation and transfer properties, and model completeness. See Sect. XII.3 for definitions of these concepts and for some sample theorems.

Categories are another general concept useful to us; they help to systematize equivalences among diverse classes of mathematical objects. Samuel Eilenberg and Saunders MacLane [**EilMa45**] formally presented this concept, although many examples were already known informally before then. What will be historically significant for us are three such categorical equivalences: Stone's duality between Boolean algebras and what are now called Boolean spaces [**Stone37**], Pierce's equivalence of the category of rings with conformal homomorphisms and the category of their sheaves [**Pier67**], and Hu's equivalence of primal varieties, mentioned earlier [**Hu69**]. See [**MacL65**] for early developments and applications of this rapidly growing subject.

Modern analysis, and indirectly general topology, has been a source of inspiration for the ideas in this monograph. Early on we have John von Neumann's paper on rings of operators [**vonN36**]. A later influence on our efforts came from functional analysis, where Melvin Henrikson and Meyer Jerison [**HenrJe65**] and Kist [**Kist69**] systematically exploited the spectrum of minimal prime ideals. Kist had moved from functional analysis to commutative ring theory.

It is easy to appreciate the significance of functional analysis for the type of theorems we are heading toward. The ring $\boldsymbol{C}(\boldsymbol{X})$ of all continuous functions from a topological space $\boldsymbol{X}$ to the real numbers $\mathbb{R}$ has the appearance of a sheaf space $\boldsymbol{X} \times \mathbb{R}$, with the obvious projection $\pi: \boldsymbol{X} \times \mathbb{R} \to \boldsymbol{X}$, where $\boldsymbol{X} \times \mathbb{R}$ is given the product topology. Thus, it is already decomposed into a subdirect product whose stalks $\mathbb{R}$ have no divisors of zero. But its similarity to a sheaf is only that, for the technical condition that π be a local homeomorphism fails; related to this failure is the fact that each stalk $\mathbb{R}$ is not discrete. (Sometimes we speak of $\boldsymbol{X} \times \mathbb{R}$ as being merely a 'bundle'.) Thus, we might seek a different factoring of $\boldsymbol{C}(\boldsymbol{X})$. Marshall Stone has some intriguing comments in [**Stone70**, p. 240] about how this interest in $\boldsymbol{C}(\boldsymbol{X})$ shifted to concerns with representing algebras by sets of functions subject to certain constraints. See also the selection of essays [**Aull85**], edited by Charles Aull, for some current views on $\boldsymbol{C}(\boldsymbol{X})$ as an algebraic object. Hofmann has additional comments in [**Hofm72**, pp. 295–296] on the influence of functional analysis on the evolution of sheaf theory towards ring theory.

As a beginning to the book, this tour through the evolution of ideas leading to sheaves is over. Now we are ready to delve into what this book covers – both known theorems and new ones.

2. Survey of Results

With the historical background of Sect. 1 in mind, we now briefly describe the principal results of this monograph, chapter by chapter, omitting minor caveats. The flow of this book as a whole is first to develop tools for constructing sheaves that are helpful in understanding the structure of general algebras, specializing as needed, with applications in the middle chapters, and finally to close with a backward glance at how some of the earlier theorems might be extended.

Chapter II lays out the traditional background needed from general algebra. In Sect. 1, there is one novelty: 'sesquimorphisms' as a substitute for congruences. The three isomorphism theorems are presented both conventionally and in terms of sesquimorphisms.

Section 2 introduces direct and categorical products, and then studies the five kinds of factor objects that may identify products internally: bands, congruences, sesquimorphisms, ideals, and elements; the last four coming in complementary pairs (Theorems II.2.5, II.2.12, II.2.19).

Chapter III outlines the concepts and theorems needed from several disciplines: equational logic, categories, topology, and Boolean algebras, including Stone's representation of these, the grandfather of many of the theorems in this book and an essential tool for proving them.

Chapter IV sets the stage for the book proper by introducing the notion of a complex and showing that it always gives a sheaf of algebras. Crucial to decomposing an algebra as a subdirect product of quotient algebras is a measure of how close or far apart the elements of the algebra are. Complexes originated in the theory of rings and modules, and they are the algebraic analog of metric spaces. A metric now becomes a binary operation from the carrier of an algebra, not necessarily going to the real numbers, but taking as values open sets in some topological space. This binary operation satisfies axioms similar to both those for a metric space and those for a congruence preserving the operations of an algebra. Complexes are an intermediate step on the way to sheaves.

Sheaf constructions next illustrate how a well-developed topological tool can shed light on a principally algebraic device. Out of each complex, one constructs a sheaf whose algebra of all global sections contains a subalgebra isomorphic to the original (Theorem IV.2.1); arguments common to this construction can now be made once and for all in the context of complexes. Sheaves have proven their value in many situations, and here they will also do so.

We also look at systems of congruences from which one obtains a subdirect product. This is proven equivalent to the notion of a complex whenever the underlying topology is T_0 and the equalizers of global sections form a subbasis (Theorem IV.2.5).

Another concept is the 'Hausdorff sheaf', where the sheaf space is T_2. A sheaf being Hausdorff is equivalent to equalizers being clopen and the base space being Hausdorff (Proposition IV.2.9). When the base space is also a Boolean space, we have the well-studied notion of a 'Boolean product'.

The constructions of this chapter are set into an adjoint situation between the categories $\mathfrak{Complex}$ and $\mathfrak{Sheaf}$ for a given algebraic type (Theorems IV.3.15 and IV.3.18). The functors and natural transformations entering into this adjoint situation will be successively specialized in subsequent chapters.

Another way to capture the separation of two elements of an algebra is through a congruence by which they are not related. The typical situation introduced in Chap. V, which will occur repeatedly throughout this monograph, is where a subset of congruences separating all elements is singled out for special attention. We want to pick a set of congruences that is appropriate to the algebra at hand and to the aspects of it we wish to study. It is noteworthy that usable sets of congruences need not be sublattices of the lattice of all congruences. Such sets need only be closed to intersection and a complementation respecting the partial ordering of inclusion on congruences. This fragment, to be called a 'Boolean subsemilattice', will be a complemented distributive lattice, in which the join operation may be greater than that in the complete lattice of all congruences.

To set things up for later ideas covered in the book, we prove Theorem V.2.1, which states that any Boolean subsemilattice of an algebra determines a complex over a Boolean space, which in turn determines a sheaf of algebras over the same space. The original algebra will be a subalgebra of the algebra of all global sections of a sheaf of quotient algebras. In general, the larger the subsemilattice, the more numerous are the stalks of the sheaf; and the larger the congruences themselves, the smaller the quotients. But the quotients do not come directly from the congruences of the Boolean subsemilattice. Instead, the essential construction behind this theorem is to look at the Boolean space of prime ideals in this selected Boolean algebra of congruences. Each prime ideal has a supremum that is again a congruence in the given algebra, although not usually in the Boolean subsemilattice. These suprema are the points of the Boolean space over which floats the sheaf space of stalks, which are the quotient algebras by the suprema. The continuous cuts through the sheaf space are the global sections. The mapping of elements of the original algebra into them is called the Gel'fand morphism.

But the converse is also true: we prove in Theorem V.2.9 that every representation of an algebra by a sheaf of algebras over a Boolean space must arise by the previous construction from some Boolean subsemilattice of congruences. As one is free to choose the Boolean subsemilattice, so one is also free to choose the nature of the quotient algebras, and thus to tailor the extent of their indecomposability. For example, discovering the right congruences will factor out divisors of zero in shells.

The patchwork, partition, and interpolation properties associated with sheaves over Boolean spaces make the global sections of such sheaves especially malleable. Even easier to work with are the more specialized Boolean products, which have been used to good advantage in universal algebra, and which are briefly looked at for the sake of comparison. Also included in Sect. V.3 are Boolean powers, Boolean extensions, and Hausdorff sheaves.

Introduced at the end of Chap. V is the category $\mathfrak{BooleBraceRed}$ of reduced Boolean braces – they consist of an algebra and a selected Boolean subsemilattice of congruences. This category forms an adjunction with the category $\mathfrak{CompBooleRed}$ of reduced complexes over Boolean spaces

(Theorem V.4.14). In turn, this last category is a full subcategory of the category of all complexes over arbitrary topological spaces, and thus, this adjunction, when composed with the adjunction of the last chapter, forms an adjunction of $\mathfrak{BooleBraceRed}$ with the category $\mathfrak{SheafBooleRed}$ of reduced sheaves over Boolean spaces, which is a full subcategory of $\mathfrak{Sheaf}$ (Theorem V.4.17). Reduction limits the number of trivial stalks in a sheaf and its related structures.

∞∞∞∞

As the set of all congruences of an algebra is a lattice, it is natural to consider sublattices. Of special interest in Chap. VI are those congruences θ having a complementary partner θ' in the sense of forming a factorization: $\boldsymbol{A} \cong (\boldsymbol{A}/\theta) \times (\boldsymbol{A}/\theta')$. Davey [**Davey73**] considered a Boolean sublattice of commuting (= permuting) factor congruences – this sublattice together with its algebra we call a 'factorial brace'. As Boolean lattices are equivalent to Boolean algebras, one has a Boolean subsemilattice, the previous situation. Thus, one obtains a sheaf over a Boolean space. But now we have an isomorphism: Theorem VI.1.8 states that the algebra of global sections of this sheaf is isomorphic to the original algebra, that is, the Gel'fand morphism is also surjective, not just injective, as in Chap. V.

This set-up is important enough to warrant a section devoted to characterizing Boolean algebras of factor congruences alternatively by factor bands and sesquimorphisms.

Comer postulated in his paper [**Comer71**] that all the factor congruences form a distributive sublattice, that is, a Boolean sublattice of the lattice of all congruences, which is described as the algebra having 'Boolean factor congruences' (BFC). For many algebras occurring in practice, such a condition is easy to check. Section 3 characterizes their sheaves in Theorem VI.3.15 as those that are 'reduced' and 'factor-transparent'. This is called the 'canonical' sheaf representation of an algebra with BFC. Historically these results of Comer came before those of Chaps. IV and V, but it is now easier to present them as a special case of those earlier chapters. But not all algebras have BFC. Theorems VI.3.2 and VI.3.9 give many conditions equivalent to an algebra having Boolean factor congruences.

As was done in the previous two chapters, the final section of this chapter recasts its achievements in categorical terms. If a Boolean brace is taken into $\mathfrak{Sheaf}$ and then back again to the algebra of global sections, then the new Boolean brace has additional properties. The new Boolean subsemilattice now has a distributive sublattice of commuting factor congruences, creating a factorial brace as examined in the previous paragraph. All reduced factorial braces constitute the category $\mathfrak{FactorBraceRed}$, which is isomorphic to a full subcategory of $\mathfrak{BooleBraceRed}$. Most importantly, $\mathfrak{FactorBraceRed}$ is categorically equivalent to $\mathfrak{SheafBooleRed}$, by Theorem VI.4.2; thus we extend Stone's representation theorem for Boolean algebras.

Additionally, if the set of *all* factor congruences is a Boolean sublattice of **Con** $\boldsymbol{A}$, then these algebras constitute the category $\mathfrak{AlgBFC}$ of algebras with Boolean factor congruences. By Theorem VI.4.5 it is isomorphic to the category of reduced and factor-transparent sheaves over Boolean spaces. Table VI.1 lists the many categories of structures and sheaves that occur in this book and the various adjunctions and equivalences that exist among them. Figure 1 summarizes in a Venn diagram the various levels of generality considered so far, as well as the shells to be discussed next.

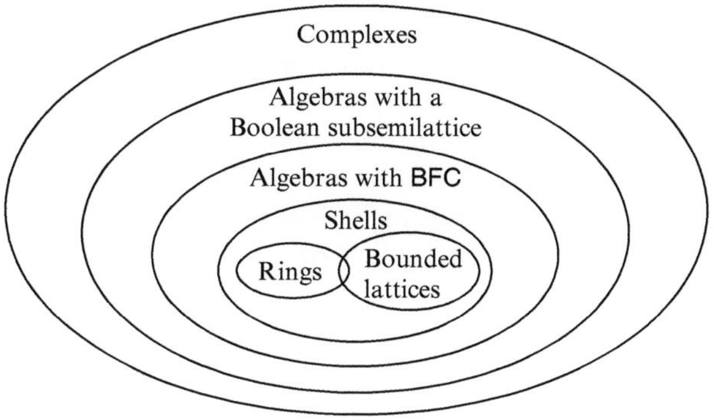

FIGURE 1. Kinds of algebras with a ready-made sheaf structure

In the heart of this monograph, Chap. VII introduces the notion of a 'unital shell', which is an algebra $\langle A; +, \times, 0, 1 \ldots, \omega, \ldots \rangle$ with two binary operations $+$ and $\times$, two constants 0 and 1, and other arbitrary operations ω as desired, in which no identities need hold other than what is expected of what we call a nullity and unity:

$$(2.1) \qquad 0 + a = a = a + 0, \qquad 0 \times a = 0, \qquad 1 \times a = a.$$

The remaining operations ω, if any, need not have any relationship to the first four. Clearly, a nullity is needed in order to talk about divisors of zero, which will appear in the Chap. VIII. Although addition is not always needed, a unity for multiplication appears to be needed to obtain our results on factor objects.

Examples abound: rings and linear algebras with a unity, Boolean algebras with operators, as well as bounded lattices and trellises, perhaps also with operators. The sparse identities of (2.1) provide all we need to mimic the factorization of rings by central idempotents. This definition is that of 'unital shell' in a strict sense; but we will also use 'shell' in a loose sense to refer to various weakenings of this definition. By omitting $+$, we prune it to 'unital half-shell'. Examples are bounded semilattices, and more generally, monoids with a nullity. Much of the theory still holds for even weaker shells, and this will now be explained in detail.

We start off this chapter with sesqui-elements and sesquishells, temporary concepts leading us to prove BFC in Theorem VII.1.10 with the weakest hypothesis possible in the context of shells, with the sheaf being reduced and factor-transparent. The slightly stronger unital half-shells also have BFC, and are studied in Sect. 2. Theorems tie factor elements to sesqui-elements. Within a unital half-shell, the set of its factor elements is now a Boolean algebra, anti-isomorphic to the Boolean algebra of factor congruences (Theorem VII.2.15).

Products were captured internally in Sect. II.2 by factor congruences, bands, and sesquimorphisms. In unital shells, we add to this list complementary pairs of factor ideals and complementary pairs of factor elements. In Sect. 3, the concept of unital shell is sufficiently stronger to support a characterization of factor elements solely in terms of factor identities, as given in Theorem VII.3.4. Better still, Theorems VII.3.7 and VII.3.14 characterize the inner direct product of ideals independently of the other factor objects, the latter theorem becoming the traditional definition of inner product in unital rings.

In Sect. 4, we explore the one-to-one correspondences between the five kinds of factor objects in two-sided unitary shells. (A unitary shell is 'two-sided' if, in addition to (2.1), the equations, $a \times 0 = 0$ and $a \times 1 = a$, also hold.) Each factor ideal is the 0-coset of a factor congruence, and the congruence is uniquely determined by its 0-coset. Each factor ideal is generated by a factor element, which serves as a relative unity, and conversely, each factor element generates a factor ideal. In a unital shell, the set of all factor ideals forms a Boolean algebra, and thus, so do the factor elements, and these Boolean algebras are isomorphic or anti-isomorphic to the Boolean algebras of the previously defined factor objects. Formulas are developed for these correspondences and the Boolean operations.

But not all algebras with Boolean factor congruences have factor elements. 'Separator algebras', generalizing shells, are introduced as a device for proving that any algebra $\boldsymbol{A}$ with BFC is embeddable in a ring or lattice, whose new factor elements capture the factorizations of $\boldsymbol{A}$ where there were none before (Theorem VII.5.5).

The category $\mathfrak{UnitShell}$ of unital shells of a given type is a full subcategory of $\mathfrak{AlgBFC}$ of the same type. Theorem VII.6.4 and its corollary establish that this new category is isomorphic to the category of reduced and factor-transparent sheaves of unital shells over Boolean spaces. The morphisms of $\mathfrak{UnitShell}$ are characterized as those homomorphisms, called 'conformal', that take factor elements into factor elements. Similar and equivalent categories also exist for the more general unital half-shells and their sheaves.

One of the high points of this monograph is the generalization to unital shells of Kist's theorem [**Kist69**] on the decomposition of commutative

Baer unital rings into a sheaf of integral domains over a Boolean space, which in turn is a generalization of the classical result that every von Neumann regular, commutative, unital ring is a subdirect product of fields. This application in Chap. VIII illustrates the power of sheaves over spaces constructed out of factor elements, and for which the previous chapters have prepared the reader. Here, the adjective 'Baer' originally defined for rings, becomes 'Baer-Stone' to include 'Stone' lattices: the annihilator of any element a is generated by a single factor element e, that is, the annihilator is a principal ideal:

$$a^{\perp} = \{b \in R \mid ab = 0\} = [e].$$

Theorem VIII.1.13 then states that every two-sided unital half-shell that is Baer-Stone has a canonical sheaf representation where the stalks are integral, that is, they have no divisors of zero. Here we apply the crucial fact that an ideal is integral[2] if it is associated with a congruence that is the supremum of a prime ideal of factor congruences.

The categorical interpretation of Chap. VII can be further specialized to this result: the category of Baer-Stone two-sided unital shells with conformal homomorphisms is categorically equivalent to the category of sheaves of integral shells ('integral' means no divisors of zero). This theorem can also be phrased outside of the language of sheaves. Each Baer-Stone shell is isomorphic to a subdirect product of integral shells.

The biregular rings of Arens and Kaplansky [**AreKa48**], and Dauns and Hofmann [**DauHo66**] present another situation that can be generalized; this was extended to near-rings by [**Szeto77**]. But we extend it further to unital half-shells; they are called biregular if every principal ideal is generated by a factor element. Theorem VIII.2.3 then reads essentially as it does for the classical case: a biregular unital half-shell is isomorphic to the half-shell coming from a reduced and factor-transparent sheaf over a Boolean space with simple stalks (providing certain technical conditions hold).

The sheaf representations of the last two chapters were all surjective, that is, each continuous section represents some element in the original algebra. Chapter IX relaxes this; an algebra now may be represented merely as some subalgebra of the algebra of all global sections of some sheaf. This means that the hypotheses of the last chapter, such as being Baer-Stone, may also be relaxed.

---

[2]This means that if a product of two elements is in the ideal, then one of the factors must be in there also. In commutative rings, this is synonymous with being a prime ideal.

In the first section of Chap. IX, to achieve any results on integrality, one must first study shells without an addition, a unity, or additional operations; I dub these new algebras 'strict half-shells'. Also, one must expect the multiplication to satisfy certain nilpotent conditionals, too involved to state here. We then prove in Theorem IX.1.3 that any such half-shell is isomorphic to a half-shell of some of the global sections of a sheaf over a Boolean space of half-shells without divisors of zero.

In the second section, consequences of this theorem are explored. We abstract its conclusion by calling a half-shell semi-integral if it is a subdirect product of half-shells without divisors of zero, and give several equivalent formulations of semi-integrality, for example, Theorem IX.2.3.

In the third section, it is further assumed that the strict half-shell has a unity and consequently every factor ideal is principal, which leads to an especially transparent form for the factor elements when the half-shell is semi-integral. Then this section returns to shells where analogous results hold. But now it is necessary to make some additional assumptions: $+$ satisfies the loop laws; $+$ is distributive over $\times$; and $\times$ satisfies the nilpotence conditions mentioned above. Theorem IX.3.8 tells us that such a strict shell is isomorphic to a subshell of the Baer-Stone shell coming from a sheaf over an extremally disconnected base space whose stalks are integral. In some detail, we trace the relationship of this result to some older theorems in ring theory.

Chapter X starts off with a new proof of the classical result that any algebra in the variety generated by a primal algebra is isomorphic to the algebra of all global sections of a sheaf over a Boolean space all of whose stalks are the primal algebra. Recall that a primal algebra is a finite nontrivial algebra whose operations lead by composition to all finitary functions on the carrier. It is natural to seek other algebras close to primality whose generated varieties will have nice sheaf representations. The preprimal algebras do not disappoint us.

An algebra is preprimal if it is one step away from being primal, that is, if any one function not composed from its operations is added to them, then the new algebra is primal. The preprimal algebras fall naturally into seven classes, identified by relations that all their operations preserve. Most of the varieties generated by them have BFC (see Table X.1). For three of the classes we find the stalks in the sheaves of their algebras: those coming from a preprimal preserving certain permutations (Theorem X.2.1); those preserving certain Abelian groups (Theorem X.3.2); and those preserving a proper subalgebra (Theorem X.4.1).

In Chap. XI, we attempt to get away from the language of shells and half-shells, and return to arbitrary algebras, in so far as possible; there are two independent sections.

The first section iterates our sheaf construction for a shell until all quotients have become directly indecomposable. In a commutative unital ring, the stalks are always directly indecomposable. As surprising as it may seem, when noncommutative, the stalks need not be directly indecomposable. Walter Burgess and William Stephenson [**BurgSt78**] took this opportunity for unital rings to iterate the construction of Pierce sheaves on the stalks themselves until it could be pushed no further. This forces the ultimate 'factors' to be directly indecomposable; but we may no longer have a sheaf, only a subdirect product. Theorem XI.1.2 adapts this iterative construction to general algebras in varieties with Boolean factor congruences.

The second section looks at the lattice of congruences as a shell and considers how its decomposition might lead to a decomposition of the algebra itself. The crucial observation upon which Theorem XI.2.3 is based is that the factor elements of the congruence lattice form a Boolean lattice; hence, we can obtain its associated Boolean space. This, in turn, induces a sheaf, and so the Gel'fand morphism maps elements of the given algebra to some of the global sections.

As this book was being written, it became clear that there were many related topics to be pursued, and many tempting trails on which to venture. The techniques introduced in this monograph might well be extended in any number of directions. In order to draw this book to a close, rather than try to develop these ideas in detail, Chap. XII outlines additional applications, without proofs; these point to five regions ripe for research, beyond what is already known. The first application wants to extend the sheaf representations in classical ring and lattice theory to shells and beyond. The second considers algebras derived from logic. The third is about model theory: preservation of properties, decidability, and model completeness. The fourth weakens the metrics of complexes and the topology of Boolean spaces. The fifth ranges over the diverse sheaves that may exist for a particular algebra.

This introductory chapter delineates the scope of this book. It could be summarized by saying that there are two approaches to decomposing algebras by sheaves: (1) take one large algebra at a time and decompose it into smaller pieces with a sheaf; and (2) take one small algebra, or a finite collection of finite algebras, and decompose all the algebras in the variety generated by them. This book concentrates mostly on (1) and only on (2) in Chap. X about preprimals.

Outlined below is what might have been covered but was not, and a few topics that are introduced but not pursued at length since there are already excellent monographs covering these.

Relational structures are not included; most of our examples are algebras, or they can be made into them, such as lattice-ordered groups. Other structures omitted are several-sorted algebras, which have more than one carrier, and infinitary algebras, which have operations with an infinite number of arguments. So the book is restricted to single-sorted, finitary algebras.

In the spirit of (2), Burris and McKenzie [**BurMc81**] characterize those algebras decomposable as Boolean products, which we mention. The book of Pinus [**Pinus93**] is mostly about Boolean products, and consequently overlaps the book at hand; but the present book starts from a broader perspective. On the other hand, Pinus's book has many more results on Boolean products than we can cover.

In another vein, Clark and Davey [**ClaDa98**] seek in their natural duality theory a unique topological representation of all algebras that are subdirect powers of a particular one. To achieve uniqueness, one must add relations to the topological spaces. By way of contrast, this book seeks a decomposition of a large algebra as a sheaf with much smaller, usually nonisomorphic stalks, hopefully indecomposable in some sense.

Borceux [**Borc94**] and Johnstone [**John82**] approach sheaves from the vantage point of pointless topology, that is locales. This has the advantage of carving out a large part of an argument without the axiom of choice. Then, when the axiom of choice is finally invoked, its role is made clear. The further abstraction to topoi is out of bounds of this book. There are other approaches starting with categories, algebraic theories, and monads, such as those by Kennison [**Kenn81**] and Lambek and Rattray [**LamRa78**].

Finally, there are many books on traditional areas of sheaf theory that we only skirt, for example, those by Tennison [**Tenn75**] and Bredon [**Bred97**]. Homological algebra in algebraic geometry and topology, presented in Cartan [**Cart49**], Grothendieck and Dieudonné [**GroDi60**], and Godement [**Gode58**], is not written about, mainly because it is restricted to modules in these books, although it is an important origin of sheaf theory. For similar reasons, not covered are sheaves over modules as cultivated by Pierce [**Pier67**], although the first part of his monograph on sheaves of rings was a prime motivation for this book; for more, see [**Prest**].

II
ALGEBRA

This chapter provides background material; it has two sections. The first briefly introduces universal algebra. The second surveys the many ways that products of algebras may be captured both externally and internally.

1. Universal Algebra

To set the stage for general algebraic systems, we quickly review the definitions, examples, and theorems that are needed to explain and exploit their decompositions by sheaves. Some of the topics covered are the isomorphism theorems, examples of lattices and shells, and the new notion of sesquimorphism. If the reader desires amplification, there are a number of good texts available; closest to our notation are those by Stanley Burris and H. P. Sankapannavar [**BurSa81**], George Grätzer [**Grät79**], and Ralph N. McKenzie, George F. McNulty and Walter F. Taylor [**McMcT87**]. Others are those by George M. Bergman [**Berg98**], Paul M. Cohn [**Cohn81**] and Richard S. Pierce [**Pier68**]. Ross Willard outlines some later work [**Will94**].

1.1. DEFINITION. An **algebra**,

$$\boldsymbol{A} = \langle A; \omega_1, \omega_2, \ldots, \omega_i, \ldots \rangle$$

is a nonempty set A, called the **carrier**, together with a sequence of multiplace functions $\omega_1, \omega_2, \ldots$ from powers of A to A, called **operations**:

$$\omega_i \colon A^{n_i} \to A.$$

A. Knoebel, *Sheaves of Algebras over Boolean Spaces*,
DOI 10.1007/978-0-8176-4642-4_II,

This sequence may be finite, infinite and even uncountable, but each operation ω_i in it must be **finitary**, that is, have a finite number n_i of arguments. Call their sequence,

$$\boldsymbol{n} = \langle n_1, n_2, \ldots, n_i, \ldots \rangle,$$

the **type** of the algebra. To evaluate an operation ω_i, write as usual

$$\omega_i(a_1, a_2, \ldots, a_{n_i}) \qquad (a_1, a_2, \ldots, a_{n_i} \in A).$$

Implicitly included in the operations of an algebra are the **projections**:

$$\pi_i^n(a_1, a_2, \ldots, a_n) = a_i,$$

which allow us to permute and identify variables. An algebra with only one element will be called **trivial**.

In practice, the type is fixed and various algebras of this type are considered simultaneously, in which case we consider ω_i to be an **operation-symbol** and write its evaluation in a particular algebra $\boldsymbol{A}$ as $\omega_i^{\boldsymbol{A}}$. However, we shall be casual about including the superscript. Familiar examples illustrating this are the operation-symbols $+$ and $\times$ acting on the integers, rational numbers or real numbers, as desired.

Nullary operations ω_i with no arguments, where $n_i = 0$, are curious but useful and necessary; each can take on only one value:

$$\omega_i(\,) = c_i.$$

Call such an ω_i a **constant** and write it simply as an element c_i of the algebra. Most algebras occurring in practice do have constants, often designated 0 and 1.

To avoid subscripts, we often write ω as the generic operation of n arguments, and the algebra simply as $\langle A; \ldots, \omega, \ldots \rangle$. To save space, we often write the sequence $a_1, a_2, \ldots, a_n$ of arguments as a vector $\vec{a}$, and the evaluation as $\omega\vec{a} = \omega(a_1, a_2, \ldots, a_n)$.

These concepts are illustrated by many examples: rings, semilattices, lattices, and Boolean algebras. The set of integers,

$$\mathbb{Z} = \{\ldots, -2, -1, 0, 1, 2, \ldots\},$$

is the carrier of the unital ring of integers,

$$\mathbf{Z} = \langle \mathbb{Z}; +, \times, 0, 1 \rangle,$$

which is an algebra of type $\langle 2, 2, 0, 0 \rangle$. Of the same type is the ring $\mathbf{Z}_n$ of the integers modulo n:

$$\mathbf{Z}_n = \langle \mathbb{Z}_n; +, \times, 0, 1 \rangle,$$

where $\mathbb{Z}_n = \langle 0, 1, 2, \ldots, n-1 \rangle$ with the operations performed modulo n. There may be divisors of zero; for example, $3 \times 4 = 0$ in $\mathbf{Z}_{12}$.

An algebra may have only one binary operation. Such are **semilattices**, $\boldsymbol{S} = \langle S; \wedge\rangle$, algebras satisfying the idempotent, commutative and associative laws for all a, b and c in S:

$$a \wedge a = a,$$
$$a \wedge b = b \wedge a, \text{ and}$$
$$a \wedge (b \wedge c) = (a \wedge b) \wedge c.$$

To each semilattice is associated a partial order: $a \le b$ iff $a \wedge b = a$. The identities imply that each pair of elements have a greatest lower bound in the partial order associated with it.

An example is $\boldsymbol{SL}_3$, a three-element semilattice $\langle\{0, a, b\}; \wedge\rangle$ of type $\langle 2\rangle$, whose binary operation is given by the table:

$\wedge$	0	a	b
0	0	0	0
a	0	a	0
b	0	0	b

It is pictured by its associated partial order presented as a Hasse diagram:

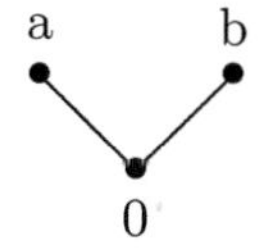

This algebra will serve as a counterexample for many conjectures.

A subset T of a semilattice $\boldsymbol{S}$ has an **infimum** if there is a largest element less than all elements of S, notated $\bigwedge T$. The **supremum** is defined dually, notated $\bigvee T$. If the infimum of each subset of a semilattice exists, then it is said to be **complete**.

A **lattice** $\langle L; \vee, \wedge\rangle$ has two binary operations, $\vee$ and $\wedge$, each a semilattice operation, such that the absorptive laws also hold for all a and b in L:

$$a \wedge (b \vee a) = a \text{ and}$$
$$a \vee (b \wedge a) = a.$$

The identities for a lattice reflect the property that each pair of elements must have both a least upper bound and a greatest lower bound in the partial orders associated with each semilattice, and that they are dually equal:

$$a \le_\wedge b \text{ iff } b \le_\vee a.$$

Traditionally, the order associated with a lattice is that for $\wedge$.

In what follows, we state only the basic notions of lattice theory that are needed in this book. For more detail see the classical text of Birkhoff [**Birk67**], the introductory text of Davey and Priestley [**DavPr02**], or the advanced text of Grätzer [**Grät98**].

By **completeness** of a lattice we mean that both semilattice operations are complete. But completeness of one operation in a lattice insures completeness of the other.

A **distributive** lattice $\boldsymbol{L}$ is one satisfying the law:

$$a \wedge (b \vee c) = (a \wedge b) \vee (a \wedge c) \qquad (a, b, c \in L).$$

The other distributive law follows from it and the other lattice identities:

$$a \vee (b \wedge c) = (a \vee b) \wedge (a \vee c) \qquad (a, b, c \in L).$$

A lattice $\boldsymbol{L}$ is **modular** if:

$$a \leq b \Rightarrow a \vee (b \wedge c) = b \wedge (a \vee c) \qquad (a, b, c \in L).$$

Any distributive lattice is modular.

When there are two elements, 0 and 1, below and above everything else, we say that the lattice is **bounded**. A bounded lattice may be written:

$$\boldsymbol{L} = \langle L; \vee, \wedge, 0, 1 \rangle \text{ with type } \langle 2, 2, 0, 0 \rangle.$$

Of necessity, finite lattices are always bounded, although the bounds may not be part of the type. These are most easily pictured by Hasse diagrams.

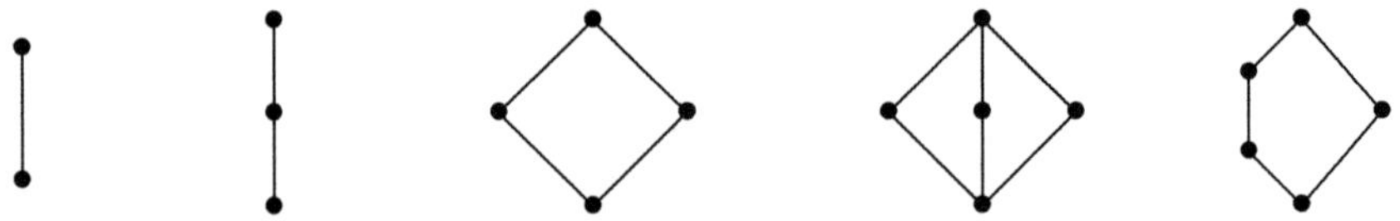

$\boldsymbol{C}_2$ $\boldsymbol{C}_3$ $\boldsymbol{M}_2$ $\boldsymbol{M}_3$ $\boldsymbol{N}_5$

Note that among these examples the first three satisfy the distributive laws, which do not generally hold in lattices. The fourth $\boldsymbol{M}_3$ is not distributive; however, it is modular. The fifth $\boldsymbol{N}_5$ satisfies neither. The last three have divisors of zero. The integers $\mathbb{Z}$ have operations that turn them into an unbounded lattice:

$$m \vee n = \max(m, n) \text{ and } m \wedge n = \min(m, n).$$

In a bounded lattice, an element a may have a **complement** a':

$$a \vee a' = 1, \quad a \wedge a' = 0.$$

Complements do not always exist, and even when they do, there may be more than one. In the chain $\boldsymbol{C}_3$, the middle element has no complement. In the example of $\boldsymbol{M}_3$, complements of any middle element are not unique. However, in $\boldsymbol{M}_2$, the complements of all elements exist and are unique.

A bounded distributive lattice in which all elements have a complement is called **Boolean**. The distributive law guarantees that complements are unique. De Morgan's laws follow:

$$(a \wedge b)' = a' \vee b'; \quad (a \vee b)' = a' \wedge b' \qquad (a, b \in L).$$

Examples are $\boldsymbol{C}_2$ and $\boldsymbol{M}_2$. Putting complementation within bounded lattices into the type gives us **Boolean algebras**:

$$\langle L; \vee, \wedge, ', 0, 1\rangle \text{ of type } \langle 2, 2, 1, 0, 0\rangle.$$

There is more about them in III.4.

Each bounded lattice has within it a Boolean sublattice called the **center**, which consists of its complemented neutral elements. As element a of a lattice $\boldsymbol{L}$ is called **neutral** if any set of three elements in L containing a generates a distributive sublattice.

Other examples of algebras are groups, rings, modules, fields, and vector spaces, with which we assume the reader is familiar. In Chap. VII, shells of various kinds generalize many of these examples.

There are several ways to relate algebras to one another; we first talk about subalgebras and homomorphisms. To be compared they must have the same type. For a **subalgebra** the new carrier is a nonempty subset of the old – closure of the new carrier to the old operations is the defining characteristic. An example is the ring of integers, which is a subalgebra of the ring of all rational numbers, which in turn is a subalgebra of the ring of all real numbers, etc. One typically writes $\boldsymbol{A} \subseteq \boldsymbol{B}$ for the subalgebra relationship.

A **homomorphism** is a function φ from the carrier of one algebra $\boldsymbol{A}$ to another $\boldsymbol{B}$ of the same type such that any operation ω of the type is preserved in going from $\boldsymbol{A}$ to $\boldsymbol{B}$:

$$\varphi\big(\omega^{\boldsymbol{A}}(a_1, \ldots, a_n)\big) = \omega^{\boldsymbol{B}}\big(\varphi(a_1), \ldots, \varphi(a_n)\big) \qquad (a_1, \ldots, a_n \in A).$$

Abbreviate this as $\varphi: \boldsymbol{A} \to \boldsymbol{B}$. Reducing integers modulo 12 is an example of a homomorphism from $\mathbb{Z}$ onto $\mathbb{Z}_{12}$. Any homomorphism $\varphi: \boldsymbol{A} \to \boldsymbol{B}$ that maps A onto B is called **surjective**. If it is one-to-one, it is called **injective**. The examples of lattices given earlier have many homomorphisms, some surjective and others injective. When a homomorphism is both injective and surjective, that is bijective, it is called an **isomorphism**, and notated $\boldsymbol{A} \cong \boldsymbol{B}$. When a homomorphism goes from an algebra back to itself, we have an **endomorphism**. An endomorphism that is also an isomorphism is an **automorphism**.

Here is a useful notion that combines subalgebras with homomorphisms. One algebra $\boldsymbol{B}$ is a **retract** of another $\boldsymbol{A}$ whenever there are two homomorphisms $\mu: \boldsymbol{A} \to \boldsymbol{B}$ and $\iota: \boldsymbol{B} \to \boldsymbol{A}$ such that their composition $\mu \circ \iota$ is the identity function on B. Of necessity, μ is surjective and ι is injective. For example, $\mathbb{Z}_4$ is a retract of $\mathbb{Z}_{12}$.

The external effect of a homomorphism, $\varphi: \boldsymbol{A} \to \boldsymbol{B}$, can be captured internally within $\boldsymbol{A}$ itself by the concept of a congruence. A **congruence** of an algebra $\boldsymbol{A}$ is an equivalence relation θ on its carrier A such that for any operation ω of $\boldsymbol{A}$:

$$\text{if } a_1\,\theta\,b_1, \ldots, a_n\,\theta\,b_n, \text{ then } \omega(a_1, \ldots, a_n)\,\theta\,\omega(b_1, \ldots, b_n).$$

An example is the congruence on $\mathbb{Z}$ of the integers modulo 12, another is the equivalence relation corresponding to the partition $\{\{a\},\{0,b\}\}$ of the semilattice $\boldsymbol{SL}_3$ given earlier. In analogy with number theory, one sometimes writes $a \equiv b \ (\theta)$ for $a \ \theta \ b$. For the set of all congruences of an algebra $\boldsymbol{A}$ write $\operatorname{Con} \boldsymbol{A}$. Also write $\theta(\boldsymbol{B})$ for the smallest congruence of $\boldsymbol{A}$ in which all elements of a subset $\boldsymbol{B}$ are related. An algebra is called **simple** if it has exactly two congruences; of necessity, these will have to be the largest and smallest congruences. A simple algebra is never trivial. Examples of simple algebras are $\mathbb{Z}_p$ for p a prime and the two-element Boolean algebra $\boldsymbol{B}_2$.

Out of each congruence θ of an algebra $\boldsymbol{A}$, a **quotient algebra** $\boldsymbol{A}/\theta$ of the same type is constructed as follows. First, designate the congruence class $\{b \in A \mid b \ \theta \ a\}$ of θ modulo an element a of A as a/θ. The carrier of $\boldsymbol{A}/\theta$ is the set A/θ of congruence classes a/θ of θ as a runs over A. The operations ω/θ on A/θ are defined by

$$\frac{\omega}{\theta}\left(\frac{a_1}{\theta}, \ldots, \frac{a_n}{\theta}\right) = \frac{\omega(a_1, \ldots, a_n)}{\theta} \qquad (a_1, \ \ldots, a_n \in A).$$

In algebras with a group operation, one can recover the whole congruence θ from only one congruence class o/θ. Rings also do not need congruences θ since every equivalence class is a coset of the ideal $0/\theta$. Thus, it suffices to work with normal subgroups or more generally ideals.

Unfortunately, as the abundance of congruences in most lattices makes clear, there is no longer such a handy one-to-one correspondence between ideals and congruences. To see this consider these two congruences on the three-element chain $\boldsymbol{C}_3$.

The upper two elements in the first lattice are related whereas in the second they are not. The bottom singleton is insufficient to determine the congruence. So, in lattice theory and in most other algebras without a group operation, the broader concept of congruence is essential, replacing normal subgroups and ideals.

Throughout we will implicitly use the three isomorphism theorems originally formulated for modules and rings by Emmy Noether [**Noet26**, p. 40], see also [**BurSa81**, Sect. II.6]. We label them the 'homomorphism', 'cancellation', and 'Noether' theorems. The first makes precise the connection between external homomorphisms and internal congruences, and captures the one-to-one correspondence between them: each homomorphism determines a unique congruence, and each congruence determines a homomorphism that is unique up to a commutative diagram.

1.2. THEOREM (Homomorphism). (a) *For any congruence θ of an algebra $\boldsymbol{A}$ there is a surjective homomorphism onto the quotient algebra, $\varphi: \boldsymbol{A} \to \boldsymbol{A}/\theta$, given by $\varphi(a) = a/\theta$.*

(b) *For any surjective homomorphism $\psi: \boldsymbol{A} \to \boldsymbol{B}$, there is a congruence θ of $\boldsymbol{A}$ such that $\boldsymbol{A}/\theta \underset{\iota}{\cong} \boldsymbol{B}$. This congruence is defined by*

$$a_1 \,\theta\, a_2 \text{ iff } \psi(a_1) = \psi(a_2) \quad (a_1, a_2 \in A)$$

and the isomorphism ι by $\iota(a/\theta) = \psi(a)$. This congruence will be called the **kernel** *of the homomorphism: $\theta = \ker \psi$.*

(c) *Starting with a congruence, passing to the surjective homomorphism, and then to its kernel yields back the original congruence.*

(d) *Starting with a surjective homomorphism $\psi: \boldsymbol{A} \to \boldsymbol{B}$, passing to its kernel, $\theta = \ker \psi$, and then to its quotient homomorphism φ yields an algebra isomorphic to $\boldsymbol{B}$ and a composition of functions equal to the original ψ, that is, this diagram commutes.*

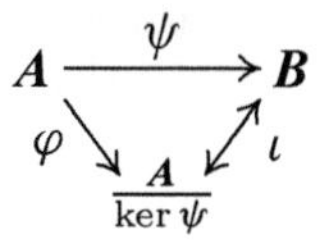

The set Con $\boldsymbol{A}$ of congruences of an algebra $\boldsymbol{A}$ has some structure, namely the partial order of inclusion, which turns it into a lattice $\mathbf{Con}\,\boldsymbol{A}$. The intersection $\theta \cap \eta$ of any two congruences θ and η is again a congruence; also for each pair of congruences there is smallest congruence, $\theta \vee \eta$, that includes both. This join is the union of all compositions of θ and η; these few suffice for the union: $\theta \circ \eta$, $\theta \circ \eta \circ \theta$, $\theta \circ \eta \circ \theta \circ \eta$, etc. These two operations create a lattice. There are always the special congruences: the smallest $0_{\mathrm{Con}\,\boldsymbol{A}}$, which is the identity relation, often called the **trivial** congruence, and the largest $1_{\mathrm{Con}\,\boldsymbol{A}}$, the **improper** congruence, the other congruences being called **proper**.

Thus, $\mathbf{Con}\,\boldsymbol{A}$ is a bounded lattice $\langle \mathrm{Con}\,\boldsymbol{A}; \vee, \wedge, 0_{\mathrm{Con}\,\boldsymbol{A}}, 1_{\mathrm{Con}\,\boldsymbol{A}} \rangle$ for any algebra $\boldsymbol{A}$. An example is $\mathbb{Z}_4$, where $\mathbf{Con}\,\mathbb{Z}_4$ is isomorphic to $\boldsymbol{C}_3$. Also, $\mathbf{Con}\ \boldsymbol{A}$ is a complete lattice.

1.3. THEOREM (Cancellation). *For any algebra $\boldsymbol{A}$ and any congruence θ, the sublattice of all congruences η between and including θ and $1_{\mathrm{Con}\,\boldsymbol{A}}$ is isomorphic to the congruence lattice $\mathbf{Con}(\boldsymbol{A}/\theta)$ of the quotient algebra. The isomorphism sends η to η/θ where η/θ is the congruence on $\boldsymbol{A}/\theta$ defined by*

$$\frac{a}{\theta}\,\frac{\eta}{\theta}\,\frac{b}{\theta} \text{ iff } a \,\eta\, b.$$

Moreover, $(\boldsymbol{A}/\theta)(\eta/\theta) \cong \boldsymbol{A}/\eta$.

As an illustration in the ring $\mathbb{Z}$ of integers, in the interval within $\mathbf{Con}\,\mathbb{Z}$ from mod 4 to mod 1, there are three congruences: mod 4, mod 2 and mod 1. By the cancellation theorem, this three-element lattice is isomorphic to $\mathbf{Con}\,\mathbb{Z}_4$, which is $\boldsymbol{C}_3$, as we already know.

The subalgebras $\boldsymbol{B}$ and congruences θ of an algebra interact in the last isomorphism theorem. For that, extend $\boldsymbol{B}$ to a larger subalgebra, the union of all those congruence classes of θ with at least one member in $\boldsymbol{B}$:

$$\theta B = \{c \mid c\ \theta\ b \text{ for some } b \text{ in } B\}.$$

1.4. THEOREM (Noether). *Let $\boldsymbol{B}$ be a subalgebra of $\boldsymbol{A}$ and θ a congruence of $\boldsymbol{A}$. Then*

$$\frac{\theta B}{\theta|(\theta B)} \cong \frac{\boldsymbol{B}}{\theta|B},$$

where $\theta|B$ is the restriction of θ to $\boldsymbol{B}$.

Classically, homomorphisms of one algebra onto another are captured by congruences, as we have just seen. But congruences are relations. Are there simpler structures that will accomplish the same?

There are several: sesquimorphisms, transversals, and ideals. The first work universally for all algebras; these are projections that replace endomorphisms in earlier investigations. Precursors of sesquimorphisms occur in [**GouGr67**] and [**HobMc88**, def. 2.1], the last as idempotent polynomials restricted to their ranges. The last two, transversals and ideals, which are subsets of an algebra, are always definable, but have nice one-to-one correspondences with congruences only for certain algebras.

We define these concepts, give examples, and relate them to each other. Whenever possible, we would like an axiomatic definition for each concept that is independent of the others. Our immediate goal is to establish the isomorphism theorems for sesquimorphisms. Their definition and some of their properties were formulated in [**Knoe07a**].

1.5. DEFINITION. *A* **sesquimorphism**[1] *is a function μ from the carrier of an algebra $\boldsymbol{A}$ to itself such that*

(i) $\mu\big(\mu(a)\big) = \mu(a) \quad (a \in A)$

(ii) $\mu\big(\omega(\mu(a_1), \ldots, \mu(a_n))\big) = \mu\big(\omega(a_1, \ldots, a_n)\big) \qquad (a_1, \ldots, a_2 \in A)$
for all operations ω of $\boldsymbol{A}$.

Although appearing a bit strange, sesquimorphisms do occur in the literature, but are rarely given a name. Here are three examples, with three more given in the next section on products. For any e in a distributive lattice, Birkhoff [1967, Sect. III. 9] displays the functions,

$$\mu\colon \boldsymbol{L} \to \boldsymbol{L} : x \mapsto e \vee x \quad \text{and} \quad \nu\colon \boldsymbol{L} \to \boldsymbol{L} : x \mapsto e \wedge x.$$

[1] Here is the reason for this name. The prefix 'sesqui' means a ratio of 3:2. When written with a sequence $\vec{a}$ of elements of A, formula (ii) becomes $\mu\big(\omega(\mu \circ \vec{a})\big) = \mu(\omega(\vec{a}))$, with three occurrences of μ as opposed to two in the formula for a homomorphism, $\omega(\mu \circ \vec{a}) = \mu(\omega(\vec{a}))$.

These are sesquimorphisms, in fact, idempotent endomorphisms. However, in a Boolean algebra these two functions are no longer endomorphisms, but they are still sesquimorphisms.

Other examples of sesquimorphisms that are not endomorphisms are found in the ring of integers $\mathbb{Z}$. For a fixed integer m define

$$\begin{aligned}\mu_m(n) &= n \bmod m \\ &= \text{that } k \text{ such that } 0 \leq k < m \text{ and } k \equiv n \pmod{m}.\end{aligned}$$

The range $\mu_m\mathbb{Z}$ is then just the finite set $\{0, 1, \ldots, m-1\}$, with the usual addition and multiplication reduced modulo m.

While any idempotent endomorphism is a sesquimorphism, an arbitrary sesquimorphism is not necessarily an endomorphism, as just shown. However, in the context of products, Proposition 2.14 will show that for many classical algebras, factor sesquimorphisms are endomorphisms.

Each sesquimorphism μ engenders a congruence θ_μ:

$$a \,\theta_\mu\, b \text{ if } \mu(a) = \mu(b),$$

its **induced congruence**. Unequal sesquimorphisms, $\mu \neq \nu$, may induce equal congruences, $\theta_\mu = \theta_\nu$.

For a precise one-to-one correspondence between sesquimorphisms and congruences, we need also the concept of a transversal. The goal is to capture a homomorphism $\varphi: \boldsymbol{A} \to \boldsymbol{B}$ internally within $\boldsymbol{A}$ with a sesquimorphism that picks an element from each congruence class of $\ker \varphi$. A **transversal** of a congruence θ is a subset T of A such that each congruence class of θ has exactly one representative in T. Overall, call T a **transversal** of an algebra $\boldsymbol{A}$ if it is the transversal of some congruence of $\boldsymbol{A}$. In the integers, the congruence mod m has the transversal $\{0, 1, \ldots, m-1\}$, which is the range of the sesquimorphism $\mu_m(n) = n \bmod m$. Generally, for a sesquimorphism μ, its range, $\mu A = \{\mu a \mid a \in A\}$, is a transversal of θ_μ. Conversely, a congruence θ and a transversal T of it produce a sesquimorphism μ:

$$\mu(a) = \text{that unique } b \text{ such that } a \,\theta\, b \text{ and } b \in T. \tag{1.1}$$

These observations may be formalized and proven.

1.6. Proposition. *The sesquimorphisms of an algebra* $\boldsymbol{A}$ *are in one-to-one correspondence with pairs consisting of a congruence and one of its transversals.*

Proof. The only difficult part might be to show that for any transversal the function μ defined by (1.1) is indeed a sesquimorphism. To that end, let $a_1, \ldots, a_n$ be any sequence of arguments for an operation ω of $\boldsymbol{A}$. Then, $a_1 \,\theta\, \mu(a_1), \ldots, a_n \,\theta\, \mu(a_n)$, and hence

$$\omega(a_1, \ldots, a_n) \,\theta\, \omega(\mu(a_1), \ldots, \mu(a_2)).$$

By the uniqueness of transversal elements in congruence classes,

$$\mu\omega(a_1, a_2, \ldots) = \mu\omega(\mu a_1, \mu a_2, \ldots). \qquad \square$$

1.7. PROBLEM. Which algebras have congruences that can be represented by transversals that are subalgebras? Vector spaces are an example.

1.8. DEFINITION. We may go further within an algebra $\boldsymbol{A}$ and turn the range μA of each sesquimorphism μ of it into an algebra. To do this, **relativize** each operation ω of $\boldsymbol{A}$ to μA:

$$\omega^{\mu A}(a_1, a_2, \ldots) = \mu\big(\omega^{\boldsymbol{A}}(a_1, a_2, \ldots)\big) \qquad (a_1, a_2, \ldots \in \mu A).$$

Thus, μA becomes an algebra $\mu\boldsymbol{A}$ of the same type as $\boldsymbol{A}$. Sometimes $\mu\boldsymbol{A}$ is called the **relativization** of $\boldsymbol{A}$ by μ.

1.9. EXERCISE. Consider the four-element Boolean algebra $\boldsymbol{B}$ on the set $\{0, a, b, 1\}$ and the unique congruence θ with transversal $\{a, 1\}$. Find the corresponding sesquimorphism, and write out the operation tables for the relativized Boolean operations on the set $\{a, 1\}$; this is a again a Boolean algebra!

That each external homomorphism casts a shadow as an internal sesquimorphism gives rise to a new version of the Homomorphism Theorem.

1.10. THEOREM (Internal Homomorphism). (a) *Any sesquimorphism* μ *of an algebra* $\boldsymbol{A}$ *is a surjective homomorphism:*

$$\mu : \boldsymbol{A} \to \mu\boldsymbol{A},$$

where $\mu\boldsymbol{A}$ *is its transversal.*

(b) *Any surjective homomorphism* $\psi : \boldsymbol{A} \to \boldsymbol{B}$ *is realized internally by a sesquimorphism* $\mu : \boldsymbol{A} \to \mu\boldsymbol{A}$ *that makes this diagram commute:*

$$\begin{array}{ccc} \boldsymbol{A} & \xrightarrow{\ \psi\ } & \boldsymbol{B} \\ & {\scriptstyle \mu}\searrow \quad \nearrow{\scriptstyle isom.} & \\ & \mu\boldsymbol{A} & \end{array}$$

Here, μ *is any sesquimorphism associated with the kernel of* ψ, *and the isomorphism from* $\mu\boldsymbol{A}$ *to* $\boldsymbol{B}$ *is* $\psi|(\mu\boldsymbol{A})$. *Thus* $\ker\psi = \ker\mu$.

To see what this has to do with the traditional Homomorphism Theorem phrased in terms of congruences, realize that

$$\frac{\boldsymbol{A}}{\ker\psi} \cong \mu\boldsymbol{A}.$$

The Cancellation Theorem, when phrased in terms of sesquimorphisms, is simpler than when phrased in terms of congruences. To state it, sesquimorphisms need to be partially ordered, using their composition.

1.11. DEFINITION. For sesquimorphisms μ and ν in an algebra $\boldsymbol{A}$,

$$\mu \le \nu \quad \text{if} \quad \mu \circ \nu = \mu = \nu \circ \mu.$$

Consequently, if $\mu \le \nu$, then

(a) $\mu \circ \nu$ and $\nu \circ \mu$ are sesquimorphisms,
(b) $\nu(a) = \nu(b)$ implies $\mu(a) = \mu(b)$ $(a, b \in A)$,
(c) $\mu(A) \subseteq \nu(A)$,
(d) $\theta_\mu \supseteq \theta_\nu$.

The reversal of ordering in (d) will fit in with the Boolean algebras of factor objects to be defined in Sect. VI.2, and in particular Theorem VI.3.2. For now, accept that we are ordering sesquimorphisms by the size of their transversals.

1.12. THEOREM (Internal Cancellation). *If μ and ν are sesquimorphims of an algebra $\boldsymbol{A}$ such that $\mu \le \nu$ and $\bar{\mu} = \mu|(\nu A)$, then $\bar{\mu}$ is a sesquimorphism of $\nu\boldsymbol{A}$ and $\mu = \bar{\mu} \circ \nu$.*

PROOF. Since $\nu(\nu(a)) = \nu(a)$ and $\mu \le \nu$, then $\bar{\mu}(\nu(a)) = \mu(a)$. □

To connect with the traditional Cancellation Theorem, usually couched in terms of two congruences, $\eta \subseteq \theta$, let ν be a sesquimorphism for η. Choose a transversal T for θ such that $T \subseteq \nu A$, thereby determining a sesquimorphism μ such that $T = \mu A$. Then $\mu \le \nu$. By the Internal Homomorphism Theorem,

$$\frac{\boldsymbol{A}}{\eta} \cong \nu\boldsymbol{A}, \quad \frac{\boldsymbol{A}/\eta}{\theta/\eta} \cong \bar{\mu}\nu\boldsymbol{A}, \quad \frac{\boldsymbol{A}}{\theta} \cong \mu\boldsymbol{A}.$$

The conclusion of the Internal Cancellation Theorem yields the traditional Cancellation Theorem.

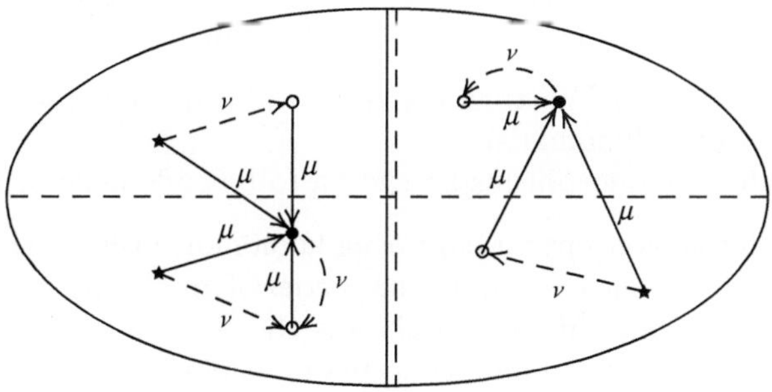

FIGURE 1. The action of sesquimorphisms.

Figure 1 illustrates the general interaction of two sesquimorphisms, μ and ν, of an algebra $\boldsymbol{A}$, with corresponding congruences, $\theta \supseteq \eta$. Their congruence classes are outlined in solid and dashed lines, respectively – two classes for θ and four for η. The symbol $\star$ represents arbitrary elements of A, the symbol $\circ$ their images under ν, and $\bullet$ their images under μ. Thus, there is one $\circ$ for each congruence class of ν, and one $\bullet$ for each class of μ.

As ν and μ are idempotent, $\circ$ and $\bullet$ go to themselves under the respective sesquimorphisms. If also $\mu \leq \nu$, as in Theorem 1.12, then $\circ$ and $\bullet$ would become one and the same in the first and third quadrants of this figure.

1.13. PROBLEM. Given an algebra $\boldsymbol{A}$, can one choose for each congruence θ a sesquimorphism μ_θ representing it ($\theta_{\mu_\theta} = \theta$) so that altogether they are compatible:

$$\theta \subseteq \eta \quad \text{if, and only if,} \quad \mu_\theta \geq \mu_\eta \qquad (\theta \in \operatorname{Con} \boldsymbol{A})?$$

The Noether Theorem becomes even simpler when viewed internally.

1.14. THEOREM (Internal Noether). *Let $\boldsymbol{B}$ be a subalgebra of $\boldsymbol{A}$. Suppose μ is a sesquimorphism of $\boldsymbol{A}$ such that $b \in B$ implies $\mu(b) \in B$. Define $\boldsymbol{C} = \mu^{-1}\boldsymbol{B}$. Then $\boldsymbol{C}$ is a subalgebra of $\boldsymbol{A}$, and*

$$\mu \boldsymbol{C} = \mu \boldsymbol{B}.$$

PROOF. To check closure of $\boldsymbol{C}$ to an n-ary operation ω of $\boldsymbol{A}$, apply it to elements $c_1, \ldots, c_n$ of $\boldsymbol{C}$. There must be elements b_i of B such that $\mu c_i = b_i$. Thus,

$$\mu\omega(c_1, \ldots, c_n) = \mu\omega(\mu b_1, \ldots, \mu b_n) = \mu\omega(b_1, \ldots, b_n) \in \boldsymbol{B},$$

since μ is a sesquimorphism. Hence, $\omega\vec{c} \in \boldsymbol{C}$.
The equality of the subalgebras is straightforward to verify. □

Again, we see simple parallels with congruences. If θ is the congruence determined by μ, then $\boldsymbol{C}$ is the subalgebra of $\boldsymbol{A}$ that is the union of all those congruence classes of θ with at least one member in $\boldsymbol{B}$, that is, $\boldsymbol{C} = \theta\boldsymbol{B}$, as defined in the external Noether theorem. Applying μ to both sides of this equation yields the quotient algebras of Theorem 1.4.

1.15. PROBLEM. (a) When can sesquimorphisms be chosen to be term-operations or polynomials?
(b) When can sesquimorphisms be chosen to be endomorphisms?

The last internal concept for capturing homomorphisms is that of 'ideal'. It depends on fixing one element of the carrier of an algebra as an 'origin' o. For a congruence θ, its **ideal** or **o-class** is the equivalence class o/θ. This notion captures several common concepts in algebra: in a group an ideal is a normal subgroup, providing the origin is the unity 1; in a ring or Boolean algebra, an ideal is the usual notion, providing that $o = 0$; and in a Boolean algebra, if $o = 1$, then we obtain 'filters'. In these cases, any other element can also serve as the origin, in the sense that its ideals uniquely determine the congruences from which they come. Ivan Chajda, Günther Eigenthaler and Helmut Länger [**ChaEL03**, p. 64] name this property: an algebra is **regular** if any congruence θ of $\boldsymbol{A}$ is determined by o/θ for any o in A, that is, for any congruences θ and η, and any element o of A, if $o/\theta = o/\eta$, then $\theta = \eta$. However, an ideal generally does not uniquely determine a congruence, as shown earlier.

Classically, the concept of an ideal as an internal determiner of homomorphisms is defined in terms of equations, as for normal subgroups. In [**ChaEL03**, p. 137], ideals are defined in this way with what are called 'ideal terms'. Their notion agrees with ours for the specific algebras mentioned in the last paragraph, but otherwise they may disagree. We do not pursue how generally equations may be used to capture the concept of an ideal since our more special interest is in complementary factor ideals, where these will be characterized in shells by somewhat different sentences at the end of Sect. VII.3.

Ideals are subalgebras when there is an element o of the algebra such that for any operation, $\omega(o, o, \ldots, o) = o$, that is, $\{o\}$ is a subalgebra. An ideal of an algebra is **maximal** if any larger ideal is the whole algebra.

This section closes with a miscellanea of useful concepts and conventions. Sometimes we need to forget about some of the operations of an algebra; this is called a **reduct**. For example, the group $\langle \mathbb{Z}; +, 0\rangle$ of integers is a reduct of the ring $\langle \mathbb{Z}; +, \times, 0, 1\rangle$. The opposite of a reduct is an **expanse**, where more operations are adjoined to an algebra.

Other times we want an operation to act on subsets of the carrier rather than on elements. This is done by mimicking the complex multiplication often found in group and ring theory:

$$\omega(A_1, A_2, \ldots, A_n) = \{\omega(a_1, a_2, \ldots a_n) \mid a_1 \in A_1, a_2 \in A_2, \ldots, a_n \in A_n\}.$$

Occasionally, as in categories and sheaves, we need to relax the totality of the operations. A **partial algebra** is an algebra in which some or all of the operations are not defined everywhere. An example is the notion of a field in which division by 0 is not allowed. We might specify the field as a partial algebra $\langle F; +, -, \times, /, 0, 1\rangle$. A composition of these operations is said to exist for particular arguments if each stage of evaluation exists. Thus, in a field, $0 + 1/(1+1)$ exists, but $0 + 1/(1-1)$ does not. An equation in which each side is a composition of partial operations is said to be **satisfied** for particular arguments if, when one side exists, so does the other, and they are equal.

The definition of homomorphism for partial algebras also needs a proviso. A **homomorphism** of partial algebras, $\varphi \colon \boldsymbol{A} \to \boldsymbol{B}$, must satisfy for each n-ary operation ω:

$$\varphi(\omega(a_1, a_2, \ldots, a_n)) = \omega\big(\varphi(a_1), \varphi(a_2), \ldots, \varphi(a_n)\big)$$

whenever $\omega(a_1, a_2, \ldots, a_n)$ exists, and thus also $\omega\big(\varphi(a_1), \varphi(a_2), \ldots, \varphi(a_n)\big)$ must exist.

The symbols 0 and 1 are used throughout this book in many contexts, most loosely connected. In a bounded partial order, $\boldsymbol{P} = \langle P; \leq\rangle$, the bottom element is designated $0_{\boldsymbol{P}}$ and the top $1_{\boldsymbol{P}}$. Thus, in the congruence lattice $\mathbf{Con}\,\boldsymbol{A}$ of an algebra $\boldsymbol{A}$, the identity relation, which is the smallest, is $0_{\mathbf{Con}\,\boldsymbol{A}}$, and the universal relation, the largest, is $1_{\mathbf{Con}\,\boldsymbol{A}}$. This goes along with notations in bounded lattices. Further 1_S represents the identity function on a set S, in other words, 1_S is the unity of the semigroup of functions on S. When obvious, these subscripts may be dropped to improve clarity. Also, this notation may be used for identity morphisms in a category; we will see in Sect. IV.3 that the identity morphism $1_{\mathcal{A}}$ of a sheaf $\mathcal{A}$ has a more complicated structure than a simple identity function has on a set. Chapter VII will use 0 and 1 in shells analogously to their use in unital rings, monoids and bounded lattices.

Here are other common terms and notations. The **power set** $\mathcal{P}(A)$ of a set A is the set of all its subsets. The cardinality of a set A is notated $|A|$.

The **composition** of two binary relations η and θ is given by:

$$\eta \circ \theta = \{\langle a,b\rangle \mid a \;\eta\; x \text{ and } x \;\theta\; b \text{ for some } x\}.$$

If $\eta \circ \theta = \theta \circ \eta$, then η and θ are said to **commute** or **permute.**[2] For example, in an algebra with a group operation, any two congruences commute. Generally, the composition of two congruences is not again a congruence; but when they commute it is.

Composition specializes to functions α and β, when they are considered as sets of ordered pairs, so that upon evaluation:

$$(\alpha \circ \beta)(a) = \alpha(\beta(a)).$$

The next exercise is about commuting sesquimorphisms.

1.16. EXERCISE. Find proofs and counterexamples.

(a) A composition of sesquimorphism is not necessarily a sesquimorphism.
(b) If two sesquimorphisms commute, then their composition is a sesquimorphism.
(c) Even if two sesquimorphisms commute, their congruences may not.
(d) Even if two congruences commute, some corresponding sesquimorphisms may not.

The anomalies of Exercise 1.16 will vanish in the next section when complementary sesquimorphisms create products.

Composition may be extended to functions of more than one argument. For an m-ary function α and an n-ary function β the **composition** $\alpha \circ \beta$ is an mn-ary function.

$$\begin{aligned}&(\alpha \circ \beta)(a_1^1, a_2^1, \ldots, a_n^1,\ a_1^2, a_2^2, \ldots, a_n^2,\ \ldots,\ a_1^m, a_2^m, \ldots, a_n^m)\\ &= \alpha\big(\beta(a_1^1, a_2^1, \ldots, a_n^1),\ \beta(a_1^2, a_2^2, \ldots, a_n^2),\ \ldots,\ \beta(a_1^m, a_2^m, \ldots, a_n^m)\big).\end{aligned}$$

It is most convenient to express this by adapting matrix notation:

$$\alpha(M\beta) = \alpha\big(\beta(a_1^1, a_2^1, \ldots, a_n^1),\ \beta(a_1^2, a_2^2, \ldots, a_n^2),\ \ldots,\ \beta(a_1^m, a_2^m, \ldots, a_n^m)\big),$$

[2]The traditional term for relations is 'permute', but the preferred term in this book is 'commute' since other related notions, such as endomorphisms, traditionally commute.

where

$$M = \begin{pmatrix} a_1^1, & a_2^1, & \dots, & a_n^1 \\ a_1^2, & a_2^2, & \dots, & a_n^2 \\ \vdots & \vdots & & \\ a_1^m, & a_2^m, & \dots, & a_n^m \end{pmatrix}.$$

Here, $M\beta$ is the column vector resulting from β operating on each row, and α operates on this column to produce a single element. Similarly, αM is the row vector resulting from α operating on each column, and β operates on this row to yield $(\alpha M)\beta$. If $\alpha(M\beta) = (\alpha M)\beta$, then α and β are said to **commute**; equivalently, if $\beta(M^T\alpha) = (\beta M^T)\alpha$, using the transpose. This is an extension of commuting functions of one variable. For example, when α and β are binary operations of an algebra, they commute iff

$$\alpha\big(\beta(a,b), \beta(c,d)\big) = \beta\big(\alpha(a,c), \alpha(b,d)\big).$$

We extend this notation to the preservation of relations by functions. Let a^i be the ith row of the matrix M and a_j its jth column. For an n-ary function φ and an m-ary relation ρ, both on the set A, we say that φ **preserves** ρ

$$\text{if } \rho(a_1), \rho(a_2), \dots, \rho(a_n), \text{ then } \rho(\varphi(a^1), \varphi(a^2), \dots, \varphi(a^m)),$$

for any m by n matrix M with entries a^i_j in A.

The product $\alpha \times \beta$ of two m-ary relations, α on A and β on B, is an m-ary relation on $A \times B$ given by:

$$\langle\langle a_1, b_1\rangle, \dots, \langle a_m, b_m\rangle\rangle \in \alpha \times \beta \text{ if } \langle a_1, \dots, a_m\rangle \in \alpha \text{ and } \langle b_1, \dots, b_m\rangle \in \alpha.$$

It should be clear how to define products of more than two relations and powers of a single relation.

Since functions are so fundamental, composition of them will often be abbreviated, $\alpha\beta = \alpha \circ \beta$, and their evaluation, $\alpha a = \alpha(a)$. Confusion might result when composition and evaluation are juxtaposed and iterated, but associative laws save the day – at least for functions of one argument:

$$\alpha \circ (\beta \circ \gamma) = \alpha\beta\gamma = (\alpha \circ \beta) \circ \gamma;$$
$$(\alpha \circ \beta)(a) = \alpha\beta a = \alpha(\beta(a)).$$

By $\Phi\ (\varphi)(\mathrm{a})$ where φ is a function and Φ is a function of functions, we mean of course $\big(\Phi\ (\varphi)\big)(\mathrm{a})$. The **domain** of a function, $\varphi\colon A \to B$, is A; and its **range** is denoted: $\operatorname{rng}\varphi = \{\varphi(a) \mid a \in A\}$. If S is a subset of A, then

$$\varphi(S) = \{\varphi(s) \mid s \in S\} \quad \text{and}$$
$$\varphi|S = \{\langle s, \varphi(s)\rangle \mid s \in S\},$$

the restriction of a function.

In analogy with the set former $\{n^2 \mid n \in \mathbb{Z}\}$ – the collection of integers that are squares – we write $\langle n^2 \mid n \in \mathbb{Z}\rangle$ for the function φ that squares integers. We may call this a **family** when the emphasis is on the range of values. This function former may also be spelled out as

$$\varphi : \mathbb{Z} \to \mathbb{Z} : n \mapsto n^2;$$

more generally,

$$\varphi : \boldsymbol{A} \to \boldsymbol{B} : a \mapsto \varepsilon(a),$$

where ε is some expression.

To make formulas easily readable we continue to juxtapose symbols when no confusion arises. For example, in shells the second binary operation-symbol may be omitted: $ab = a \times b$ in rings, and $ab = a \wedge b$ in lattices. These implicit operations have the greatest cohesiveness in groupings; in lattices for example,

$$a(bc \vee de) \vee f = \Big(a \wedge \big((b \wedge c) \vee (d \wedge e)\big)\Big) \vee f.$$

For algebras with a binary operation such as $\times$, we may combine the convention of juxtaposition with complex multiplication:

$$eA = e \times A = \{e \times a \mid a \in A\} = \{ea \mid a \in A\}.$$

The repeated argument o of a function ω is written

$$\omega(\dot{o}) = \omega(o, o, \ldots, o).$$

Here, $\dot{o}$ is o repeated enough times to fill out ω. It is also convenient in long derivations to abbreviate the sequence of arguments $a_1, a_2, \ldots, a_n$ of a function ω as $\vec{a}$. Thus,

$$\omega(\vec{a}) = \omega(a_1, a_2, \ldots, a_n).$$

We continue the convention adopted in model theory of distinguishing between a structure and its carrier: a bold letter for the algebra and an ordinary font for the carrier, and similarly for a topological space. This convention breaks down when both operations and a topology are present on the carrier and the definition of each takes several stages, as in the definition of sheaves in Sect. IV.1.

This convention continues over functions. Thus, when $\varphi: \boldsymbol{A} \to \boldsymbol{B}$ is a homomorphism, $\varphi(A)$ is a *set* that is the range of φ. But $\varphi(\boldsymbol{A})$ is the image of φ as a *subalgebra* of $\boldsymbol{B}$. A similar comment applies to topological spaces.

Operators, such as Con and Γ, need different conventions. Thus, Con $\boldsymbol{A}$ is the *set* of congruences on an algebra $\boldsymbol{A}$, whereas **Con** $\boldsymbol{A}$ is the *lattice* of congruences. Similarly, $\Gamma(\mathcal{A})$ will be the *set* of global sections of the sheaf on an algebra $\boldsymbol{A}$, whereas $\boldsymbol{\Gamma}(\mathcal{A})$ will be the algebra of such sections.

A few words about emptiness. In general we have excluded empty algebras, the conventional stand, although there are good reasons for including them. Among these are that each variety would then have a free algebra on zero generators; this free algebra would be the initial object of the variety when viewed as a category. Also, there would be no need to distinguish between subalgebras (ordinarily nonempty) and subuniverses (possibly empty). When nullary operations are present, empty algebras do not exist: so in this case, these points are moot. And this is so for most of our applications. To have included empty algebras in this book would have meant adding extra clauses and some ad hoc constructions to the definitions involving specific categories.

2. Products and Factor Objects

Homomorphic images and subalgebras of an algebra are smaller algebras. We turn our attention now to constructions that create larger algebras: direct products, subdirect products, and disjoint unions. Decomposing an algebra into various products of smaller and more manageable pieces is the foundation for constructing the Pierce sheaf. This will be achieved in Chap. VI by refining a family of factor congruences into a subdirect product indexed by a topological space.

This section studies how to factor an arbitrary algebra into a product, both externally and internally. Relating outer direct products to inner direct products is natural and well developed in classical systems. Jónsson and Tarski [**JónTa47**] extended this correspondence to more general algebras (JT-algebras) when they proved uniqueness of direct decompositions of their finite algebras.

We exhibit the many ways in which factorizations may be characterized. The external ways are the outer direct product and the categorical product. There are up to five ways to recreate these outer products internally. Complementary factor congruences are well known, and factor bands less so; there are also complementary factor ideals and elements, the analog of central idempotents in rings. Complementary factor sesquimorphisms have been defined up to now in the literature only for special algebras, and often they are just endomorphisms.

In unital rings and bounded lattices, as well as in their common generalization, unital shells, all external and internal concepts are equivalent. But in general, among all the ways to express factorizations in arbitrary algebras, there is a bijective correspondence between only some of these: outer direct products, categorical products, bands, complementary congruences and complementary sesquimorphisms. Equivalence with the remaining two, elements and ideals, requires something like a weak sum or multiplication in Chap. VII. This section concludes with subdirect products and disjoint unions of algebras.

2.1. DEFINITION. The **outer direct product** $\boldsymbol{P}$, or just **product**, of two algebras $\boldsymbol{A}$ and $\boldsymbol{A}'$ of the same type has as a carrier the Cartesian product $A \times A'$, with the operations defined on it coordinate-wise:

$$\omega^{\boldsymbol{P}}\big(\langle a_1, a_1'\rangle, \ldots, \langle a_n, a_n'\rangle\big) = \langle \omega^{\boldsymbol{A}}(a_1, \ldots, a_n), \omega^{\boldsymbol{A}'}(a_1', \ldots, a_n')\rangle,$$

for any n-ary operation ω of the given type with a_i in A and a'_i in A'. Associated with it are the **projections**: $\pi: \boldsymbol{P} \to \boldsymbol{A}: \langle a, a' \rangle \mapsto a$ and $\pi': \boldsymbol{P} \to \boldsymbol{A}': \langle a, a' \rangle \mapsto a'$. One writes $\boldsymbol{P} = \boldsymbol{A} \times \boldsymbol{A}'$, and omits the words modifying 'product' when clear. For the same factor repeated n times, we have the power $\boldsymbol{A}^n$.

Carried throughout this discussion will be the example of a product of cyclic rings: $\mathbb{Z}_{12} \cong \mathbb{Z}_3 \times \mathbb{Z}_4$, where the isomorphism is given by projections: $m \mapsto \langle m \bmod 3, m \bmod 4 \rangle$.

Power sets, viewed as Boolean algebras, provide more products:

$$\mathcal{P}(A \cup B) \times \mathcal{P}(A \cap B) \cong \mathcal{P}(A) \times \mathcal{P}(B).$$

2.2. DEFINITION. An algebra $\boldsymbol{P}$ is said to be a **categorical product** of algebras $\boldsymbol{A}$ and $\boldsymbol{A}'$ if there are homomorphisms $\pi: \boldsymbol{P} \to \boldsymbol{A}$ and $\pi': \boldsymbol{P} \to \boldsymbol{A}'$ such that for any other algebra $\boldsymbol{Q}$ and homomorphisms $\rho: \boldsymbol{Q} \to \boldsymbol{A}$ and $\rho': \boldsymbol{Q} \to \boldsymbol{A}'$ there is a unique homomorphism $\chi: \boldsymbol{Q} \to \boldsymbol{P}$ for which $\rho = \pi \circ \chi$ and $\rho' = \pi' \circ \chi$; that is, this diagram commutes:

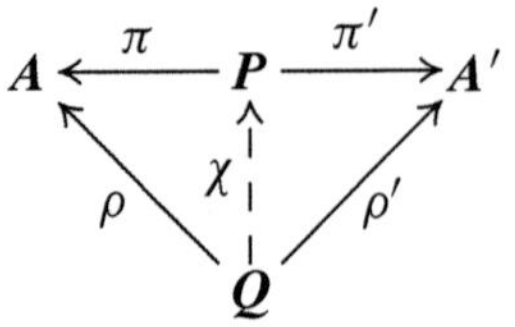

Categorical notions are defined relative to a class of objects and mappings between them. For now, it suffices to consider all algebras of a given type, and all homomorphisms between them. In this case, the outer direct product $\boldsymbol{A} \times \boldsymbol{A}'$ defined earlier will also be a categorical product. Conversely, for any categorical product $\boldsymbol{P}$ as notated above, there is the isomorphism, $\chi : \boldsymbol{P} \cong \boldsymbol{A} \times \boldsymbol{A}'$ (in the defining diagram take $\boldsymbol{Q} = \boldsymbol{A} \times \boldsymbol{A}'$ with ρ and ρ' the Cartesian projections).

One says 'a' product since categorically products are defined only up to isomorphism. For example, in the category of sets, a product of $\{0, 1, 2\}$ and $\{0, 1\}$ may be any six-element set. Note also that the projections are considered an integral part of a categorical product $\langle \boldsymbol{P}, \pi, \pi' \rangle$. Two categorical products on the same algebra, $\langle \boldsymbol{P}; \pi, \pi' \rangle$ of $\boldsymbol{A}$ and $\boldsymbol{A}'$, and $\langle \boldsymbol{P}; \rho, \rho' \rangle$ of $\boldsymbol{B}$ and $\boldsymbol{B}'$, are said to be **isomorphic** if there are isomorphisms making this diagram commute.

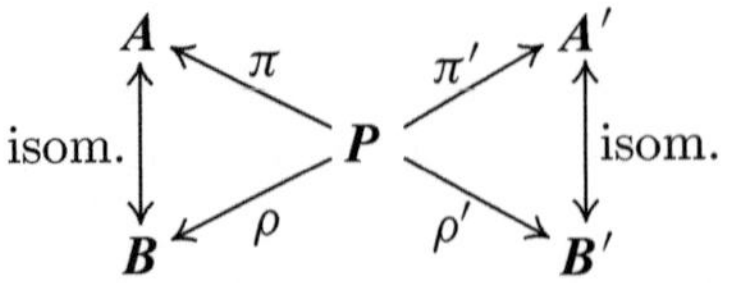

The internal ways of factoring are five in number. If a unital ring factors, $\boldsymbol{R} = \boldsymbol{S} \times \boldsymbol{T}$, then this product may be captured within $\boldsymbol{R}$ as follows:

(a) By complementary central idempotents, $e = \langle 1, 0 \rangle$ and $e' = \langle 0, 1 \rangle$

(b) By complementary ideals, $I = eR$ and $I' = e'R$
(c) By complementary endomorphisms onto these ideals, $\pi(r) = er$ and $\pi'(r) = e'r$
(d) By complementary congruences, θ and θ', where $r\ \theta\ s$ iff $er = es$, and $r\ \theta\ 's$ iff $e'r = e's$
(e) By a band, $\beta(r, s) = er + e's$.

In addition one can talk about an inner direct product, $\boldsymbol{R} = e\boldsymbol{R} + e'\boldsymbol{R}$, associated with this outer direct product.

In analogy with analytic geometry, Fig. 2 illustrates these concepts with the product of rings, $\mathbb{Z}_{12} \cong \mathbb{Z}_3 \times \mathbb{Z}_4$. Think of the bottom row, the ideal

$9 \sim \langle 0,1\rangle = e'$	$5 \sim \langle 2,1\rangle$	$1 \sim \langle 1,1\rangle$
$6 \sim \langle 0,2\rangle$	$2 \sim \langle 2,2\rangle$	$10 \sim \langle 1,2\rangle$
$3 \sim \langle 0,3\rangle$	$11 \sim \langle 2,3\rangle$	$7 \sim \langle 1,3\rangle$
$0 \sim \langle 0,0\rangle$	$8 \sim \langle 2,0\rangle$	$4 \sim \langle 1,0\rangle = e$

FIGURE 2. Representing the ring $\mathbb{Z}_{12}$ as $\mathbb{Z}_3 \times \mathbb{Z}_4$

$4\mathbb{Z}_{12}$, as the X-axis – this is isomorphic to $\mathbb{Z}_3$; and the left-hand column, the ideal $3\mathbb{Z}_{12}$, as the Y- or X'-axis – this is isomorphic to $\mathbb{Z}_4$. An X-coordinate x at the bottom points up and an X'-coordinate x' at the left points to the right, giving us an entry in the body that is $\beta(x, x')$. Endomorphisms do the opposite: π projects down to the X-axis and π' projects left to the X'-axis. More generally, $\beta(r, s)$ is that entry that is in same column as r and the same row as s. The corresponding congruences are $\theta = \bmod 3$, and $\theta' = \bmod 4$. As for their congruence classes, the columns are the θ-classes and the rows the θ'-classes.

We now define these internal notions within any general algebra, calling them collectively **factor objects**, starting with bands and ending with elements, prefixing the adjective 'factor' to each to indicate their origin in a product. But to fully capture products by all these notions we will have to wait till additional assumptions on the algebra are added.

For any product of algebras, $\boldsymbol{P} = \boldsymbol{A} \times \boldsymbol{A}'$, there is the homomorphism, $\beta: \boldsymbol{P} \times \boldsymbol{P} \to \boldsymbol{P}$, given by $\beta(\langle a, a'\rangle, \langle b, b'\rangle) = \langle a, b'\rangle$. A product given merely as an isomorphism, $\boldsymbol{P} \cong \boldsymbol{A} \times \boldsymbol{A}'$, will carry this binary function back to $\boldsymbol{P}$. In any case, it satisfies these equations for elements in $\boldsymbol{P}$:

$$\beta(p, p) = p, \tag{2.1}$$

$$\beta(\beta(p, q), \beta(s, t)) = \beta(p, t), \tag{2.2}$$

$$\beta(\omega(p_1, \ldots, p_n), \omega(q_1, \ldots, q_n)) = \omega(\beta(p_1, q_1), \ldots, \beta(p_n, q_n)), \tag{2.3}$$

where ω is any n-ary operation of the given type. We call such a binary function β on any algebra a **factor band** ; it is also called a **decomposition operation** (see [**McMcT87**, vol. 1, pp. 162 ...]). The last equation amounts to the commutativity of β and ω, expressed more compactly: $\beta \circ \omega = \omega \circ \beta$. In the example of the ring $\mathbb{Z}_{12}$, which is isomorphic to $\mathbb{Z}_3 \times \mathbb{Z}_4$, one has that $\beta(r,s) = 4r + 9s$, the coefficients to be found by the Chinese Remainder Theorem. It follows from its defining identities that a factor band is always associative:

$$\beta(p, \beta(q,r)) = \beta(\beta(p,p), \beta(q,r)) = \beta(p,r) = \ldots = \beta(\beta(p,q),r);$$

hence, the name 'band', as it is applied to any idempotent and associative binary operation. There are the trivial factor bands, $\beta(a,b) = a$ and $\beta(a,b) = b$ for all a and b in A, corresponding to factoring an algebra $\boldsymbol{A}$ as a product of itself with a one-element algebra.

Factor bands may also be obtained directly from the categorical definition. Let $\boldsymbol{P}$ be the outer direct product $\boldsymbol{A} \times \boldsymbol{A}'$ with projections π and π'. As $\boldsymbol{P} \times \boldsymbol{P}$ is also an outer direct product, there are two projections associated with it: $\Pi, \Pi' \colon \boldsymbol{P} \times \boldsymbol{P} \to \boldsymbol{P}$. If we view $\boldsymbol{P}$ as a categorical product, there must be a unique $\beta \colon \boldsymbol{P} \times \boldsymbol{P} \to \boldsymbol{P}$ for which $\pi \circ \beta = \pi \circ \Pi$ and $\pi' \circ \beta = \pi' \circ \Pi'$. One easily checks that β is the previously defined factor band. A commutative diagram illustrates this construction.

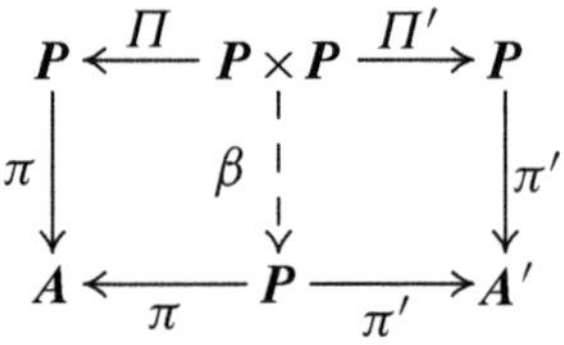

2.3. PROPOSITION. *Two categorical products, $\langle \boldsymbol{P} : \pi, \pi' \rangle$ and $\langle \boldsymbol{P} : \rho, \rho' \rangle$ of a common algebra $\boldsymbol{P}$, are isomorphic if, and only if, their respective factor bands β and γ are equal.*

PROOF. From the diagram above the factor bands are defined by

$$\beta(p,q) = \chi\langle \pi p, \pi' q \rangle,$$
$$\gamma(p,q) = \psi\langle \rho p, \rho' q \rangle.$$

In the first formula, the isomorphism χ comes from inserting the outer direct product $\boldsymbol{A} \times \boldsymbol{A}'$ into the definition of categorical product:

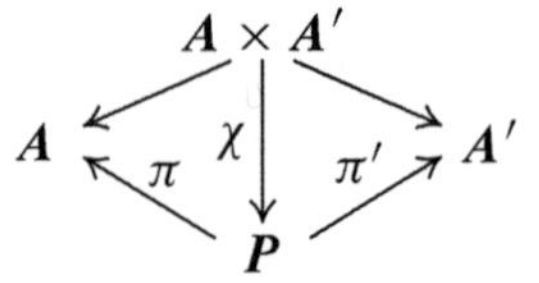

In the second formula, ψ comes from a similar diagram for $\boldsymbol{B} \times \boldsymbol{B}'$.

The two directions of logical implication are proven by developing the next commutative diagram in two different ways.

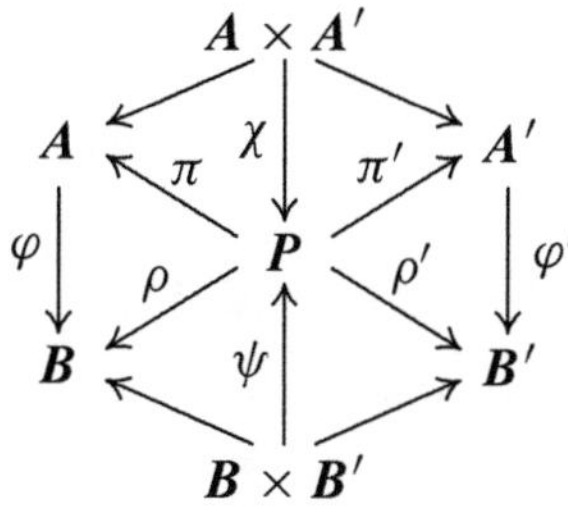

$\Rightarrow$. Create this diagram from two earlier diagrams for categorical products and their isomorphism. Clearly, $\chi = \psi \circ \langle \varphi, \varphi' \rangle$. Hence, for all p and q in P,

$$\gamma pq = \psi \langle \rho p, \rho' q \rangle = \psi \langle \varphi \pi p, \varphi' \pi' q \rangle = \chi \langle \pi p, \pi' q \rangle = \beta pq.$$

$\Leftarrow$. Create this diagram again by defining $\langle \varphi, \varphi' \rangle$ as $\psi^{-1} \circ \chi$. □

The four remaining inner factor objects come in pairs.

2.4. DEFINITION. Two congruences θ and θ' of an algebra $\boldsymbol{A}$ are **complementary factor** congruences if

$$\theta \cap \theta' = 0_{\mathrm{Con}\,\boldsymbol{A}}, \tag{2.4}$$

$$\theta \circ \theta' = 1_{\mathrm{Con}\,\boldsymbol{A}}. \tag{2.5}$$

Pierce [**Pier68**, p. 88] calls these **decomposition** congruences .

It follows from (2.5), by taking the converse of each side, that also $\theta' \circ \theta = 1$, and hence for their join that $\theta \vee \theta' = 1$. (More generally, for any two congruences θ and η of an algebra $\boldsymbol{A}$, $\theta \vee \eta = \theta \circ \eta$ iff $\theta \circ \eta = \eta \circ \theta$.)

Conditions (2.4) and (2.5) have useful interpretations. Let θ and θ' be arbitrary congruences of an algebra $\boldsymbol{A}$, not necessarily complementary. There is always the **canonical** homomorphism,

$$\varphi \colon \boldsymbol{A} \to \frac{\boldsymbol{A}}{\theta} \times \frac{\boldsymbol{A}}{\theta'} : a \mapsto \left\langle \frac{a}{\theta}, \frac{a}{\theta'} \right\rangle.$$

Now (2.4) holds just when φ is injective; and (2.5) just when φ is surjective.

Another viewpoint considers, for any a and b in A, the possible solutions x to the system of congruences:

$$\begin{cases} x \equiv a \ (\theta), \\ x \equiv b \ (\theta'). \end{cases}$$

Condition (2.5) insures that solutions exist, and (2.4) promises uniqueness.

Up to isomorphism there is a one-to-one correspondence between outer direct products, factor bands and complementary factor congruences, as stated next.

2.5. THEOREM. (a) *An algebra decomposes as a product,* $\boldsymbol{P} \cong \boldsymbol{A} \times \boldsymbol{A}'$, *if, and only if,* $\boldsymbol{P}$ *has a pair of complementary factor congruences* θ *and* θ' *such that*

$$\boldsymbol{A} \cong \frac{\boldsymbol{P}}{\theta} \text{ and } \boldsymbol{A}' \cong \frac{\boldsymbol{P}}{\theta'}.$$

(b) *In any algebra, via its product decompositions in part (a), factor bands* β *correspond one-to-one to pairs* $\langle \theta, \theta' \rangle$ *of complementary factor congruences:*
(1) $\beta(a,b) = c$ *if, and only if,* $a \,\theta\, c$ *and* $c \,\theta'\, b$;
(2) $a \,\theta\, b$ *if, and only if,* $\beta(a,b) = b$; $a \,\theta'\, b$ *if, and only if,* $\beta(a,b) = a$.

PROOF. See [**McMcT87**, Theorem 4.33]. □

Perhaps the meaning of this proposition should be amplified: if a factor band β comes from a pair of complementary factor congruences θ and θ' by (b1) and new factor congruences $\widehat{\theta}$ and $\widehat{\theta}'$ are subsequently defined by (b2), then $\widehat{\theta} = \theta$ and $\widehat{\theta}' = \theta'$. And vice versa, starting with a factor band and going full circle via (b2) on through (b1) gives back the original band. Note that the trivial band β in which $\beta(a,b) = a$ for all a and b corresponds to the trivial congruences 0 and 1.

Observe that $\mathbf{Con}(\mathbb{Z}_2 \times \mathbb{Z}_2)$ depends on whether we are talking about groups or rings. For groups it is isomorphic to the lattice $\boldsymbol{M}_3$, whereas for rings it is isomorphic to $\boldsymbol{M}_2$ (defined in Sect. 1). This seemingly innocuous discrepancy will be significant in later chapters.

2.6. EXERCISE. For any factor band β and its corresponding complementary factor congruences θ and θ' in an algebra $\boldsymbol{A}$, prove that

$$\beta(a,b) = \beta(c,d) \text{ if, and only if, } a \,\theta\, c \text{ and } b \,\theta'\, d \qquad (a,b,c,d \in A)$$

Interpret this as showing that the operation table of any factor band has a characteristic appearance. Namely, it breaks up into rectangular blocks, not necessarily contiguous, with each block containing one element of A and each element of A appearing in one block.

2.7. EXERCISE. Show that any rectangular band β on a set A is Abelian:

$$\beta(a,c) = \beta(a,d) \Rightarrow \beta(b,c) = \beta(b,d),$$
$$\beta(a,c) = \beta(b,c) \Rightarrow \beta(a,d) = \beta(b,d).$$

But find an Abelian binary function that is not a rectangular band. (For the general meaning of 'Abelian', see Definition V.3.10).

2.8. PROBLEM. When are factor bands term-operations or polynomials?

As there are outer projections, so there are inner ones, $\mu, \mu': A \to A$ within an algebra $\boldsymbol{A}$, to be called 'factor sesquimorphisms'; but only in special cases are these endomorphisms, such as in rings, groups and lattices. We may create them out of a factor band β of $\boldsymbol{A}$ and an element o:

$$\mu(a) = \beta(a, o) \qquad (a \in A), \tag{2.6}$$

$$\mu'(a) = \beta(o, a) \qquad (a \in A). \tag{2.7}$$

Here, o is a fixed element of A. From the properties of a factor bands, one easily verifies the following properties of μ and μ'.

2.9. Proposition. *For a factor band β of an algebra $\boldsymbol{A}$ with the sesquimorphisms μ and μ' as defined by (2.6) and (2.7):*

(a) $\mu(\mu(a)) = \mu(a)$, *and* $\mu'(\mu'(a)) = \mu'(a)$;
(b) $\mu(\mu'(a)) = \mu'(\mu(a))$;
(c) *for any n-ary operation ω of $\boldsymbol{A}$,*
  (1) $\mu(\omega(\mu(a_1), \ldots, \mu(a_n)) = \mu(\omega(a_1, \ldots, a_n))$ *and*
  (2) $\mu'(\omega(\mu'(a_1), \ldots, \mu'(a_n))) = \mu'(\omega(a_1, \ldots, a_n))$;
(d) *if $\mu(a) = \mu(b)$ and $\mu'(a) = \mu'(b)$, then $a = b$;*
(e) *for all a and b there is an x such that $\mu(x) = \mu(a)$ and $\mu'(x) = \mu'(b)$.*

2.10. Definition. Any two functions, $\mu, \mu': A \to A$, satisfying conditions (a)–(e) of this proposition will be called **complementary factor sesquimorphisms** of an algebra $\boldsymbol{A}$. And any function μ on the carrier of an algebra for which there exists another function μ' that satisfies these properties will be called a **factor sesquimorphism**. (See the previous section for an introduction to sesquimorphisms.)

Here are three examples of factor sesquimorphisms – three other examples of sesquimorphisms were given after Definition 1.5. Jónsson and Tarski algebras (Definition VII.3.17) exhibit factor sesquimorphisms that are idempotent endomorphisms (see [**JónTa47**] and [**McMcT87**, p. 283]). In a categorical setting, [**Hofm72**, p. 323] has factor sesquimorphisms, which are also endomorphisms. In the context of identities, factor sesquimorphisms are to be found in [**Knoe73, Knoe82**]; these are not necessarily endomorphisms.

Note that the conditions (a)–(e) in Proposition 2.9 do not presuppose a fixed element o. But they do yield such an element.

2.11. Proposition. *In an algebra $\boldsymbol{A}$ with complementary factor sesquimorphisms, μ and μ', there is an element o of A such that:*

$$\mu(\mu'(a)) = o = \mu(o) = \mu'(o) \qquad (a \in A).$$

Additionally,

$$\mu(\mu'(a)) = \mu(\mu'(b)) = \mu'(\mu(a)) = o \qquad (a, b \in A).$$

PROOF. To see that $\mu\mu'a = \mu\mu'b$ for any a and b in A, realize by property (e) that there exists an x such that $\mu x = \mu a$ and $\mu' x = \mu' b$, and hence

$$\mu\mu'a = \mu'\mu a = \mu'\mu x = \mu\mu' x = \mu\mu' b.$$

Call this fixed element o. Then $\mu o = \mu\mu\mu' a = \mu\mu' a = o$, and likewise $\mu' o = o$. □

Call this o the **origin** of this particular pair of sesquimorphisms. It is convenient to have an origin common to all the sesquimorphisms chosen to factor an algebra. To that end, in analogy with pointed spaces in topology, define a **pointed algebra** to be a pair $\langle \boldsymbol{A}, o \rangle$ where $\boldsymbol{A}$ is an algebra and o is any element of A, called the **origin**. An origin will also be needed to define factor ideals and elements. Its choice is arbitrary in general but for specific systems such as rings it is best to choose the nullity, and for groups the unity. Choosing the origin of $\mathbb{Z}_{12}$ to be 0 in our running example, we find from β that $\mu(m) = 4m$ and $\mu'(m) = 9m$, so that multiplication effects the action of a sesquimorphism. Factor sesquimorphisms correlate well with other factor objects; collate the next theorem with Theorem 2.5.

2.12. THEOREM. *Let $\langle \boldsymbol{A}, o \rangle$ be a pointed algebra. There is a one-to-one correspondence between factor bands β and pairs $\langle \mu, \mu' \rangle$ of complementary factor sesquimorphisms with origin o, and another one between these $\langle \mu, \mu' \rangle$ and pairs $\langle \theta, \theta' \rangle$ of complementary factor congruences of $\boldsymbol{A}$. This means that, with formulas (a)–(d) below, factor objects on the left may be defined uniquely in terms of those on the right. Going full circle returns us to the original factor objects. For example, going from a factor band to a pair of sequimorphisms by (b) and then returning by (a) gives back the same band:*

(a) *$\beta(a, b) = c$ if, and only if, $\mu(c) = \mu(a)$ and $\mu'(c) = \mu'(b)$;*
(b) *$\mu(a) = \beta(a, o)$ and $\mu'(a) = \beta(o, a)$;*
(c) *$a\, \theta\, b$ iff $\mu(a) = \mu(b)$ and $a\, \theta'\, b$ iff $\mu'(a) = \mu'(b)$;*
(d) *$\mu(a)$ is that unique x such that $x\, \theta\, a$ and $x\, \theta'\, o$, and $\mu'(a)$ is that unique y such that $y\, \theta'\, a$ and $y\, \theta\, o$.*

Also for a factor band β with corresponding sesquimorphisms μ and μ':

(e) *$\mu\beta(a, b) = \mu(a)$ and $\mu'\beta(a, b) = \mu'(b)$;*
(f) *$\beta(\mu(a), \mu'(a)) = a$.*

PROOF. It is long but elementary, and so it is not given. Hint: it is easiest to prove (e) and (f) first. □

This theorem is in contrast to the last section where a transversal as a well as a congruence was needed in order to uniquely determine a sesquimorphism. Note that in a pointed algebra $\langle \boldsymbol{A}, o \rangle$ the trivial band β, in which $\beta(a, b) = a$ for all a and b, has the trivial sesquimorphisms, $\mu(a) = a$ and $\mu'(a) = o$.

2.13. EXERCISE. In the group $(\mathbb{Z}_2)^2$ there are three nontrivial decompositions as a product. Find the corresponding factor bands and factor sesquimorphisms, assuming $\langle 0, 0 \rangle$ is the origin.

In JT-algebras [**JónTa47**] (see Definition VII.3.17), factor sesquimorphisms are endomorphisms. Generally, this is not the case; for example, complementation in Boolean algebras complicates the situation. But when the origin is a one-element subalgebra, this is so, and conversely, as shown next.

2.14. PROPOSITION. *For the factor sesquimorphisms of a pointed algebra* $\boldsymbol{A}$ *with origin* o*, the following are equivalent:*

(a) *The singleton* $\{o\}$ *is a subalgebra of* $\boldsymbol{A}$*.*
(b) *All factor sesquimorphisms are endomorphisms of* $\boldsymbol{A}$*.*
(c) *At least one pair of complementary factor sesquimorphisms are endomorphisms of* $\boldsymbol{A}$*.*
(d) *The constant sesquimorphism,* $x \mapsto o$*, is an endomorphism of* $\boldsymbol{A}$*.*

PROOF. (a) $\Rightarrow$ (b). Use (c) and (d) of Proposition 2.9 to show that $\mu\omega\vec{a} = \omega\mu\vec{a}$. One also needs Proposition 2.11.

(b) $\Rightarrow$ (d). Obvious

(d) $\Rightarrow$ (c). The complement of the constant sesquimorphism is the identity map, and it is obviously an endomorphism.

(c) $\Rightarrow$ (a). That $\mu\mu' a = o$ for any a implies, for a generic operation ω with the repeated argument o and endomorphisms μ and μ', that

$$\omega\dot{o} = \omega(\mu'\mu o, \mu'\mu o, \ldots, \mu'\mu o) = \mu'\mu\omega\dot{o} = o. \qquad \square$$

Rings and groups have an obvious origin in the one-element subalgebra 0; hence all factor sesquimorphisms are endomorphisms. Lattices are another good example of this proposition, where any element may be taken as the origin, proving that all factor sesquimorphisms are endomorphisms.

2.15. EXERCISE. Consider the 4-element Boolean algebra $\boldsymbol{B}$ on the set $\{0, a, b, 1\}$, and write out the operation table for a factor band β corresponding to the nontrivial factorization of $\boldsymbol{B}$. Be perverse by choosing a to be the origin o, and describe the corresponding factor sesquimorphisms. Compare with Exercise 1.9.

2.16. EXERCISE. Do an exercise similar to the preceding for the ring $\mathbb{Z}_6$ with 2 taken as the origin o; write out the operation tables for the relativized ring on $\{2, 5\}$.

2.17. EXERCISE. Let β be the factor band for a product $\boldsymbol{A} \times \boldsymbol{B}$, and choose an origin $\langle a_o, b_o \rangle$. For the corresponding sesquimorphisms μ and μ' show that $\mu(\langle a, b \rangle = \langle a, b_o \rangle$ and $\mu(\langle a, b \rangle = \langle a_o, b \rangle$.

2.18. PROBLEM. Which of the axioms for complementary factor sesquimorphisms are independent of the rest?

As a projection, what does a factor sesquimorphism project onto? Answer: its image can be turned into an algebra isomorphic to a factor. To do this, as in Definition 1.8, confine each operation ω of an algebra $\boldsymbol{A}$ to the subset, $\mu(A) = \{\mu(a) \mid a \in A\}$, and evaluate it by restricting it:

$$\omega^{\mu(A)}(a_1, \ldots, a_n) = \mu(\omega^{\boldsymbol{A}}(a_1, \ldots, a_n)) \quad (a_1, \ldots, a_n \in \mu(A));$$

designate the resulting algebra $\mu(\boldsymbol{A})$. Then call the pair, $\mu(\boldsymbol{A})$ and $\mu'(\boldsymbol{A})$, **complementary factor ideals** of the algebra $\boldsymbol{A}$ whenever μ and μ' are complementary factor sesquimorphisms. That these are ideals in the sense of Sect. 1 will surface in the next proposition. An axiomatic definition, independent of sesquimorphisms, will have to wait till there is more structure on $\boldsymbol{A}$, as in Chap. VII. Also $x \in \mu(A)$ iff $\mu(x) = x$, since μ is idempotent.

In our running example where $\mathbb{Z}_{12} \cong \mathbb{Z}_3 \times \mathbb{Z}_4$ and $\beta(r, s) = 4r + 9s$, we have that $\mu(\mathbb{Z}_{12}) = \{0, 4, 8\}$ and $\mu'(\mathbb{Z}_{12}) = \{0, 3, 6, 9\}$. The ring operations of $+$ and $\times$, when confined to these factor ideals, do not change, thanks to the ring identities. For the constants, 0 stays the same, but 1 becomes 4 in $\mu(\mathbb{Z}_{12})$ and 9 in $\mu'(\mathbb{Z}_{12})$. See Fig. 2.

Classically, in the presence of a group operation, factor ideals are known as 'direct summands'. Thus, inner direct products might just as well be called sums, and often are, but we have chosen a term consistent with the previously defined outer direct products of algebras.

The next two results use the commutativity of congruences to connect complementary factor ideals, μA and $\mu' A$, with products and with the transversals defined in the last section, thereby showing that μA and $\mu' A$ are truly ideals in the sense of that section. The proof of the first result is straightforward.

2.19. THEOREM. *Let μ and μ' be complementary factor sesquimorphisms corresponding to the factor band β of a pointed algebra $\boldsymbol{A}$ with an origin o. Then, μ and μ' are homomorphisms from $\boldsymbol{A}$ onto $\mu(\boldsymbol{A})$ and $\mu'(\boldsymbol{A})$, respectively; and $\boldsymbol{A} \cong \mu(\boldsymbol{A}) \times \mu'(\boldsymbol{A})$ by the isomorphism $a \mapsto \langle \mu(a), \mu'(a) \rangle$, with inverse β. Further, for the corresponding factor congruences θ and θ',*

$$\frac{\boldsymbol{A}}{\theta} \cong \mu(\boldsymbol{A}) \text{ with } \mu(A) = \frac{o}{\theta'}; \text{ and } \frac{\boldsymbol{A}}{\theta'} \cong \mu'(\boldsymbol{A}) \text{ with } \mu'(A) = \frac{o}{\theta}.$$

2.20. PROPOSITION. *Let $\boldsymbol{A}$ be a pointed algebra with an origin o. Two congruences, θ and η, are complementary factor congruences if, and only if,*

(a) $\theta \circ \eta = \eta \circ \theta$,
(b) $\theta \cap \eta = 0_{\mathbf{Con}\,\boldsymbol{A}}$,
(c) *o/θ is a transversal of η,*
(d) *o/η is a transversal of θ.*

PROOF. $\Rightarrow$. Clear.

$\Leftarrow$. We need only to prove that $\theta \circ \eta = 1_{\mathbf{Con}\,\boldsymbol{A}}$. Let μ be the sesquimorphism coming from the congruence θ and the transversal o/η, and let λ come from η and o/θ, as in Sect. II.1. For arbitrary a and b in A,

$$a \ \theta \ \mu(a) \ \eta \ o \ \theta \ \lambda(b) \ \eta \ b;$$

thus, $a \ (\theta \vee \eta) \ b$. By commutativity, $\theta \vee \eta = \theta \circ \eta$; hence $\theta \circ \eta = 1_{\mathbf{Con}\,\boldsymbol{A}}$. □

In general, condition (a) of Proposition 2.20 is necessary; the semilattice $\boldsymbol{SL}_3$ on three elements that is not a chain demonstrates this. Of course, (a) is automatically fulfilled in varieties with commuting congruences.

2.21. EXERCISE. Groups and vector spaces, which have natural origins, are a good place to observe factor bands and other factor objects. In their common generalization, groups with operators (see Sect. 1), factor sesquimorphisms become endomorphisms of a special kind and factor ideals become normal subgroups.

(a) Interpret our language and propositions into that of groups with operators. For example, two normal subgroups M and N of a group $\langle G; \times, 1, \ldots, \omega, \ldots \rangle$ with operators form an inner direct product if, and only if,
   (1) $M \times N = G$,
   (2) $M \cap N = \{1\}$.
(b) Prove that two functions $\mu, \mu' \colon V \to V$ on a vector space $\boldsymbol{V}$ are complementary factor sesquimorphisms if, and only if, μ and μ' are idempotent linear transformations such that for all v in V:
   (1) $\mu(v) + \mu'(v) = v$, and
   (2) $\mu(\mu'(v)) = \mu'(\mu(v))$.
(c) Show that any factor band β of a vector space $\boldsymbol{V}$ is of the form:

$$\beta(v, w) = Mv + Nw \qquad (v, w \in V)$$

for linear transformations M and N such that $M^2 = M$, $N^2 = N$, $M + N = 1$, and $MN = 0 = NM$. And conversely, show that any binary function of this form on a vector space is a factor band.

Although $\mu(\boldsymbol{A})$ and $\mu'(\boldsymbol{A})$ appear to form what might be called an 'inner product', there is no longer, in general, a one-to-one correspondence between pairs of complementary factor ideals and the previous factor objects, even up to isomorphism or a choice of the origin. To illustrate, consider the set $\{0, 1, \ldots, 5\}$ as an algebra $\boldsymbol{A}$ with no operations but with an origin 0. It may be factored in two different ways yielding the same factor ideals, $\{0, 1\}$ and $\{0, 2, 4\}$.

1	3	5
0	2	4

and

1	5	3
0	2	4

In view of this unfortunate insufficiency of factor ideals to capture inner products uniquely, it seems reasonable to define this concept with sesquimorphisms.

2.22. DEFINITION. An **inner direct product** in an arbitrary algebra is a pair of complementary factor sesquimorphisms. Their images become a pair of algebras, although not subalgebras, whose product is isomorphic to the original algebra.

This is as close as we may get to the classical notion of inner product in groups and rings as a pair of subalgebras satisfying certain conditions. Of interest here are JT-algebras [**JónTa47**], where pairs of complementary factor ideals, now subalgebras, do correspond one-to-one to inner products. By Theorem 2.12, any pair of complementary factor congruences θ and θ' in a pointed algebra $\langle \boldsymbol{A}, o\rangle$ gives an inner direct product in which their ideals o/θ and o/θ' are the images of the corresponding sesquimorphisms. In unital shells and half-shells, to be studied intensively later, their sets of factor congruences are Boolean algebras. More generally, we will show in Sect. VI.2 that, whenever the factor congruences form a Boolean algebra, their factor ideals can give back the congruences from which they came. Inner products relate to outer products as follows.

2.23. PROPOSITION. *Let $\langle \boldsymbol{P}, \pi, \pi'\rangle$ be the outer product of the algebras $\boldsymbol{A}$ and $\boldsymbol{A}'$, that is, $\boldsymbol{P} = \boldsymbol{A} \times \boldsymbol{A}'$. Assume that $\boldsymbol{P}$ is pointed with origin o. Further, assume that β is the induced factor band of $\boldsymbol{P}$. Write μ and μ' for the complementary factor sesquimorphisms created by β via the origin. There are isomorphisms, $\boldsymbol{A} \cong \mu(\boldsymbol{P})$ and $\boldsymbol{A}' \cong \mu(\boldsymbol{P}')$, such that the outer product is related to the inner product by this commutative diagram.*

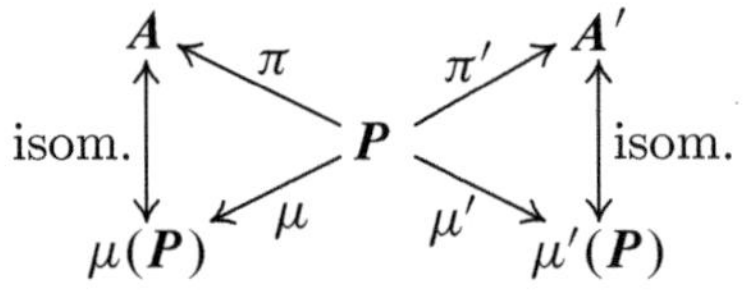

Factor elements are the last kind of factor object to be investigated; these are the counterpart in universal algebra to central idempotents in ring theory. Like factor ideals they apparently need to be defined in terms of factor bands or sesquimorphisms, at least in general.

In order to identify factor elements another fixed element is needed. Call a triple $\langle \boldsymbol{A}, o, t\rangle$ a **doubly pointed algebra** when o and t are any elements of an algebra $\boldsymbol{A}$ – these need not be constant operations; name o the **origin**, as before, and t the **terminus**. Although there are no restrictions on the choice of o and t for now, when we do come to unital rings $\langle R; +, \times, 0, 1\rangle$ and their generalizations, it will be most advantages to assume that o is 0 and t is 1.

There are at least three equivalent ways to define factor elements: through a factor band, through its factor sesquimorphisms, or through the corresponding factor congruences.

2.24. DEFINITION. Call two elements e and e' **complementary factor elements** of a doubly pointed algebra $\langle \boldsymbol{A}, o, t\rangle$ if there is a factor band β of $\boldsymbol{A}$ for which

$$\beta(t,o) = e, \text{ and } \beta(o,t) = e'.$$

With the corresponding sesquimorphisms, $\mu(t) = e$ and $\mu'(t) = e'$. A **factor element** is one of such a complementary pair.[3]

2.25. PROPOSITION. *Let e and e' be complementary factor elements that come from a factor band β in a doubly pointed algebra $\langle \boldsymbol{A}, o, t\rangle$. The factor congruences corresponding to β uniquely define e and e' by the relationships:*

$$o\theta' e \text{ and } e\theta t,$$
$$o\theta e' \text{ and } e'\theta' t.$$

By choosing the terminus to be 1 and the origin 0 in the running example of $\mathbb{Z}_{12}$ with $\beta(a,b) = 4a+9b$, this definition and proposition are illustrated by computing in several ways its complementary factor elements: $e = 4$ and $e' = 9$. In unital rings, factor elements are central idempotents, which generate the corresponding factor ideals as principal ideals. Significantly for the future unfolding of the theory, $\mu(a) = e \times a$ and $\mu'(a) = e' \times a$.

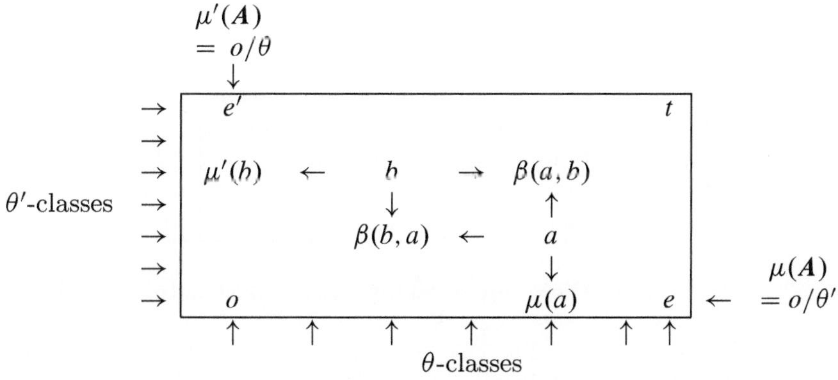

FIGURE 3. An algebra decomposed as a product

Figure 3 is an abstraction to general algebras of the earlier Fig. 2 for a product of two unital rings. The box encloses the elements of a doubly pointed algebra $\boldsymbol{A}$ with origin o and terminus t, sorted into the congruences classes of two complementary congruences, θ and θ' with the θ-classes being the columns and the θ'-classes the rows. In this picture analogous to the

[3]Swamy and Murti [**SwaMu81a**] discuss factor elements in semigroups, where they call them 'central' elements. Central elements are more generally defined as sequences of elements in [**VagSá04**] and [**SánVa09**]. There the origin and terminus become sequences of unary operations.

Cartesian plane, the algebra $\mu\boldsymbol{A}$ ($= o/\theta'$ qua ideal) is the X-axis and $\mu'\boldsymbol{A}$ ($= o/\theta$) the Y-axis. The sesquimorphisms μ, μ' applied to an element a give its 'co-ordinates' and β recovers a by the formula $\beta\big(\mu(a), \mu'(a)\big) = a$. Two arbitrary elements a and b go to $\beta(a, b)$ and $\beta(b, a)$ as indicated. One should take this figure with a few grains of salt, as there may be some equalities and collapsing among o, e, t and e'.

Although we have defined factor elements, they may be useless in general. An example illustrates how there may be no choice for the origin and the terminus so that each pair of complementary factor elements determines the factor band from which they came. Consider the lattice of all finite subsets of an infinite set, with union and intersection being the lattice operations. Then one can show that, for any choice of the origin and terminus, there are an infinite number of factor bands yielding the same pair of complementary factor elements.

One virtue of unital rings is that each of the five kinds of factor objects uniquely characterize products internally. Especially easy to use are factor elements. This will also be true of two-sided unital shells. But, in broader classes of algebras more complicated set-theoretical structures are needed, such as congruence relations or bands. The less structure needed the better: elements are better than subsets, such as ideals, and subsets are better than relations, etc., as shown in Fig. 4.

	Factor object	as a Structure	
Most	band	binary operation	↓
widely	congruence	binary relation	↓
applicable	sesquimorphism	unary function	Most
↑	ideal	set	desirable
↑	element	element	to use

FIGURE 4. Hierarchy of factor objects as structures

2.26. EXERCISE. Can an origin and a terminus be chosen in each vector space so as to create pairs of complementary factor elements via factor bands such that these pairs uniquely determine the factor bands whence they came? Hint: use Exercise 2.21. This was originally a problem, subsequently settled by Diego Vaggione [**Vagg10**].

2.27. PROBLEM. What properties should an algebra or variety possess in order to have complementary factor ideals or elements from which all products may be reconstructed? The shells of Chap. VII are a partial answer. Pedro Sánchez Terraf and Diego Vaggione [**SánVa09**] have a more extensive answer; see Theorem VI.3.11.

Although factor elements play a pivotal role in unital rings, and in shells in subsequent chapters, there is no more to say about them at this point.

Products of more than two factors are possible. Infinite products lead to the important subjects of refinements and subdirect products, discussed only briefly here. As in the case with two factors there are both outer and inner characterizations. Taking to heart remarks made a page ago, we might define inner direct products in terms of sesquimorphisms. However, in anticipation of conventional definitions of refinement to come in Sect. VI.2, we present them equivalently in terms of congruences, leaving their development with sesquimorphisms and infinitary bands to the reader.

2.28. DEFINITION. The **outer direct product** or simply 'direct product' $\prod_{i\in I} \boldsymbol{B}_i$, of the algebras $\boldsymbol{B}_i$ indexed by i in I is an algebra $\boldsymbol{P}$ with carrier $\prod_{i\in I} B_i$ and each n-ary operation ω defined coordinate-wise, as previously done with two factors:

$$\omega^{\boldsymbol{P}}(\langle a_1^i \mid i \in I\rangle, \ldots, \langle a_n^i \mid i \in I\rangle) = \langle \omega^{\boldsymbol{B}_i}(a_1^i, \ldots, a_n^i) \mid i \in I\rangle \qquad (a_j^i \in B_i).$$

If the index set I is empty, then the product has just one element. For two factors (or just a few) we write $\boldsymbol{A} \times \boldsymbol{B}$, etc. If α is a congruence of $\boldsymbol{A}$ and β is a congruence of $\boldsymbol{B}$, their product $\alpha \times \beta$ in $\boldsymbol{A} \times \boldsymbol{B}$ is defined

$$\langle a_1, b_1\rangle \, (\alpha \times \beta) \, \langle a_2, b_2\rangle \text{ if } a_1 \, \alpha \, a_2 \text{ and } b_1 \, \beta \, b_2 \qquad (a_i \in A,\ b_i \in B).$$

A nontrivial algebra is called **directly indecomposable** if it is not isomorphic to a product of two nontrivial algebras.

In order to illuminate the inner factoring of congruences, we develop some definitions and propositions.

2.29. DEFINITION. For a family of homomorphisms, $\varphi_i\colon \boldsymbol{A} \to \boldsymbol{B}_i$ with i in I, with common domain, there is the **canonical homomorphism** φ to the outer direct product:

$$\varphi : \boldsymbol{A} \underset{\text{can.}}{\to} \prod_{i\in I} \boldsymbol{B}_i\colon a \mapsto \langle \varphi_i(a) \mid i \in I\rangle.$$

More generally, assume that $\eta \in \operatorname{Con} \boldsymbol{A}$ and $\eta \subseteq \bigcap \Theta$ for a collection of congruences in $\boldsymbol{A}$ with product, $\boldsymbol{P} = \prod_{\theta\in\Theta} \boldsymbol{A}/\theta$. Its **canonical homomorphism**, $\psi : \boldsymbol{A}/\eta \underset{\text{can.}}{\to} \boldsymbol{P}$, maps $a/\eta \mapsto \langle a/\theta \mid \theta \in \Theta\rangle$.

2.30. PROPOSITION. *Assume $\theta, \eta \in \operatorname{Con} \boldsymbol{A}$. Then, θ and η are complementary factor congruences if, and only if, the canonical homomorphism,*

$$\varphi : \boldsymbol{A} \underset{\text{can.}}{\to} \frac{\boldsymbol{A}}{\theta} \times \frac{\boldsymbol{A}}{\eta},$$

is an isomorphism.

2.31. DEFINITION. An **inner direct product** of a congruence η of an algebra $\boldsymbol{A}$ is a nonempty collection Θ of congruences of $\boldsymbol{A}$ such that $\eta = \bigcap \Theta$ and the canonical homomorphism of quotients is an isomorphism:

$$\frac{\boldsymbol{A}}{\eta} \underset{\text{can.}}{\cong} \prod_{\theta\in\Theta} \frac{\boldsymbol{A}}{\theta}.$$

Abbreviate this as $\eta = \bigsqcap \Theta$. With only two factors, write $\eta = \theta_1 \sqcap \theta_2$. An **inner direct product** of congruences of an algebra $\boldsymbol{A}$ is an inner direct product of $0_{\mathbf{Con}\,\boldsymbol{A}}$.

For the moment we neglect to have an origin, which would return from complementary θ and θ' the sesquimorphisms of a true inner product in the sense of Definition 2.22. The cancellation isomorphism theorem of Sect. 1 allows us to pass easily between inner products of algebras and their congruences. Also, any congruence θ of an inner direct product of an algebra is a factor congruence, that is, θ and $\bigcap(\Theta{\sim}\{\theta\})$ are complementary factor congruences.

2.32. PROPOSITION. *A congruence η of an algebra $\boldsymbol{A}$ is an inner direct product Θ if, and only if, for any family $\langle a_\theta \mid \theta \in \Theta \rangle$ of elements of A, the family of congruences*

$$x \equiv a_\theta \;(\theta) \qquad (\theta \in \Theta)$$

has a unique solution x modulo η; that is, if y is another solution, then $x \equiv y \;(\eta)$.

The next proposition will prove to be useful in Sect. VI.2. Atomic and complete Boolean algebras are defined near the end of Sect. III.4.

2.33. PROPOSITION. *Let Θ be an inner direct product of congruences of an algebra $\boldsymbol{A}$, and consider the collection of all intersections of them:*

$$\overline{\Theta} = \left\{ \bigcap H \mid H \subseteq \Theta \right\}.$$

(a) *Then, $\overline{\Theta}$ is a complete and atomic Boolean lattice of commuting factor congruences of $\boldsymbol{A}$.*
(b) *If $1_{\mathbf{Con}\,\boldsymbol{A}} \notin \Theta$, then $\overline{\Theta}$ is anti-isomorphic to the set of all subsets of Θ; this is given by the correspondence:*

$$\psi\colon \mathcal{P}\Theta \to \overline{\Theta} : H \mapsto \bigcap H,$$

where the lattice operations are transformed:

$$\psi(Z \cup H) = \psi(Z) \cap \psi(H) \qquad (Z, H \subseteq \Theta), \tag{2.8}$$

$$\psi(Z \cap H) = \psi(Z) \vee \psi(H) \qquad (Z, H \subseteq \Theta), \tag{2.9}$$

$$\psi(\varnothing) = 1_{\mathbf{Con}\,\boldsymbol{A}}, \tag{2.10}$$

$$\psi(\Theta) = 0_{\mathbf{Con}\,\boldsymbol{A}}. \tag{2.11}$$

PROOF. (a) That $\overline{\Theta}$ is a complete, atomic Boolean algebra follows from (b) unless $1 \in \Theta$. But this makes no difference, since 1 is already in $\overline{\Theta}$ as $\bigcap\varnothing$.

View this as a problem in the corresponding outer direct products. Any partition of Θ, say H and $\Theta{\sim}H$, produces a product:

$$\prod_{\theta \in \Theta} \frac{\boldsymbol{A}}{\theta} \cong \frac{\boldsymbol{A}}{\bigcap H} \times \frac{\boldsymbol{A}}{\bigcap(\Theta{\sim}H)}.$$

Hence, $\bigcap H$ and $\bigcap(\Theta{\sim}H)$ are complementary factor congruences.

To prove that the congruences of $\overline{\Theta}$ commute, assume for some a and b in A that $a\,(\bigcap Z \circ \bigcap H)\,b$, where $Z, H \subseteq \Theta$; one needs to show that $a\,(\bigcap H \circ \bigcap Z)\,b$. Now there exists an x in A such that $a \bigcap Z\, x$ and $x \bigcap H\, b$. So $a\,\zeta\,x$ when $\zeta \in Z$, and $x\,\eta\,b$ when $\eta \in H$. Because Θ is an inner product, there is a solution y to the system of congruences:

$$y \equiv \begin{cases} b\ (\zeta) & (\zeta \in Z), \\ a\ (\eta) & (\eta \in H{\sim}Z). \end{cases}$$

For any η in $H \cap Z$, then $y\,\eta\,b\,\eta\,x\,\eta\,a$. Consequently, $a \bigcap H\, y \bigcap Z\, b$.

(b) To show that ψ is one-to-one, assume that there are two unequal subsets H and Z of Θ; the object is to prove that $\bigcap H \neq \bigcap N$. Without loss of generality, assume that there is an η_0 in H but not in Z. Since $\eta_0 \neq 1_{\mathbf{Con}\,\boldsymbol{A}}$, there are a and b in A not related by η_0. By solvability in an inner product, there is an x in A such that

$$x \equiv \begin{cases} a\ (\zeta) & (\zeta \in Z), \\ b\ (\eta_0). \end{cases}$$

So $x \equiv a\ (\bigcap Z)$ but $x \not\equiv a\ (\bigcap H)$.

The passage of the lattice operations through ψ is clear except for the second one transforming $\cap$ into $\vee$. An argument similar to but simpler than the one for commutation will do the trick. □

2.34. EXERCISE. Phrase and prove a converse to this proposition.

2.35. PROBLEM. Go beyond factor bands on just two arguments and fashion a theory of products on arbitrary index sets that includes as many as possible of the remaining factor objects.

Throughout this section and the previous one, one sees a general philosophy at work: for any outer concept find a corresponding inner concept. This will reappear in the notion of 'refinement' in Sect. VI.3, where the outer and inner notions diverge.

Decomposing an algebra into a product of other nontrivial algebras is not always possible, even though the algebra may break up in some other way. It might be only the subalgebra of a direct product, as developed next.

2.36. DEFINITION. *An algebra* $\boldsymbol{A}$ *is an* **outer subdirect product** *of a family* $\langle A_i \mid i \in I\rangle$ *of algebras all of the same type if* $\boldsymbol{A}$ *is a subalgebra of the direct product of the family and each projection* π_i *of* $\boldsymbol{A}$ *to each factor* $\boldsymbol{A}_i$ *is surjective. In notation,*

$$\boldsymbol{A} \underset{s.d.}{\subseteq} \prod_{i \in I} \boldsymbol{A}_i .$$

Or more briefly, call this a 'subdirect product' or even just 'subproduct'.

The group $\mathbb{Z}$ is a case in point; it is isomorphic to a subproduct of all groups of prime order:

$$\mathbb{Z} \underset{s.d.}{\subseteq} \mathbb{Z}_2 \times \mathbb{Z}_3 \times \mathbb{Z}_5 \times \cdots .$$

In this example, one maps $\mathbb{Z}$ to $\mathbb{Z}_p$ by projecting each integer z to it: $\pi_p(z) = z \bmod p$. Since each factor is fully utilized, these projections are surjective and we have a subproduct. They have an internal characterization.

2.37. PROPOSITION. *An algebra* $\boldsymbol{A}$ *is an outer subdirect product of a family* $\langle \boldsymbol{A}_i \mid i \in I \rangle$ *of algebras if, and only if, there are congruences* θ_i *of* $\boldsymbol{A}$ *such that*

$$\boldsymbol{A}_i \cong \boldsymbol{A}/\theta_i \quad (i \in I),$$

$$\bigcap \{\theta_i : i \in I\} = 0_{\mathbf{Con}\,\boldsymbol{A}}.$$

This proposition suggests a definition.

2.38. DEFINITION. An **inner subdirect product** in an algebra $\boldsymbol{A}$ is a collection Θ of congruences of $\boldsymbol{A}$ such that their intersection is the identity relation:

$$\bigcap \Theta = 0_{\mathbf{Con}\,\boldsymbol{A}}.$$

2.39. DEFINITION. A nontrivial algebra $\boldsymbol{A}$ is called **subdirectly irreducible** if, for any inner subdirect product in $\boldsymbol{A}$, at least one of the congruences θ_i is already the identity relation. Likewise, a congruence θ of an algebra $\boldsymbol{A}$ is **subdirectly irreducible** if $\boldsymbol{A}/\theta$ is.

This concept is needed to state a well-known and much used result due to Birkhoff [**Birk44**]; see also [**BurSa81**, Sect. II.8].

2.40. THEOREM. *Every nontrivial algebra is a subdirect product of subdirectly irreducible algebras.*

Thus subdirect decompositions have the advantage that they always exist. However, the quotient algebras may be too small for a particular purpose; also it may be impossible to conveniently specify the subalgebra of the product. Direct products on the other hand may be too coarse. For these reasons we seek an intermediate path where we take as initial ingredients a Boolean algebra of congruences, and then take suprema of maximal ideals of these; these suprema are again congruences, but usually not factor congruences. The quotient algebras coming from these suprema have a trivial intersection, thereby yielding a subdirect product, and even better, these stalks, as they will be called, will bind together topologically to form a sheaf in Theorem V.2.1.

A Venn diagram of algebras relates different kinds of factors.

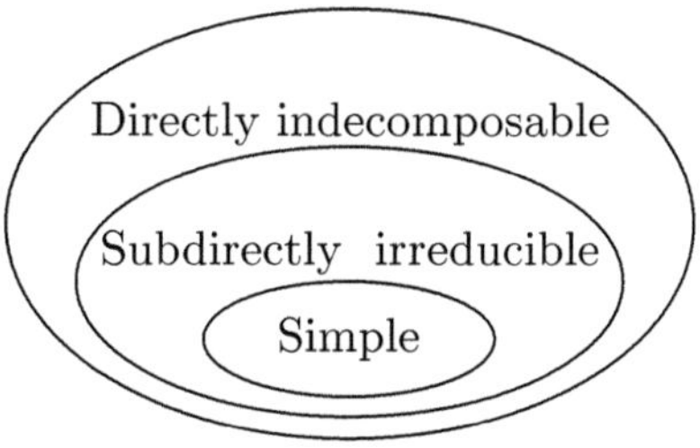

These inclusions are all proper. The three-element semilattice $\boldsymbol{SL}_3$ that is not a chain is directly indecomposable as a product but it is subdirectly reducible. And the group $\mathbb{Z}_4$ is subdirectly irreducible but not simple.

2.41. PROBLEM. In a subdirect product of two algebras, a factor band may be approached as for a product, but it is now a partial function, not defined for all pairs of arguments. Carefully define, if possible, this new concept of a 'subdirect band' so that it corresponds one-to-one to subdirect products of any number of algebras. Likewise, can these subdirect definitions and propositions be rephrased in terms of sesquimorphisms?

There is one last construction needed for making sheaves in Chap. IV.

2.42. DEFINITION. For a family $\{\boldsymbol{A}_x \mid x \in X\}$ of algebras of the same type, consider the disjoint union of their carriers: $\mathcal{A} = \biguplus_{x\in X} A_x$. It is not necessary for the components to be disjoint to start with. To make them disjoint, employ the set construction:

$$\biguplus_{x\in X} A_x = \{\langle x, a\rangle \mid x \in X \text{ and } a \in A_x\}.$$

This **disjoint union** has the natural structure of a partial algebra of the same type, $\boldsymbol{\mathcal{A}} = \biguplus_{x\in X} \boldsymbol{A}_x$: for a generic operation ω with n arguments, when $a_1, \ldots, a_n \in A_x$ for some x in X,

$$\omega^{\boldsymbol{\mathcal{A}}}(\langle x, a_1\rangle, \ldots, \langle x, a_n\rangle) = \langle x, \omega^{\boldsymbol{A}_x}(a_1, \ldots, a_n)\rangle;$$

otherwise, it is undefined.

Nullary operations, that is, constants in the type, pose a problem. By fiat, we postulate that they are undefined in the disjoint union. We also allow the index set X to be empty, whence the disjoint union is also empty; so partial algebras in this case may have an empty carrier. This trivial allowance will be useful in capturing the one-element algebra by a sheaf.

In Chap. VII much more will be said about all the internal factor objects in the context of shells with binary operations and constants. Inner products, as sums of ideals, will capture external products uniquely. Each kind of factor object will be defined independently of the others, and directly in terms of the operations of the shell.

Recall that out of the operation-symbols ω_i of a given type we can build **terms** or what are sometimes called 'words' or 'polynomial symbols'; we designate these compositions as $t(x_1, x_2, \ldots)$, being casual sometimes about the number of arguments, except that crucially it is always finite.

1.1. DEFINITION. An **identity** is a pair of terms, t_1 and t_2, each with the same variables, and denoted suggestively by $t_1 \approx t_2$.

Formulas in the language are built from identities, the Boolean connectives, and universal and existential quantifiers. For example, in the theory of fields, a typical sentence defines solvability:

$$\forall a, b \; \big(a \neq 0 \Rightarrow \exists x \; (a \times x = b)\big).$$

Deduction is should be familiar to the reader. An important part of it is equational deduction, in which there are five rules: reflexivity, symmetry, transitivity, substitution of terms for variables, and replacement of a term by one equated to it already by an identity.

We assume that the reader knows the recursive definition of the interpretation $t^{\boldsymbol{A}}$ of terms t within an algebra $\boldsymbol{A}$. The interpretation $t^{\boldsymbol{A}}$ is called a **term-operation** (often the superscript $\boldsymbol{A}$ will be dropped). An m-ary **polynomial** p of an algebra $\boldsymbol{A}$ is an n-ary term-operation t with $m \leq n$ in which the last variables beyond x_m are replaced by elements of $\boldsymbol{A}$:

$$p(x_1, \ldots, x_m) = t(x_1, \ldots, x_m, a_{m+1}, \ldots, a_n).$$

The set of all term-operations is the **clone** of $\boldsymbol{A}$, designated $\operatorname{Clo} \boldsymbol{A}$. The algebra $\boldsymbol{A}$ **satisfies** an identity $t_1 \approx t_2$ on n variables if $t_1^{\boldsymbol{A}} = t_2^{\boldsymbol{A}}$, that is,

$$t_1(a_1, a_2, \ldots, a_n) = t_2(a_1, a_2, \ldots, a_n) \qquad (a_1, a_2, \ldots, a_n \in A);$$

this may be written as

$$\boldsymbol{A} \models t_1 \approx t_2. \tag{1.1}$$

This relation $\models$ is a polarity between algebras and identities, and the two closure operators obtained from it are significant [**Birk67**].

The set of all identities satisfied by an algebra $\boldsymbol{A}$ is denoted

$$\operatorname{Id} \boldsymbol{A} = \{t_1 \approx t_2 \mid \boldsymbol{A} \models t_1 \approx t_2\}.$$

In analogy with modular arithmetic, we also write

$$t_1 \equiv t_2 \quad (\operatorname{Id} \boldsymbol{A}),$$

for (1.1). Similarly, a set $\mathfrak{A}$ of algebras satisfies an identity if all algebras in it do so, and $\operatorname{Id} \mathfrak{A}$ designates the set of all such:

$$\operatorname{Id} \mathfrak{A} = \{t_1 \approx t_2 \mid \boldsymbol{A} \models t_1 \approx t_2 \text{ when } \boldsymbol{A} \in \mathfrak{A}\}.$$

An algebra is considered to be a **model** of a set I of identities if the algebra satisfies all of them. Write $\operatorname{Mod} I$ for the set of all models of I, called a **variety** or **equational class**. The two-fold expression, $\operatorname{Id} \operatorname{Mod} I$, turns out to be the equational closure of the set I of identities. The **equational** or

deductive closure of I is the smallest set of identities in the given type that is closed to the five rules of equational deduction.

Turned around, $\mathrm{Mod}\,\mathrm{Id}\,\mathfrak{A}$, for any set $\mathfrak{A}$ of algebras, is most importantly $\mathsf{HSP}\mathfrak{A}$, the **semantical closure** of $\mathfrak{A}$.[1] Abbreviate this as $\mathrm{Var}\,\mathfrak{A}$, the **variety** generated by $\mathfrak{A}$, that is, the smallest set of algebras containing $\mathfrak{A}$ and closed to taking homomorphic images H, subalgebras S, and products P. A variety $\mathfrak{V}$ is **finitely generated** if it is generated by a finite set $\mathfrak{A}$ of finite algebras, that is, $\mathfrak{V} = \mathsf{HSP}\mathfrak{A}$. Occasionally, we also need to close up a set $\mathfrak{A}$ by adding all isomorphic images $\mathsf{I}\mathfrak{A}$.

Note that identities are preserved by products:

$$\mathrm{Id}\,(\boldsymbol{A} \times \boldsymbol{B}) = \mathrm{Id}\,\boldsymbol{A} \cap \mathrm{Id}\,\boldsymbol{B}.$$

This is also true for infinite products. If $\boldsymbol{A}$ is a homomorphic image of an algebra $\boldsymbol{B}$, then $\mathrm{Id}\,\boldsymbol{A} \supseteq \mathrm{Id}\,\boldsymbol{B}$. Also, if $\boldsymbol{A}$ is a subalgebra of $\boldsymbol{B}$, then $\mathrm{Id}\,\boldsymbol{A} \supseteq \mathrm{Id}\,\boldsymbol{B}$. Most useful for us is preservation of identities under subdirect products.

1.2. PROPOSITION. *If $\boldsymbol{A}$ is a subdirect product of $\boldsymbol{B}_i$ $(i \in I)$, then*

$$\mathrm{Id}\,\boldsymbol{A} = \bigcap_{i \in I} \mathrm{Id}\,\boldsymbol{B}_i.$$

In sheaves over Boolean spaces, sentences more general than identities are preserved (see Proposition V.2.7).

As an example of these notions, we say that two algebras $\boldsymbol{A}$ and $\boldsymbol{B}$ of the same type are **independent** if there is a binary term φ in their type such that

$$\varphi(x, y) \equiv \begin{cases} x & (\mathrm{Id}\,\boldsymbol{A}); \\ y & (\mathrm{Id}\,\boldsymbol{B}). \end{cases} \tag{1.2}$$

The rings $\mathbb{Z}_3$ and $\mathbb{Z}_4$ are independent, as witnessed by

$$\varphi(x, y) = 4x + 9y.$$

This is a factor band, which satisfies (2.1–2.3) of Chap. II. It suggests the following.

1.3. PROPOSITION. *Two algebras $\boldsymbol{A}$ and $\boldsymbol{B}$ are independent by a binary term φ if, and only if, the factor band β for their product is a term-operation. In either direction of implication, $\varphi^{\boldsymbol{A}\times\boldsymbol{B}} = \beta$.*

PROOF. $\Rightarrow$. We assume (1.2) for the product of $\boldsymbol{A}$ and $\boldsymbol{B}$. Recall the coordinate-wise definition of operations in a product:

$$\varphi^{\boldsymbol{A}\times\boldsymbol{B}}(\langle a_1, b_1\rangle, \langle a_2, b_2\rangle) = \langle \varphi^{\boldsymbol{A}}(a_1, a_2), \varphi^{\boldsymbol{B}}(b_1, b_2)\rangle. \tag{1.3}$$

By independence, the right side is $\langle a_1, b_2\rangle$. So $\varphi^{\boldsymbol{A}\times\boldsymbol{B}}$ is the factor band for $\boldsymbol{A} \times \boldsymbol{B}$.

---

[1] See [**Birk35**] and [**BurSa81**, Sect. II.11].

$\Leftarrow$. The factor band β satisfies (1.3) when φ is replaced by β. Realizing that the left side is just $\langle a_1, b_2\rangle$ by virtue of β being a factor band, and separating the coordinates, we see that $a_1 = \beta^{\boldsymbol{A}}(a_1, a_2)$ and $b_2 = \beta^{\boldsymbol{B}}(b_1, b_2)$. Hence, β witnesses independence. □

Independence leads to the product of varieties.

1.4. DEFINITION. Let $\mathfrak{V}$, $\mathfrak{V}_1$ and $\mathfrak{V}_2$ be three varieties of the same type. Their **join**, $\mathfrak{V} = \mathfrak{V}_1 \vee \mathfrak{V}_2$, is the smallest variety including both. It is seen to be $\operatorname{Mod}(\operatorname{Id}\mathfrak{V}_1 \cap \operatorname{Id}\mathfrak{V}_2)$. We say that $\mathfrak{V}_1$ and $\mathfrak{V}_2$ are **independent** if there is a binary term φ such that

$$\varphi(x, y) \equiv \begin{cases} x & (\operatorname{Id}\mathfrak{V}_1), \\ y & (\operatorname{Id}\mathfrak{V}_2). \end{cases} \tag{1.4}$$

If $\mathfrak{V}_1$ and $\mathfrak{V}_2$ are independent and their join is $\mathfrak{V}$, we say that $\mathfrak{V}$ is a **varietal product** of $\mathfrak{V}_1$ and $\mathfrak{V}_2$, and write $\mathfrak{V} = \mathfrak{V}_1 \otimes \mathfrak{V}_2$. It follows that $\varphi^{\boldsymbol{A}}$ is a factor band decomposing each algebra $\boldsymbol{A}$ of $\mathfrak{V}$:

$$\boldsymbol{A} \cong \boldsymbol{A}_1 \times \boldsymbol{A}_2 \qquad (\boldsymbol{A}_1 \in \mathfrak{V}_1 \text{ and } \boldsymbol{A}_2 \in \mathfrak{V}_2).$$

Mal'cev conditions are another example of the interplay between terms and models. Without defining them in general, we give two illustrations sufficient for our purposes (for more, see [**BurSa81**, Sects. 2.12] and [**McMcT87**, 4.12]).

1.5. DEFINITION. A variety $\mathfrak{V}$ has **commuting congruences** (or is **congruence-commutable**) if for all algebras $\boldsymbol{A}$ in $\mathfrak{V}$,

$$\eta \circ \theta = \theta \circ \eta \qquad (\eta, \theta \in \operatorname{Con}\boldsymbol{A}).$$

A variety $\mathfrak{V}$ has **distributive congruences** if for all algebras $\boldsymbol{A}$ in $\mathfrak{V}$,

$$\zeta \cap (\eta \vee \theta) = (\zeta \vee \eta) \vee (\zeta \vee \theta) \qquad (\zeta, \eta, \theta \in \operatorname{Con}\boldsymbol{A}).$$

1.6. PROPOSITION. *Let $\mathfrak{V}$ be a variety of algebras.*

(a) *$\mathfrak{V}$ has commuting congruences if, and only if, there is a ternary term t such that*

$$t(x, x, y) \approx y \approx t(y, x, x) \quad (\operatorname{Id}\mathfrak{V}).$$

(b) *$\mathfrak{V}$ has distributive congruences if there is a ternary term t such that*

$$t(x, x, y) \approx y \approx t(x, y, x) \approx t(y, x, x) \quad (\operatorname{Id}\mathfrak{V}).$$

Groups and rings illustrate (a) and lattices (b).

Two algebras, $\boldsymbol{A}$ and $\boldsymbol{B}$, with the same carrier but not necessarily of the same type, are said to be **term-equivalent** or **equationally interdefinable** if for each operation ω of $\boldsymbol{A}$ there is a term-operation t of $\boldsymbol{B}$ such that $\omega^{\boldsymbol{A}} = t^{\boldsymbol{B}}$, and vice versa. Two varieties $\mathfrak{V}$ and $\mathfrak{W}$ are **term-equivalent** if there is a one-to-one correspondence between the algebras of $\mathfrak{V}$ and $\mathfrak{W}$ such that corresponding algebras are term-equivalent with the choice of terms corresponding to operation symbols being uniform over all algebras in the varieties. Two algebras, on the same carrier A but perhaps dissimilar in type, are said to be **polynomially equivalent** if they are term-equivalent when all elements of A are adjoined as constant operations in both algebras; in other words, they have the same polynomials. Less than term-equivalence is term-reduction: $\boldsymbol{A}$ is a **term-reduct** of $\boldsymbol{B}$ if every operation of $\boldsymbol{A}$ is a term-operation of $\boldsymbol{B}$.

Our underlying set theory is Zermelo–Fraenkel set theory with the axiom of choice, together with Alexander Grothendieck's axiom of universes, which asserts the existence of ever larger universes [**Gabr62**, p. 328]). This rather powerful axiom has the virtue of formalizing the informal sequence of successively larger entities: sets, classes, conglomerates, cartels – each an element of the next, as suggested by Horst Herrlich and George Strecker [**HerSt79**, Appendix]. Very importantly, it avoids the need to prove anew elementary results in the customary proper classes and beyond that are already known and proven for sets, since now every thing is a set. For example, traditional varieties may be intersected and joined, thereby creating a 'lattice'; this lattice would have been a conglomerate, but now ordinary theorems about lattices hold without further ado. Consult the books of Bergman [**Berg98**, Sect. 6.4] and Arthur Kruse [**Krus69**] for fuller expositions.

A slight disadvantage of Grothendieck's axiom is that it makes ambiguous the phrase 'variety of groups'. Does this mean the variety of all groups in some universe, or does it mean a subvariety of this, that is, a collection of groups satisfying certain identities? We sidestep this question by using '*a* variety of groups' to mean the latter, and the former is phrased '*the* variety of all groups in a universe', but where we will usually drop 'in a universe'. We have already used this convention in defining varieties.

Jan Mycielski [**Myci06**] proposes a significantly different but powerful set theory, which could also be used for the foundation of this book.

The axiom of choice will typically be used in the form of Zorn's lemma.

1.7. LEMMA. *If each chain in a partially ordered set has an upper bound, then there is a maximal member in the set.*

A medley of frequently used conventions and notations in set theory is found near the end of Sect. II.1.

2. Category Theory

There is a well-known story about the origin of category theory. Historically, natural transformations were discovered first, such as the isomorphism of a vector space with its second dual. However, to define natural transformations in general, functors were needed and so invented. In turn, to define functors rigorously, categories were created. But we reverse the historical origins [**EilMa45**], and present this subject axiomatically. This section is mainly a listing of the concepts, notations, and theorems needed in the sequel, with few proofs. We head for the concept of an adjunction between two categories, and even better for later, their equivalence. A reference is [**HerSt79**]. (See Sect. II.1 for a definition of partial algebras.)

2.1. DEFINITION. A **category** $\mathfrak{C}$ is a partial algebra $\langle \mathfrak{C}; \circ, \mathrm{dom}, \mathrm{cod} \rangle$ of type $\langle 2, 1, 1 \rangle$ satisfying these axioms for all a, b, c in $\mathfrak{C}$:

(i) $a \circ (b \circ c) = (a \circ b) \circ c$,
(ii) $a \circ (\mathrm{dom}\, a) = a$ and $(\mathrm{cod}\, a) \circ a = a$,
(iii) $a \circ b$ exists if, and only if, $\mathrm{dom}\, a = \mathrm{cod}\, b$.

These equations are to be read with the understanding for partial algebras that if one side of an equation exists, then so does the other side. Thus, by (ii), the operations 'dom' and 'cod' (to be read *domain* and *codomain*) must be total, that is, defined everywhere. The binary operation $\circ$ is called *composition*. The elements of a category are often called *morphisms*. This is the *nonobjective* definition of a category since there are no explicit 'objects'. As will soon become apparent, these will be those elements that may be written as $\mathrm{dom}\, a$ and $\mathrm{cod}\, a$, serving as identity elements, which will be called 'unities'.

Just as groups model permutations and monoids model functions on a single set, so categories model functions with domains and codomains that differ according to the function. In fact, monoids are examples of categories in which $\mathrm{dom}\, a = \mathrm{cod}\, b$ for all a, b in the carrier, that is, there is only one unity; call it 1. This forces composition to be a total operation. Groups are monoids in which all elements are invertible: for all a there is a b such that

$$a \circ b = 1 \quad \text{and} \quad b \circ a = 1.$$

Another example gathers these together and puts them into the category of all groups in a universe, $\mathfrak{Groups} = \langle \mathfrak{Groups}; \circ, \mathrm{dom}, \mathrm{cod} \rangle$. Nonobjectively, $\mathfrak{Groups}$ would be the set of all homomorphisms between groups:

$$h: \boldsymbol{G} \to \boldsymbol{H};$$

with $\circ$ being the composition of homomorphisms; and $\mathrm{dom}\, h = 1_{\boldsymbol{G}}$, $\mathrm{cod}\, h = 1_{\boldsymbol{H}}$, where $1_{\boldsymbol{G}}: \boldsymbol{G} \to \boldsymbol{G}$ is the identity function on $\boldsymbol{G}$. Notice how the domain and codomain are an integral part of a morphism, in contrast to the use of functions in set theory, where there is the range of a function but no codomain. More generally, any variety of algebras may be turned into the category of all homomorphisms between members.

2.2. EXERCISE. Show that any category is isomorphic to a category of partial one-place functions on a single set. Hint: use left translations or right ones, as you like.

As our definition of category is novel, a series of propositions should convince the reader that its three cryptic axioms do indeed define the traditional concept of a category.

2.3. PROPOSITION. *For all a in the carrier of a category:*

(a) $\operatorname{dom}\operatorname{dom} a = \operatorname{dom} a$,
(b) $\operatorname{cod}\operatorname{dom} a = \operatorname{dom} a$,
(c) $\operatorname{dom}\operatorname{cod} a = \operatorname{cod} a$,
(d) $\operatorname{cod}\operatorname{cod} a = \operatorname{cod} a$.

PROOF. We first prove (b). Axiom (ii) and the convention on existence imply that $a \circ (\operatorname{dom} a)$ exists. Then axiom (iii) implies (b). Dual to (b) is the proof of (c). From these parts we obtain (a) (and dually (d)):

$$\operatorname{dom} a = \operatorname{cod}\operatorname{dom} a = \operatorname{dom}\operatorname{cod}\operatorname{dom} a = \operatorname{dom}\operatorname{dom} a. \square$$

Let us agree that dom and cod are more cohesive than $\circ$; so '$\operatorname{cod} a \circ a$' means $(\operatorname{cod} a) \circ a$ and not $\operatorname{cod}(a \circ a)$. We adopt the convention of denoting composition by juxtaposition, thus $ab = a \circ b$.

2.4. DEFINITION. An element u in a category is a **unity** if for all a in it,

$$au = a \quad \text{whenever the product exists, and}$$
$$ua = a \quad \text{whenever the product exists.}$$

In the literature, unities may also be called *units*, *unit elements*, *identities*, *identity maps*, *identity morphisms*, or even *objects*.

2.5. PROPOSITION. *These five statements are equivalent in a category:*

(a) *u is a unity,*
(b) $u = \operatorname{dom} u$,
(c) $u = \operatorname{dom} a$ *for some a,*
(d) $u = \operatorname{cod} u$,
(e) $u = \operatorname{cod} a$ *for some a.*

PROOF. (a) $\Rightarrow$ (b) By axiom (ii) and the definition of unity,

$$u = u \operatorname{dom} u = \operatorname{dom} u.$$

(b) $\Rightarrow$ (c) Trivial.

(c) $\Rightarrow$ (a) Let $u = \operatorname{dom} b$ and suppose au exists. Then $a \operatorname{dom} b$ exists, and by axiom (iii) and the previous proposition,

$$\operatorname{dom} a = \operatorname{cod}\operatorname{dom} b = \operatorname{dom} b.$$

So by axiom (ii),

$$au = a \operatorname{dom} b = a \operatorname{dom} a = a.$$

Similarly, if ua exists, then $ua = a$. Thus, u is a unity.

Parts (d) and (e) have dual proofs. $\square$

As a consequence one may write $\operatorname{dom} \mathfrak{C}$ for the set of all unities in a category $\mathfrak{C}$. Functional notation is adapted to morphisms in a category $\mathfrak{C}$. By $f: u \to v$ one means an element f of $\mathfrak{C}$ such that $u = \operatorname{dom} f$ and $v = \operatorname{cod} f$; colloquially the morphism f goes from u to v.

There is the dual notion of the **coproduct** of two unities u and v; it reads just like the definition for product except all arrows are reversed:

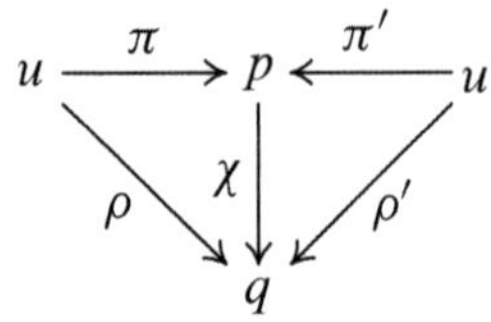

One of the good things about varieties of (total) algebras is that products and coproducts always exist in them, but this bald statement may mask considerable effort behind the scenes. The existence and construction of products is fairly straight forward; it is guaranteed by the Cartesian construction in Sect. II.2. The construction of coproducts is more elaborate, resembling the construction of free algebras; for that reason they are sometimes called *free sums.* Even more significant is that the product of two algebras is independent of the variety in which they live, whereas the coproduct definitely depends on their variety. For example, in the category of Abelian groups, the coproduct of $\mathbb{Z}$ with itself is $(\mathbb{Z})^2$; but in the category of all groups it is the free (non-Abelian) group on two generators.

Passing from categories coming from varieties to categories modeling algebras in their own right, we may use products and coproducts to define a lattice as any category satisfying these three properties:

(i) There exist products and coproducts of all pairs of unities.
(ii) There is at most one morphism between any pair of unities.
(iii) (antisymmetry) if $f: u \to v$ and $g: v \to u$, then $u = f = g = v$.

As the reader may have guessed, the set of unities is the carrier of a lattice in the usual sense; the product and coproduct are the usual meet and join; and a morphism $f: u \to v$ means simply that $u \leq v$.

This is a good place to formally define categorical duality, which we have already seen automatically gives us the notion of coproduct from that of product. The dual of a category is often described as the 'reversal' of arrows; this forces the order of composition to reverse.

2.11. DEFINITION. The **dual** or **opposite** of a category, $\mathfrak{C} = \langle \mathfrak{C}; \circ, \mathrm{dom}, \mathrm{cod} \rangle$, is the category:

$$\mathfrak{C}^{\mathrm{op}} = \langle \mathfrak{C}; \circ^{\mathrm{op}}, \mathrm{dom}^{\mathrm{op}}, \mathrm{cod}^{\mathrm{op}} \rangle,$$

where, for all a and b in $\mathfrak{C}$,

$$\begin{aligned} a \circ^{\mathrm{op}} b &= b \circ a, \\ \mathrm{dom}^{\mathrm{op}}\, a &= \mathrm{cod}\, a, \\ \mathrm{cod}^{\mathrm{op}}\, a &= \mathrm{dom}\, a. \end{aligned}$$

This notion resembles that of opposite rings, which also reverses the order of multiplication. But in Boolean algebras, which are commutative, nothing new is obtained. To get Boolean duality, one must look at Boolean algebras qua categories, as we did with lattices. Then the dual $\boldsymbol{B}^{\mathrm{op}}$ of a Boolean algebra $\boldsymbol{B}$ interchanges meet and join as well as 0 and 1.

Duality has the potential to double the number of categorical concepts and the theorems about them. Formally, let Σ be a sentence in the language of categories (with equality), and let Σ^{op} be its dual, that is, the arguments in any occurrence of $\circ$ are reversed, and simultaneously dom is replaced by cod and cod by dom. Then, for any theorem asserting that a sentence Σ is true in all categories, there is a dual theorem Σ^{op} that is also true in all categories. Notice that this is about *all* categories; without this 'all', this metatheorem is false. For example, in the single category of all finite groups, any product of two groups exist, but this is no longer true for coproducts. However, in the category of all finite Abelian groups both exist, for there the product and coproduct of two groups coincide.

Such theorems may have fragments from which one wants to abstract a dualizable definition. This may be formalized by considering a first-order categorical formula $\Sigma(x)$ with just one free variable. We say that a is such a 'gadget' in a category $\mathfrak{C}$ if $\Sigma(a)$ is true in $\mathfrak{C}$. As an example, a **monomorphism** is defined by cancellation on the left; the formula $\Sigma(x)$ is:

$$\forall a, b[xa = xb \Rightarrow a = b].$$

In varieties, monomorphisms agree with injections, that is, homomorphisms that are one-to-one; thus one has captured categorically the notion of an injective homomorphism. A useful convention here is to give categorical notions a Greek name, and set-theoretical ones a Latin-derived name.

It is harder to capture the concept of a surjective homomorphism within a category coming from a variety, but it can be done. The difficulty is that there are **epimorphisms**, the categorical dual of monomorphisms, that are not surjective, that is, not onto: this happens in the variety of rings. One may check that a homomorphism h is surjective in a variety whenever the condition $h = fg$ for some injective f implies that f is invertible. This statement may be rephrased in categorical language, thus capturing surjectivity.

An isomorphism in a variety is a homomorphism that is both injective and surjective; this is equivalent in categorical language to being invertible. There are many other concepts that can be defined categorically [**McKe96**]. But not all common concepts in general algebra are categorical; for example, freedom is not so definable [**Knoe83**]). (At least not with only one category of algebras; one also needs the forgetful functor taking an algebra to its carrier in the category of sets.)

We pass to concepts involving more than one category.

2.12. DEFINITION. One category $\mathfrak{B}$ is a **subcategory** of another, say $\mathfrak{C}$, if $B \subseteq C$ and B is closed to dom, cod and $\circ$, whenever the latter exists. It is called **full** if $\hom_{\mathfrak{B}}(u, v) = \hom_{\mathfrak{C}}(u, v)$ for all u and v in dom $\mathfrak{B}$.

Homomorphisms between categories are called *functors*. *Isomorphisms* are functors that are invertible or, alternatively, are both injective and surjective. These concepts are defined in detail as follows.

2.13. DEFINITION. (a) A **functor** from one category $\mathfrak{C}$ to another $\mathfrak{D}$ is a function $\Phi: \mathfrak{C} \to \mathfrak{D}$ such that, for all a, b in $\mathfrak{C}$,

(i) $\Phi(a \circ b) = \Phi(a) \circ \Phi(b)$ whenever $a \circ b$ exists;

(ii) $\Phi(\operatorname{dom} a) = \operatorname{dom} \Phi(a)$;

(iii) $\Phi(\operatorname{cod} a) = \operatorname{cod} \Phi(a)$.

(b) A functor $\Phi: \mathfrak{C} \to \mathfrak{D}$ is an **isomorphism** if:

(i) When $\Phi(a) = \Phi(b)$ then $a = b$; and

(ii) For all b in $\mathfrak{D}$ there is an a in $\mathfrak{C}$ such that $\Phi(a) = b$.

(c) Two categories $\mathfrak{C}$ and $\mathfrak{D}$ are **isomorphic** ($\mathfrak{C} \cong \mathfrak{D}$) when there is an isomorphism $\Phi: \mathfrak{C} \to \mathfrak{D}$.

The similarity of categories of diverse natures is more often recognized by *equivalence* rather than the more demanding isomorphism. To define this notion we need *skeletons*.

2.14. DEFINITION. A category is a **skeleton** if any two unities that are isomorphic are in fact equal. A **skeleton** of a category $\mathfrak{C}$ is any full subcategory of $\mathfrak{C}$ that is a skeleton and that is maximal with respect to this property. An **isomorphism class** of a category is a set U of all unities u isomorphic to a given unity u_0 of the category.

In the next proposition, the axiom of choice finds a skeleton in any category by shrinking its isomorphism classes to singletons. The notion of *retract*, given in Sect. II.1, is adapted to functors.

2.15. PROPOSITION. *Any category $\mathfrak{C}$ has a skeleton. This skeleton is a retract of $\mathfrak{C}$ and is unique up to isomorphism.*

PROOF. Let U, V, W, etc. be the isomorphism classes of $\mathfrak{C}$; and from each of these pick a particular unity u_o to represent it: $u_0 \in U, v_0 \in V, w_0 \in W, \ldots$. Set $\mathfrak{S} = \{u_0, v_0, w_o, \ldots\}$. Further, let $\mathfrak{S}$ be the full subcategory of $\mathfrak{C}$ on $\mathfrak{S}$. Clearly $\mathfrak{S}$ is a skeleton.

Define a function $\Phi: \mathfrak{C} \to \mathfrak{S}$ as follows. For each unity u of U pick a particular isomorphism p_{uu_0} from u to u_0, and similarly for the other unities in their respective isomorphism classes. For any morphism $f: u \to v$ in $\mathfrak{C}$, define

$$\Phi(f) = p_{vv_0} \circ f \circ p_{uu_0}^{-1}.$$

One verifies that Φ is a functor from $\mathfrak{C}$ to $\mathfrak{S}$, and that $\Phi(s) = s$ if $s \in \mathfrak{S}$. □

2.16. DEFINITION. (a) Two categories $\mathfrak{C}$ and $\mathfrak{D}$ are **equivalent** ($\mathfrak{C} \simeq \mathfrak{D}$) when their skeletons are isomorphic.

(b) Categories $\mathfrak{C}$ and $\mathfrak{D}$ are **dually equivalent** ($\mathfrak{C} \simeq^{\mathrm{op}} \mathfrak{D}$) if $\mathfrak{C}$ and $\mathfrak{D}^{\mathrm{op}}$ are equivalent.

The classical example is the dual equivalence of the category of Boolean algebras with the category of Boolean spaces (see Sect. 4). The next theorem puts the equivalence of categories into a different but perhaps more intuitive setting, analogous to isomorphism. To that end, for unities u and v in a category $\mathfrak{C}$ with a functor $\Phi\colon \mathfrak{C} \to \mathfrak{D}$, we define $\Phi_{u,v}$ to be Φ restricted to $\hom_{\mathfrak{C}}(u, v)$.

2.17. THEOREM. *Two categories* $\mathfrak{C}$ *and* $\mathfrak{D}$ *are equivalent if, and only if, there is a functor* $\Phi\colon \mathfrak{C} \to \mathfrak{D}$ *such that:*

(a) Φ *is* **faithful**: $\Phi_{u,v}$ *is injective, that is, it maps* $\hom_{\mathfrak{C}}(u, v)$ *one-to-one into* $\hom_{\mathfrak{D}}(\Phi(u), \Phi(v))$ *for all unities* u *and* v *in* $\mathfrak{C}$;
(b) Φ *is* **full**: $\Phi_{u,v}$ *is surjective, that is, for all unities* u *and* v *in* $\mathfrak{C}$, $\hom_{\mathfrak{C}}(u, v)$ *maps onto* $\hom_{\mathfrak{D}}(\Phi(u), \Phi(v))$;
(c) Φ *is* **dense**: *for every unity* v *in* $\mathfrak{D}$ *there is a unity* u *in* $\mathfrak{C}$ *such that* $\Phi(u) \cong v$.

A natural transformations passes from one functor to another. The concept is needed to define adjoint functors, which in turn give another characterization of equivalence. Much of the subsequent discussion may be found in [**HerSt79**, Sects. 13, 14, 26 and 27], especially the missing proofs.

2.18. DEFINITION. Let $\Phi\colon \mathfrak{C} \to \mathfrak{D}$ and $\Gamma\colon \mathfrak{C} \to \mathfrak{D}$ be two functors from the category $\mathfrak{C}$ to the category $\mathfrak{D}$. A **natural transformation** from Φ to Γ is a function $\eta\colon \operatorname{dom} \mathfrak{C} \to \mathfrak{D}$ yielding, for each u in $\operatorname{dom} \mathfrak{C}$, a morphism $\eta(u)\colon \Phi(u) \to \Gamma(u)$ in $\mathfrak{D}$ such that, for each morphism $f\colon u \to v$ in $\mathfrak{C}$,

$$\Gamma(f) \circ \eta(u) = \eta(v) \circ \Phi(f).$$

In symbols, this diagram commutes:

$$\begin{array}{ccc}
\Phi(u) & \xrightarrow{\ \eta(u)\ } & \Gamma(u) \\
{\scriptstyle \Phi(f)}\big\downarrow & & \big\downarrow{\scriptstyle \Gamma(f)} \\
\Phi(v) & \xrightarrow[\ \eta(v)\ ]{} & \Gamma(v)
\end{array}$$

Moreover, this natural transformation is a **natural isomorphism** if $\eta(u)$ is an isomorphism in $\mathfrak{D}$ for all u in $\operatorname{dom} \mathfrak{C}$.

A trivial example of a natural transformation is the embedding of integral domains into their fields of fractions. Let $\mathfrak{C}$ and $\mathfrak{D}$ both be the category of all monomorphisms between integral domains. Let Φ be the identity functor and Γ the functor taking a monomorphism of integral domains to the induced monomorphism of their fields of fractions. A natural transformation η from Φ to Γ takes an integral domain (or more correctly, its identity morphism) to the monomorphism embedding it in its field of fractions.

Free algebras on a set of generators motivate the concept of a Γ-universal map. The characterizing property of a free group $\boldsymbol{F}$ on a set of generators is that any other group $\boldsymbol{G}$ generated by this set is a homomorphic image of $\boldsymbol{F}$. When this is put into the language of the next definition, $\mathfrak{C}$ becomes the category of all groups and $\mathfrak{D}$ the category of sets. The functor $\Gamma:\mathfrak{C}\to\mathfrak{D}$ is the 'forgetful' functor taking a group to its carrier and taking a homomorphism of groups to the underlying function on their carriers.

2.19. DEFINITION. Let $\Gamma:\mathfrak{C}\to\mathfrak{D}$ be a functor and $v\in\operatorname{dom}\mathfrak{D}$. A pair $\langle\eta,u\rangle$ with $\eta: v\to\Gamma(u)$ and $u\in\operatorname{dom}\mathfrak{C}$ is called a Γ-**universal map** for v if, for each u' in $\operatorname{dom}\mathfrak{C}$ and each morphism $g: v\to\Gamma(u')$ of $\mathfrak{D}$, there exists a unique morphism $h: u\to u'$ of $\mathfrak{C}$ such that this triangle commutes:

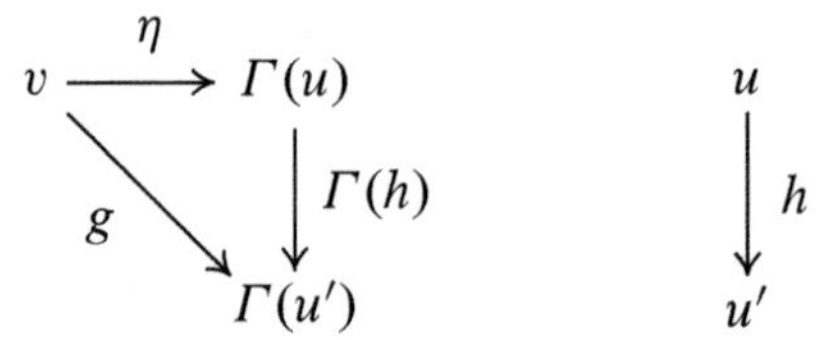

The following consequences of universal maps are tantamount to an adjunction, which will be defined after them. These are in [**HerSt79**, Theorem 26.11].

2.20. THEOREM. *Let $\Gamma:\mathfrak{C}\to\mathfrak{D}$ be a functor such that there exists a Γ-universal map $\langle\eta_v,u_v\rangle$ for each v in* $\operatorname{dom}\mathfrak{D}$. *Let η be the function $\langle\eta_v\mid v\in\operatorname{dom}\mathfrak{D}\rangle$. Two things follow:*

(a) *There exists a unique functor $\Phi:\mathfrak{D}\to\mathfrak{C}$ such that*
 (1) *$\Phi(v)=u_v$ for each v in* $\operatorname{dom}\mathfrak{D}$, *and*
 (2) *$\eta: 1_{\mathfrak{D}}\to\Gamma\circ\Phi$ is a natural transformation.*

(b) *There is a unique natural transformation $\varepsilon:\Phi\circ\Gamma\to 1_{\mathfrak{C}}$ such that*
 (1) $\Gamma\big(\varepsilon(u)\big)\circ\eta\big(\Gamma(u)\big)=1_{\Gamma(u)}$ *($u\in\operatorname{dom}\mathfrak{C}$), and*
 (2) $\varepsilon\big(\Phi(v)\big)\circ\Phi\big(\eta(v)\big)=1_{\Phi(v)}$ *($v\in\operatorname{dom}\mathfrak{D}$).*

The natural transformations and their compositions with the functors occurring in Theorem 2.20 lead into the definition of adjunction.

2.21. DEFINITION. An **adjunction** consists of two categories $\mathfrak{C}$ and $\mathfrak{D}$, two functors Φ and Γ, and two natural transformations η and ε such that:

(i) $\Gamma:\mathfrak{C}\to\mathfrak{D}$ and $\Phi:\mathfrak{D}\to\mathfrak{C}$;
(ii) $\eta: 1_{\mathfrak{D}}\to\Gamma\circ\Phi$ and $\varepsilon:\Phi\circ\Gamma\to 1_{\mathfrak{C}}$;
(iii) (1) $\Gamma\big(\varepsilon(u)\big)\circ\eta\big(\Gamma(u)\big)=1_{\Gamma(u)}$ ($u\in\operatorname{dom}\mathfrak{C}$), and
 (2) $\varepsilon\big(\Phi(v)\big)\circ\Phi\big(\eta(v)\big)=1_{\Phi(v)}$ ($v\in\operatorname{dom}\mathfrak{D}$).

Denote this relationship by

$$\langle\eta,\varepsilon\rangle:\Phi\dashv\Gamma:\langle\mathfrak{C},\mathfrak{D}\rangle.$$

Call Φ the **left adjoint** of Γ, and Γ the **right adjoint** of Φ.

Adjunctions capture categorically many useful Galois-like connections in diverse areas, and for us, they will relate certain varieties of algebras to their classes of sheaves. Some adjunctions will become equivalences in later chapters.

Here are several examples of adjunctions. The first embellishes the universal mapping property of free groups studied earlier. The natural transformation $\eta(S): S \to (\Gamma \circ \Phi)(S)$ embeds a set S of generators or variables into the carrier of the free group of all terms created on them. The natural transformation $\varepsilon(\boldsymbol{G}): (\Phi \circ \Gamma)(\boldsymbol{G}) \to \boldsymbol{G}$ is a quotient morphism: for a given group $\boldsymbol{G}$, it is the canonical mapping of the free group created on the carrier of $\boldsymbol{G}$ (as generators) onto $\boldsymbol{G}$ itself with the generators going to themselves. This example of free groups may be phrased more generally for the free algebras in all varieties.

Another example shifts the viewpoint in the previous example of integral domains: $\mathfrak{D}$ is still their category, but now $\mathfrak{C}$ is the category of all monomorphisms of fields. The functor Φ creates a field of fractions from an integral domain, and Γ is the inclusion functor placing a field into $\mathfrak{D}$.

In a third example, Φ divides any group by its commutator subgroup to obtain an Abelian group. Again Γ is the inclusion functor. A Galois connection between partially ordered sets is a further example of an adjunction. A final example is the category of completions of uniform spaces, which is adjoint to their inclusion in the category of all uniform spaces.

Some of these examples may be phrased more naturally in terms of universal maps. So the next theorem [**HerSt79**, Theorem 27.3]) compresses Theorem 2.20 and completely connects universal maps with adjunctions.

2.22. THEOREM. (a) *For a functor* $\Gamma: \mathfrak{C} \to \mathfrak{D}$, *if each* v *in* dom $\mathfrak{D}$ *has a* Γ*-universal map* $\langle \eta_v, u_v \rangle$, *then there is exists a unique adjunction* $\langle \eta, \varepsilon \rangle : \Phi \dashv \Gamma : \langle \mathfrak{C}, \mathfrak{D} \rangle$ *where* $\eta = \langle \eta_v \mid v \in \mathrm{dom}\, \mathfrak{D} \rangle$ *and* $\Phi(v) = u_v$.

(b) *Conversely, for an adjunction* $\langle \eta, \varepsilon \rangle : \Phi \dashv \Gamma : \langle \mathfrak{C}, \mathfrak{D} \rangle$ *and for each unity* v *in* $\mathfrak{D}$, *there is a* Γ*-universal map* $\langle \eta(v), \Phi(v) \rangle$.

Here is an alternative form of equivalence needed in the sequel. It comes from Theorems 14.11, 14.15 and Proposition 27.2 of [**HerSt79**].

2.23. THEOREM. *Two categories* $\mathfrak{C}$ *and* $\mathfrak{D}$ *are equivalent if, and only if, there is an adjunction* $\langle \eta, \varepsilon \rangle : \Phi \dashv \Gamma : \langle \mathfrak{C}, \mathfrak{D} \rangle$ *such that*

(a) η *is a natural isomorphism and*
(b) ε *is a natural isomorphism.*

Two adjunctions may be composed to give a third, and this will be applied in Sect. V.4. To explain and justify, let two adjunctions be given:

$$\langle \eta_1, \varepsilon_1 \rangle : \Phi_1 \dashv \Gamma_1 : \langle \mathfrak{B}, \mathfrak{C} \rangle, \qquad \langle \eta_2, \varepsilon_2 \rangle : \Phi_2 \dashv \Gamma_2 : \langle \mathfrak{C}, \mathfrak{D} \rangle.$$

The functors are composed: $\Phi_1 \circ \Phi_2 : \mathfrak{D} \to \mathfrak{B}$ and $\Gamma_2 \circ \Gamma_1 : \mathfrak{B} \to \mathfrak{D}$; they are abbreviated as

$$\Phi = \Phi_1 \circ \Phi_2 \text{ and } \Gamma = \Gamma_2 \circ \Gamma_1.$$

The new natural transformations associated with the composite adjunction are denoted: $\eta\colon 1_{\mathfrak{D}} \to \Gamma \circ \Phi$ and $\varepsilon\colon \Phi \circ \Gamma \to 1_{\mathfrak{B}}$; they operate as

$$\eta(w)\colon w \to \Gamma\big(\Phi(w)\big) \qquad (w \in \operatorname{dom}\mathfrak{D}),$$
$$\varepsilon(u)\colon \Phi\big(\Gamma(u)\big) \to u \qquad (u \in \operatorname{dom}\mathfrak{B}).$$

Specifically, for any w in dom $\mathfrak{D}$ and any u in dom $\mathfrak{B}$,

$$\eta(w) = \Gamma_2\big(\eta_1(\Phi_2(w))\big) \circ \eta_2(w), \tag{2.1}$$
$$\varepsilon(u) = \varepsilon_1(u) \circ \Phi_1\big(\varepsilon_2(\Gamma_1(u))\big). \tag{2.2}$$

2.24. THEOREM. *The functors,*

$$\Phi\colon \mathfrak{D} \rightleftarrows \mathfrak{B}\colon \Gamma,$$

together with their natural transformations,

$$\eta\colon 1_{\mathfrak{D}} \to \Gamma \circ \Phi \quad \textit{and} \quad \varepsilon\colon \Phi \circ \Gamma \to 1_{\mathfrak{B}},$$

as just given, form an adjunction

$$\langle \eta, \varepsilon\rangle\colon \Phi \dashv \Gamma\colon \langle \mathfrak{B}, \mathfrak{D}\rangle.$$

PROOF. The three parts of Definition 2.21 must be verified. Part (i) is clear.

For part (ii), the naturality of η follows from that of η_1 and η_2. That is, for any morphism $f\colon w \to w'$ in $\mathfrak{D}$, the desired commutativity of the square

$$\begin{array}{ccc}
w & \xrightarrow{\ \eta(w)\ } & \Gamma\Phi(w) \\
{\scriptstyle f}\downarrow & & \downarrow{\scriptstyle \Gamma\Phi(f)} \\
w' & \xrightarrow[\ \eta(w')\ ]{} & \Gamma\Phi(w')
\end{array}$$

follows from those in the interior of the diagram

$$\begin{array}{ccccc}
w & \xrightarrow{\ \eta_2(w)\ } & \Gamma_2\Phi_2(w) & \xrightarrow{\ \Gamma_2\eta_1\Phi_2(w)\ } & \Gamma_2\Gamma_1\Phi_1\Phi_2(w) \\
{\scriptstyle f}\downarrow & & \downarrow{\scriptstyle \Gamma_2\Phi_2(f)} & & \downarrow{\scriptstyle \Gamma_2\Gamma_1\Phi_1\Phi_2(f)} \\
w' & \xrightarrow[\ \eta_2(w')\ ]{} & \Gamma_2\Phi_2(w') & \xrightarrow[\ \Gamma_2\eta_1\Phi_2(w')\ ]{} & \Gamma_2\Gamma_1\Phi_1\Phi_2(w')
\end{array}$$

Commutativity of the second square follows from that for η_1 by composing it with Γ_2 and Φ_2, fore and aft. (Most parentheses are omitted.)

Part (iii.1) of Definition 2.21 is the perimeter in Fig. 1. Its commutativity follows from that of the inner figures. That of the triangles comes from part (iii.1) for the component adjunctions. And the square comes from the naturality of η_1 or ε_2. Part (iii.2) is proven by a similar diagram. □

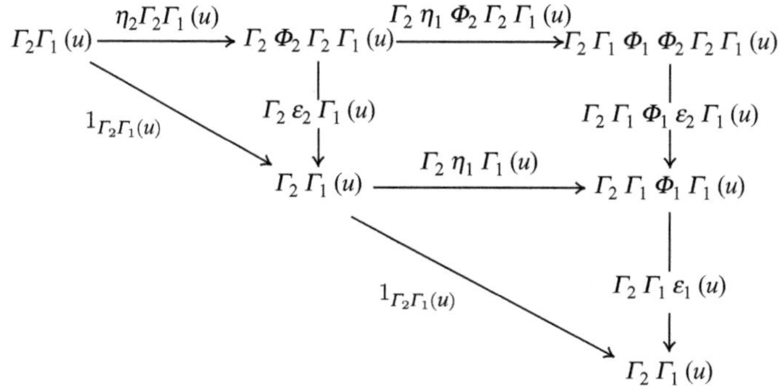

FIGURE 1. Composite Adjunction, Proof of (iii.1)

For another exposition of such compositions see the book of Herrlich and Strecker [**HerSt79**]. They follow an indirect strategy for proving our Theorem 2.24, which is their proposition 27.8.

Hu's theorem that any two varieties generated by primal algebras are categorically equivalent suggests a syntactical connection between different but categorically similar algebras [**Hu69**]. McKenzie's theorem exhibits this relationship and breaks it into three parts: term-equivalence, equivalence by a matrix power, and equivalence by an invertible, idempotent term [**McKe96**]. We will briefly describe this. Hu's theorem is an example of categorical equivalence by an invertible, idempotent term. Another example is the term-equivalence of Boolean algebras and Boolean rings.

2.25. DEFINITION. Two algebras $\boldsymbol{A}$ and $\boldsymbol{B}$ are **categorically equivalent** if their generated varieties are categorically equivalent by a functor Φ from $\operatorname{Var}\boldsymbol{A}$ to $\operatorname{Var}\boldsymbol{B}$ such that $\Phi(\boldsymbol{A}) \cong \boldsymbol{B}$.

2.26. DEFINITION. The kth **matrix power** $\boldsymbol{A}^{[k]}$ of an algebra $\boldsymbol{A}$ has carrier A^k, whose elements we now write as column vectors. For each nonnegative integer n we construct its n-ary operations as follows. From k terms $t^1, t^2, \ldots, t^k$ of the type τ of $\boldsymbol{A}$, each of (kn)-arity, define the sequence, $\vec{t} = \langle t^1, t^2, \ldots, t^k \rangle$, to be an n-ary operation-symbol of the new type $\tau^{[k]}$. It is evaluated in $\boldsymbol{A}^{[k]}$ for any k by n matrix, $a = \{a^i_j\}$, with elements a^i_j in A, and rows a^i $(1 \leq i \leq k)$ and columns a_j $(1 \leq j \leq n)$, by the formula:

$$\vec{t}(a_1, a_2, \ldots, a_n) = \langle t^1(a), t^2(a), \ldots, t^k(a) \rangle.$$

This extends to any variety $\mathfrak{A}$. The variety $\mathfrak{A}^{[k]}$ is the set of all matrix powers $\boldsymbol{A}^{[k]}$ of algebras $\boldsymbol{A}$ in $\mathfrak{A}$, and their isomorphic copies to insure that it is a variety.

An example of a matrix power will be the varieties of Chap. X coming from elementary Abelian p-groups.

2.27. DEFINITION. A unary term s of the type τ of an algebra, $\boldsymbol{A} = \langle A; \ldots, \omega, \ldots \rangle$, is **idempotent** if $s \circ s \equiv s\,(\mathrm{Id}\,\boldsymbol{A})$. It creates a new algebra $s(\boldsymbol{A})$, as was done for sesquimorphisms, by restricting its carrier to $s(A)$ and relativizing its operations to $\omega^s(\vec{a}) = s(\omega(\vec{a}))$. It is **invertible** if for some n there are unary terms $t_1, t_2, \ldots, t_n$ and an n-ary term t such that

$$t\big(s(t_1(x)), s(t_2(x)), \ldots, s(t_n(x))\big) \equiv x \quad (\mathrm{Id}\,\boldsymbol{A}).$$

This also extends to any variety $\mathfrak{A}$. The variety $s(\mathfrak{A})$ is the set of all $s(\boldsymbol{A})$ relativized from the $\boldsymbol{A}$ in $\mathfrak{A}$.

Term-equivalence was defined in Sect. II.1.

2.28. THEOREM. [**McKe96**]

(a) *Two varieties $\mathfrak{A}$ and $\mathfrak{D}$ are categorically equivalent, if and only if, there are varieties $\mathfrak{B}$ and $\mathfrak{C}$ such that $\mathfrak{B} = \mathfrak{A}^{[n]}$ for some positive integer n, and $\mathfrak{C} = s(\mathfrak{B})$ for some invertible and idempotent term s, and $\mathfrak{D}$ is term-equivalent to $\mathfrak{C}$.*

(b) *Two algebras $\boldsymbol{A}$ and $\boldsymbol{D}$ are categorically equivalent, if and only if, there are algebras $\boldsymbol{B}$ and $\boldsymbol{C}$ such that $\boldsymbol{B} = \boldsymbol{A}^{[n]}$ for some positive integer n, and $\boldsymbol{C} = s(\boldsymbol{B})$ for some invertible and idempotent term s, and $\boldsymbol{D}$ is term-equivalent to $\boldsymbol{C}$.*

This theorem will be used to prove Theorems VI.3.22 and X.3.2.

3. Topology

This section describes various kinds of topological spaces, including Boolean spaces, the most widely used in this book. New spaces are built out of old: subspaces, quotients, and products. Last to be introduced is the important glue holding sheaves together: continuous functions. For details see the texts of John Kelly [**Kell55**] and Paul Halmos [**Halm63**].

3.1. DEFINITION. A **topological space**, $\boldsymbol{X} = \langle X, \mathcal{T} \rangle$, is a set X together with a collection $\mathcal{T}$ of subsets of it that is closed to finite intersections and arbitrary unions, and contains X and the empty set $\varnothing$. The set X is called the **space**, the collection $\mathcal{T}$ its **topology** and members of $\mathcal{T}$ **open** sets. The complement of an open set X in $\mathcal{T}$ is said to be **closed**. Elements of X are often called **points**. A **neighborhood** of a point x is any subset of X containing an open subset to which x belongs.

Note that spaces may be empty, in contrast to algebras. Throughout this section let $\boldsymbol{X}$ be a topological space $\langle X, \mathcal{T} \rangle$.

A subset of X is said to be **clopen** if it is both open and closed. Clopen sets are rather rare in classical topological spaces such as the real line $\mathbb{R}$, where the only such sets are $\varnothing$ and $\mathbb{R}$ itself. But in this book they will be quite common, and exploited. Let Clop $\boldsymbol{X}$ be the set of all clopen subsets of a topological space $\boldsymbol{X}$.

The notion of topology is characterized by the closure axioms of Kuratowski [**Kell55**, p. 43]. In a topological space, define the **closure** $\overline{U}$ of a subset U of X as the smallest closed set containing it. It satisfies the four axioms:

$$\begin{aligned} \overline{\varnothing} &= \varnothing; \\ \overline{U} &\supseteq U \qquad (U \subseteq X); \\ \overline{\overline{U}} &= \overline{U} \qquad (U \subseteq X); \\ \overline{U \cup V} &= \overline{U} \cup \overline{V} \qquad (U, V \subseteq X). \end{aligned}$$

3.2. DEFINITION. A topological space $\boldsymbol{X}$ is $\mathbf{T}_0$ if for every two points in X there is an open set containing one and not the other. It is $\mathbf{T}_1$ if for every point x, its singleton set $\{x\}$ is closed. And it is $\mathbf{T}_2$, or **Hausdorff**, if for every two points in X there are two disjoint open sets containing the respective points. The topology of a topological space $\langle X, \mathcal{T} \rangle$ is **discrete** if every subset is open: $\mathcal{T} = \mathcal{P}X$. The antonym is **indiscrete**: $\mathcal{T} = \{\varnothing, X\}$. The **interior** of a subset S of a topological space $\boldsymbol{X}$ is the set Int S of points x for which there is an open set U of $\boldsymbol{X}$ such that $x \in U$ and $U \subseteq S$.

Recall that T_2 implies T_1, which in turn implies T_0. Spaces that are T_0 but not T_1 are common in algebraic geometry, as explained in [**Berg73**]. We proceed to put together the definition of a Boolean space.

3.3. DEFINITION. A **cover** of a set U in a topological space $\langle X, \mathcal{T} \rangle$ is a collection of subsets of X whose union includes U; it is *open* or *clopen* as all the subsets are. A topological space $\boldsymbol{X}$ is **compact** if every open cover of X has a finite subcover of X, that is, if for every collection $\mathcal{C}$ of open subsets of X such that $\bigcup \mathcal{C} = X$, there is finite subcollection $\mathcal{F}$ of $\mathcal{C}$ such that also $\bigcup \mathcal{F} = X$. It is **locally compact** if every point resides in an open set that is compact. A topological space is **totally disconnected** if every open set of it is the union of those clopen sets contained in it. A topological space is a **Boolean** space if it is Hausdorff, compact, and totally disconnected.

3.4. DEFINITION. A **basis** for a topology $\mathcal{T}$ is a subcollection $\mathcal{B}$ of $\mathcal{T}$ such that every member of $\mathcal{T}$ is a union of some of them. A **subbasis** for $\mathcal{T}$ is a subcollection $\mathcal{B}$ of $\mathcal{T}$ such that the collection of all finite intersections of members of $\mathcal{B}$ is a basis for $\mathcal{T}$.

A basis for a Boolean space is its collection of clopen sets.

The purpose of topology is to capture continuity, defined next.

3.5. DEFINITION. Let $\langle X_1, \mathcal{T}_1 \rangle$ and $\langle X_2, \mathcal{T}_2 \rangle$ be topological spaces, and φ a function from X_1 to X_2. Then φ is said to be **continuous** if $\varphi^{-1}(U) \in \mathcal{T}_1$ whenever $U \in \mathcal{T}_2$, and **open** if $\varphi(U) \in \mathcal{T}_2$ whenever $U \in \mathcal{T}_1$. A **homeomorphism** is when φ is bijective and it is both continuous and open; it is on each side are of the same kind. But, as we wish to deal also with non-Boolean rings, where ideals do not have a counterpart in filters, we have chosen ideals.

In order to endow Spec $\boldsymbol{B}$ with a topology, define for each element b of a Boolean algebra $\boldsymbol{B}$ a set U_b of prime ideals:

$$U_b = \{P \in \operatorname{Spec} \boldsymbol{B} \mid b \in P\}. \tag{4.2}$$

From the nature of prime ideals (4.1), one shows for any b and c in B that

$$U_b \cup U_c = U_{b\wedge c}; \quad U_b \cap U_c = U_{b\vee c}; \quad \operatorname{Spec} \boldsymbol{B} \sim U_b = U_{b'}.$$

Since $U_b \cap U_c = U_{b\vee c}$, the set of all U_b is closed to finite intersections; it can serve as a basis of open sets for a topology, which is traditionally called the **Stone** topology after Stone's paper [**Stone36**]. Each open set is a union of basis sets U_b, which are clopen by (4.1). Thus, prime ideals P of the Boolean algebra $\boldsymbol{B}$ become points P of a topological space **Spec** $\boldsymbol{B}$.

It is also called the **hull-kernel** topology, and here is why. Consider the membership relation, $b \in P$, between elements b of B and prime ideals P of $\boldsymbol{B}$. As does any such binary relation, '$\in$' gives rise to a polarity between subsets of B and subsets of Spec $\boldsymbol{B}$, and any polarity creates closure operators on each of these domains [**Birk67**, p.122]. For a subset C of B, its polar $h(C)$ is known as the **hull** and is given by

$$h(C) = \{P \in \operatorname{Spec} \boldsymbol{B} \mid \forall b \in C\ (b \in P)\}.$$

For a set Y of prime ideals, its polar $k(Y)$, the **kernel**, is given by

$$k(Y) = \{b \in B \mid \forall P \in Y\ (b \in P)\}.$$

These polars satisfy (when $C, C_i \subseteq B$ and $Y, Y_i \subseteq \operatorname{Spec} \boldsymbol{B}$):

$$C \subseteq k\big(h(C)\big); \qquad \text{if } C_1 \subseteq C_2 \text{ then } h(C_1) \supseteq h(C_2);$$
$$Y \subseteq h\big(k(Y)\big); \qquad \text{if } Y_1 \subseteq Y_2 \text{ then } k(Y_1) \supseteq k(Y_2).$$

It follows that

$$h\big(k(h(C))\big) = h(C) \text{ and } k\big(h(k(Y))\big) = k(Y).$$

Composition is designated:

$$\overline{Y} = h\big(k(Y)\big) \qquad (Y \subseteq \operatorname{Spec} \boldsymbol{B}).$$

This is a closure operator on Spec $\boldsymbol{B}$:

$$Y \subseteq \overline{Y}, \quad \overline{Y} = \overline{\overline{Y}}, \quad \text{and if } Y_1 \subseteq Y_2 \text{ then } \overline{Y_1} \subseteq \overline{Y_2}.$$

It can be verified that it is 'additive': $\overline{Y_1 \cup Y_2} = \overline{Y_1} \cup \overline{Y_2}$. Also $\bar{\varnothing} = \varnothing$, so it satisfies Kuratowski's axioms and is a topological closure operator (Sect. 3). Thus, this is called the hull-kernel topology[2] and coincides with the Stone topology **Spec** $\boldsymbol{B}$, as may be proven with the help of [**Birk67**, pp. 111–112, 116–117, and 122–126].

---

[2]This breakdown of the topological closure into hull and kernel parts is patterned after Kist's treatment of the prime spectrum for a commutative semigroup [**Kist63**].

As is well known, **Spec** $\boldsymbol{B}$ is a Boolean space when $\boldsymbol{B}$ is a Boolean algebra. By this is meant that **Spec** $\boldsymbol{B}$ is Hausdorff, compact and totally disconnected (see Sect. 3 and [**Halm63**, p.$\tilde{7}$3]).

From any topological space $\boldsymbol{X}$ comes the Boolean algebra, $\mathbf{Clop}\,\boldsymbol{X} = \langle \mathrm{Clop}\,\boldsymbol{X}; \cup, \cap, ', \varnothing, X \rangle$, of its clopen subsets, those that are both closed and open. As these are closed to union, intersection, and complementation, they form a Boolean algebra, which leads to the representation theorem discovered by Stone [**Stone36**].

4.1. Theorem. *For any Boolean algebra* $\boldsymbol{B}$ *there is the isomorphism:*

$$\boldsymbol{B} \cong \mathbf{Clop}\,\mathbf{Spec}\,\boldsymbol{B}.$$

For any Boolean space $\boldsymbol{X}$ *there is the homeomorphism:*

$$\boldsymbol{X} \cong \mathbf{Spec}\,\mathbf{Clop}\,\boldsymbol{X}.$$

Most importantly for us is the dual equivalence of these algebras and spaces, due also to Stone [**Stone37**], although not phrased in categorical language. This extends the correspondence between objects to morphisms among them. A homomorphism φ from one Boolean algebra $\boldsymbol{A}$ to another $\boldsymbol{B}$ turns into a continuous map χ from **Spec** $\boldsymbol{B}$ to **Spec** $\boldsymbol{A}$, given by

$$\chi(P) = \varphi^{-1}(P) \qquad (P \in \mathrm{Spec}\,\boldsymbol{B}).$$

A continuous map χ from one Boolean space $\boldsymbol{X}$ to another $\boldsymbol{Y}$ turns into a homomorphism φ from **Clop** $\boldsymbol{Y}$ to **Clop** $\boldsymbol{X}$, given by

$$\varphi(U) = \chi^{-1}(U) \qquad (U \in \mathrm{Clop}\,\boldsymbol{Y}).$$

4.2. Theorem. *The category* $\mathfrak{BooleAlg}$ *of Boolean algebras is dually equivalent to the category* $\mathfrak{BooleSpace}$ *of Boolean spaces:*

$$\mathfrak{BooleAlg} \simeq^{\mathrm{op}} \mathfrak{BooleSpace}.$$

For a proof see [**Halm63**, Sect. 18]. Consequently, to any property of Boolean algebras corresponds a property of its dual space. The next definition gives such a pair [**Halm63**, p. 90].

4.3. Definition. A Boolean algebra is **complete** if for an arbitrary subset S of A its join $\bigvee S$ and meet $\bigwedge S$ exist. Its dual, an **extremally disconnected** Boolean space, is defined by Stone as the paradoxical property that the closure of any open set is again open.

We close this section with several notions needed elsewhere. An **atom** of a Boolean algebra $\boldsymbol{A}$ is any non-zero element with no element between it and 0. A Boolean algebra is **atomic** if every non-zero, non-atomic element is greater than some atom. The **dual** of a Boolean algebra $\langle B; \vee, \wedge, ', 0, 1 \rangle$ is the Boolean algebra $\langle B; \wedge, \vee, ', 1, 0 \rangle$ with operations interchanged. Two Boolean algebras, $\boldsymbol{B}_1$ and $\boldsymbol{B}_2$, are **anti-isomorphic** if $\boldsymbol{B}_1$ is isomorphic to the dual of $\boldsymbol{B}_2$, written $\boldsymbol{B}_1 \overset{\text{anti}}{\cong} \boldsymbol{B}_2$.

IV

Complexes and their Sheaves

With a little thought one easily adapts the notion of complex, known for some time in ring and module theory, to universal algebra, where it is new. The analogy with metric spaces is apt. The preliminary notion of precomplex then corresponds to that of premetric space. We give many examples of complexes. Their value lies in the sheaves they generate.

To represent an algebra is to decompose it, that is, to turn its elements into sequences whose components are in simpler algebras. But generally not all sequences are required in a subdirect product, and so we need a criterion for determining whether a sequence is to be in the representation or not. This leads to a sheaf, adding topology. Its advantage is that the admissible sequences are certain continuous functions, the global sections.

In the second section, there is the Gel'fand morphism from a precomplex to the algebra of all global sections of the corresponding sheaf. This is injective when we start with a complex. It is surjective – highly desirable but elusive – only when much stronger hypotheses are assumed, as in later chapters.

Closely related to complexes are systems of congruences, which we define and compare with complexes. There also is a proposition characterizing Hausdorff sheaves.

The passage between complexes and sheaves may be put formally into a categorical setting. For each type of algebra, the third section defines a pair of categories, $\mathfrak{Complex}$ and $\mathfrak{Sheaf}$. Two functors go back and forth between them, taking morphisms of one category into the other. Together

A. Knoebel, *Sheaves of Algebras over Boolean Spaces*,
DOI 10.1007/978-0-8176-4642-4_IV,

with a couple of natural transformations, these create an adjoint situation. For particular classes of algebras, this adjunction will later become a categorical equivalence.

1. Concepts

We define complexes and sheaves, and develop their properties. Important is the notion of a global section, which represents an element of an algebra.

1.1. DEFINITION. A **precomplex** $\langle \boldsymbol{A}, :, \boldsymbol{X}\rangle$ consists of an algebra, $\boldsymbol{A} = \langle A; \ldots, \omega, \ldots\rangle$, and a topological space, $\boldsymbol{X} = \langle X, \mathcal{T}\rangle$, together with a binary function ":" from the carrier A to the set $\mathcal{T}$ of open sets on X satisfying these postulates:

$$a : a = X \qquad (a \in A); \tag{1.1}$$

$$a : b = b : a \qquad (a, b \in A); \tag{1.2}$$

$$a : b \cap b : c \subseteq a : c \qquad (a, b, c \in A); \tag{1.3}$$

$$a_1 : b_1 \cap a_2 : b_2 \cap \cdots \cap a_n : b_n \subseteq \omega(a_1, a_2, \ldots, a_n) : \omega(b_1, b_2, \ldots, b_n). \tag{1.4}$$

The last inclusion holds for each operation ω of $\boldsymbol{A}$ with n arguments and for elements $a_1, a_2, \ldots, a_n$, $b_1, b_2, \ldots, b_n$ in A. If, in addition,

$$a : b = X \text{ implies } a = b \qquad (a, b \in A), \tag{1.5}$$

then we have a **complex**. Anticipating when a and b will be elements of a subdirect product indexed by X, we call $a : b$ the **equalizer** of a and b, and its complement, $a ; b = X \sim a : b$, the **inequalizer** or **difference** of a and b.

Other notations used for $a : b$ and $a ; b$, at least in the context of subdirect products, are $E(a, b)$ and $D(a, b)$, as well as the suggestive solecisms $[\![a = b]\!]$ and $[\![a \neq b]\!]$. We might also call $a ; b$ the 'support' of a and b, but this would conflict with the convention in analysis that the support of a function is the *closure* of the set of points where it is different from zero.

Here are three examples of precomplexes in modules, rings and general algebras. For the first, let $\boldsymbol{M}$ be a module over a commutative ring $\boldsymbol{R}$, and X the set of all maximal submodules $\boldsymbol{S}$ of $\boldsymbol{M}$. Define a function, $a \mapsto |a|$, from M to subsets of X by the formula:

$$|a| = \{\boldsymbol{S} \in X \mid a \in S\}.$$

It is easy to verify for all a, b in M and r in R that:

$$|0| = X, \quad |a| \subseteq |ra|, \quad \text{and } |a| \cap |b| \subseteq |a + b|.$$

The set,

$$B = \{|a| \mid a \in M\},$$

may be chosen as a subbasis for a topology $\mathcal{T}$ on X. Historically, complexes were first defined in the context of modules, as in the lecture notes of Armand Borel [**Borel64**, 1964, p. II-3] after Jean Leray [**Leray50**].

To connect this unary function, $|*|: M \to B$, with our more general concept, introduce a binary function ":" from M to B by defining

$$a : b = |a - b|.$$

It is easy to show that : satisfies the axioms for a precomplex $\langle \boldsymbol{M}, :, \boldsymbol{X} \rangle$.

Ring theory supplies the next example. Let $\boldsymbol{R}$ be a commutative ring that is regular in the sense of von Neumann: for all a in it there is an x such that $axa = a$. Here, it is assumed that $\boldsymbol{R}$ has a unity. This time, take X to be the set of all prime ideals $\boldsymbol{P}$ of $\boldsymbol{R}$. Introduce a function from R to subsets of X by the formula, $a \mapsto |a|$,

$$|a| = \{\boldsymbol{P} \in X \mid a \in P\}.$$

To get a precomplex, define a binary operation from R to the these sets of prime ideals by the same formula as for modules. To show that we again get a complex requires some effort. Generalizing the subsequent sheaf representation to shells motivated this book (Theorem VIII.2.1).

For other examples in ring theory, see the manuscript of Kist [**Kist69a**] where the spectrums of maximal ideals and minimal ideals are turned into complexes. See also Borceux and van den Bossche [**BorVa91**] for a locale-like presentation of complexes in ring theory. Also, the global subdirect products of Krauss and Clark [**KraCl79**] and the lattice products of Volker Weispfenning [**Weis70**] could be put into the framework of complexes, although in the latter case the underlying topological spaces would become arbitrary bounded lattices.

Creating complexes in arbitrary algebras is more general and encompasses the previous two examples as well as all applications to come. Consider an algebra $\boldsymbol{A}$ and a collection X of congruences of $\boldsymbol{A}$. Define a binary function ":" from A to subsets of X by

$$a : b = \{\theta \in X \mid a \,\theta\, b\}.$$

The aggregate of all such sets $a:b$ of congruences serves as a subbasis for a topology $\mathcal{T}$ on X. The properties defining a congruence readily imply that $\langle \boldsymbol{A}, :, \boldsymbol{X} \rangle$ is a precomplex. Turning this into a complex will require strong hypotheses on the algebra and its congruences.

As a specific example of this process, consider a **semilattice**, $\boldsymbol{S} = \langle S; \wedge \rangle$, with its associated partial order $\leq$ (for their definition, see Sect. II.1). For each a in A define the principal filter $F_a = \{b \in S \mid a \leq b\}$; and let θ_a be the corresponding congruence of $\boldsymbol{S}$ that has congruence classes: F_a and $S {\sim} F_a$ (if nonempty). Equivalently θ_a could have been defined by the clauses:

$$b \,\theta_a\, c \text{ iff } \left\{ \begin{array}{llll} \text{both} & a \leq b & \text{and} & a \leq c, \text{ or} \\ \text{both} & a \not\leq b & \text{and} & a \not\leq c. \end{array} \right.$$

Then X, the set of all such congruences θ_a, when endowed with the topology described in the previous paragraph, not only yields a precomplex, but also a complex. The implication (1.5) holds since $b:c = X$ implies that $b\,\theta_a\,c$ for all a in S; hence $b = b \wedge c = c$ when a is set in turn to b and c. It will follow from Theorem 2.1 in the next section that every semilattice is a subsemilattice of a sheaf of two-element semilattices.

The task ahead in these notes is to pick appropriate complexes by choosing the family X to be a useful set of factor congruences of $\boldsymbol{A}$. Such complexes will yield significant representation theorems.

To that end we now turn to the definition of a sheaf by first recalling the disjoint union,

$$\mathcal{A} = \biguplus_{x \in X} \boldsymbol{A}_x,$$

of algebras (see Definition II.2.42). The non-constant operations of $\mathcal{A}$ are defined within each summand, but not across them. To find the index of an element of a disjoint union, there is the projection,

$$\pi : \mathcal{A} \to X : \langle x, a \rangle \mapsto x.$$

To make a sheaf, $\mathcal{A}$ and X are given topologies.

1.2. DEFINITION. A **sheaf** $\langle \mathcal{A}, \pi, X \rangle$ of algebras has three parts:

(1) a topological partial algebra $\mathcal{A}$ given by a disjoint union of algebras $\boldsymbol{A}_x$ of the same type and indexed by X,
(2) the projection $\pi : \mathcal{A} \to X : \langle x, a \rangle \mapsto x$, and
(3) a topology $\mathcal{T}$ on X;

that altogether satisfy these three conditions:

(i) within the topology of $\mathcal{A}$ each partial operation of $\mathcal{A}$ of at least one argument is continuous;
(ii) for a nullary operation, $\omega = c$, the set $\{\langle x, c^{A_x} \rangle \mid x \in X\}$ is open in $\mathcal{A}$;
(iii) the projection $\pi : \mathcal{A} \to X$ is a local homeomorphism (Definition III.3.5).

Perhaps condition (i) requires explanation. A generic operation ω is a partial function from $\mathcal{A}^n$ to $\mathcal{A}$. Thus, to say that ω is continuous is to say that it is continuous with respect to the product topology on $\mathcal{A}^n$; in other words, $\omega^{-1}(U)$ is open in $\mathcal{A}^n$ whenever U is open in $\mathcal{A}$. Condition (iii) means that locally π has a continuous inverse. A consequence is that π is an open map, that is, the image of any open set of $\mathcal{A}$ is open in $\boldsymbol{X}$. Thus π induces the quotient topology on X, its original topology: it is the largest topology for X such that π is continuous. A curious consequence of (iii) is that each component $\boldsymbol{A}_x$, considered as a subspace of $\mathcal{A}$, has the discrete topology.

Here is some standard vocabulary for the parts of a sheaf $\langle \mathcal{A}, \pi, X \rangle$. The topological partial algebra $\mathcal{A}$ is called the **sheaf space**, a component A_x a **stalk** or **fiber**. Any element of a stalk is a **germ**. The topological space X is called the **index** or **base space**, and π the **projection**.[1]

As a technical aside, we should point out how all the other information that some writers build into the definition of a sheaf can be recovered from ours. In our definition of a sheaf, we are assuming that X is both the base space and the index set for $\mathcal{A}$. Note that π and X are redundant in $\langle \mathcal{A}, \pi, X \rangle$ since the sheaf space $\mathcal{A}$ completely determines the other two parts. This is because the projection π is inherent in the disjoint union, when both are viewed as a sets of ordered pairs, as at the end of Sect. II.1. As already noted, this continuous projection being open yields the quotient topology on the base space. Hence, both π and X can be recovered from $\mathcal{A}$, and there is no need to explicitly mention them. From now on we need refer only to $\mathcal{A}$ as the sheaf.

As another aside, in ring theory Pierce [**Pier67**, p. 10] and others use the term 'ringed space' for $\langle X, \mathcal{R} \rangle$, coming from algebraic geometry, for a topologized disjoint union $\mathcal{R}$ of rings together with its continuous projection onto the base space X. For general algebras this term might be extended to 'algebraic space' for $\langle X, \mathcal{A} \rangle$. As this pair contains the same information as does $\mathcal{A}$ by itself, at least as we have defined it, we forego this suggestive term.

A **global section** of a sheaf $\mathcal{A}$ is any continuous function σ from the base space X to the sheaf space $\mathcal{A}$ that is a right inverse to the projection π, in symbols $\pi \circ \sigma = 1_X$. More generally, a **section** is any continuous function σ from an open subset U of X to $\mathcal{A}$ such that $\pi \circ \sigma = 1_U$. The **equalizer** of two sections σ and τ is the set of indices in their common domain where they are equal:

$$\sigma : \tau = \{x \in X \mid \sigma(x) = \tau(x)\}.$$

An alternative formulation is

$$\sigma : \tau = \pi(\text{rng } \sigma \cap \text{rng } \tau).$$

We state now four properties of sheaves that will be used repeatedly — some have already been mentioned.

1.3. PROPOSITION. *Let $\mathcal{A}$ be a sheaf of algebras with the continuous projection $\pi : \mathcal{A} \to X$ onto the base space X.*

(a) *The projection π is an open function.*
(b) *The range of any global section σ is an open subset of $\mathcal{A}$.*
(c) *Any global section σ is an open function, and consequently* rng σ *is homeomorphic to X.*
(d) *The equalizer $\sigma : \tau$ of two global sections is open in X.*

---

[1] One should not confuse this projection with the n-ary projections of Sect. II.1, implicit in the term-operations of any algebra.

PROOF. (a) Let a be in the open subset $\mathcal{U}$ of $\mathcal{A}$. Since π is a local homeomorphism, there is an open subset $\mathcal{U}_a$ of $\mathcal{U}$ containing a that is homeomorphic to an open subset of $\boldsymbol{X}$; and the union of these $\mathcal{U}_a$ is all of $\mathcal{U}$. So the union of their images $\pi(\mathcal{U}_a)$ is open in $\boldsymbol{X}$.

(b) Let a be in $\operatorname{rng}\sigma$. Desired is a subset $\mathcal{U}$ of $\operatorname{rng}\sigma$ that is open in $\mathcal{A}$ and contains a. Use the local homeomorphism π to find an open subset $\mathcal{V}$ of $\mathcal{A}$ containing a that is homeomorphic to an open subset of $\boldsymbol{X}$. By the continuity of σ and π the desired open set $\mathcal{U}$ is $\mathcal{V} \cap \pi^{-1}\sigma^{-1}\mathcal{V}$.

(c) If U is open in $\boldsymbol{X}$, then $\sigma U = \operatorname{rng}\sigma \cap \pi^{-1}U$, and hence it is open in $\mathcal{A}$ by (b). That σ is injective follows from $\pi\sigma = 1$.

(d) True by (a), (b), and the alternative formulation of equalizer. □

Several overlapping sections may be patched together, as proven next. Disjoint patching is possible when the base space is Boolean, as will be proven in Proposition V.2.4.

1.4. PROPOSITION (Overlapping patchwork). *Let $\mathcal{A}$ be a sheaf of algebras over the base space $\boldsymbol{X}$. Suppose that $\langle U_i \mid i \in I\rangle$ is an open cover of an open set U in $\boldsymbol{X}$, with a corresponding family $\langle \sigma_i \mid i \in I\rangle$ of sections such that each σ_i is defined over the corresponding U_i. Further assume that these sections agree on their overlaps; in other words, for all i and j in I,*

$$\sigma_i(x) = \sigma_j(x) \qquad (x \in U_i \cap U_j).$$

Then there is a unique section σ defined over U such that

$$\sigma(x) = \sigma_i(x) \qquad (i \in I \text{ and } x \in U_i).$$

PROOF. As each σ_i is continuous, so σ is continuous. □

1.5. DEFINITION. We designate by $\Gamma(\mathcal{A})$ the set of all global sections of a sheaf $\mathcal{A}$. This is indeed an algebra $\boldsymbol{\Gamma}(\mathcal{A})$ of the same type as the component algebras $\boldsymbol{A}_x$; the operations are defined pointwise. In fact, $\boldsymbol{\Gamma}(\mathcal{A})$ can always be naturally identified with a subalgebra of the product $\prod_{x\in X} \boldsymbol{A}_x$.

The terminology of sheaf theory comes from harvesting wheat, where vertical stalks are cut horizontally by a scythe as in Fig. 1. A continuous sweep of its sharp blade across the stalks cuts out a global section. Another swipe, another global section, and so on. Altogether, these global sections make up $\Gamma(\mathcal{A})$. See the expository articles of Arthur Seebach, Linda Seebach and Lynn Steen [**SeeSS70**] and Christopher Mulvey [**Mulv79**] for several classical examples of sheaves in algebraic geometry, complex analysis, differential forms and ring theory, including the influential sheaves of local rings of Grothendieck [**GroDi60**].

Fiber bundles are a rich source of examples of sheaf spaces in algebraic topology. For example, a Möbius band may be created as a product with a twist. Cross the interval $[0, 2\pi]$ with another interval $[0, 1]$, and identify points at the ends: $\langle 0, a\rangle$ with $\langle 2\pi, 1-a\rangle$ $(a \in [0, 1]$. With the product

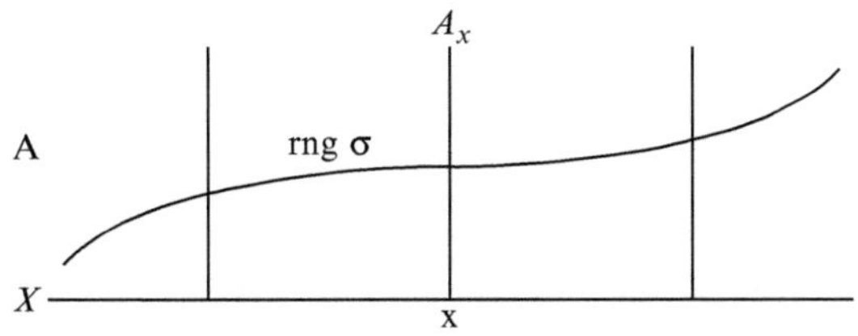

FIGURE 1. The sweep of a scythe, cutting out a global section σ, where $\boldsymbol{X}$ is the base space, $\boldsymbol{A}_x$ is a stalk over x, and $\mathcal{A}$ is the sheaf space, the union of all the disjoint stalks.

topology this is the sheaf space $\mathcal{A}$. The base space $\boldsymbol{X}$ is the half-open interval $[0, 2\pi)$, and the stalks are $[0, 1]$. If one wishes this to be a sheaf of algebras, then one adds operations to the stalks; for example,

$$\mu(a, b, c) = \text{the median of } a, b \text{ and } c,$$

$$\nu(a) = 1 - a,$$

for a, b, c in $[0, 1]$. The global sections are all continuous loops going once around the band.

There are two ways to present sheaves; as espaces étalés and through presheaves. Espaces étalés, the sheaves defined in Definition 1.2, are most natural for algebra, and were the original form given by Leray, but many of the other early expositions, mentioned in Sect. I.1, were with presheaves.

Davey [**Davey73**], using presheaves, gives an equivalent but categorical, and in some ways smoother, approach to $\boldsymbol{\Gamma}(\mathcal{A})$ via morphisms, functors, direct limits, and sections that are not global. See also [**Hofm72**], [**Romo94**], [**SeeSS70**], and [**Tenn75**] for expositions of presheaves on classical algebraic structures. For comparison, we sketch this approach, softening the role of categories.

1.6. DEFINITION. Given a topological space $\boldsymbol{X}$ and an algebraic type τ, a **presheaf** on $\boldsymbol{X}$ has data of two kinds:

(i) a function φ from open sets of $\boldsymbol{X}$ to algebras of type τ;
(ii) a family of homomorphisms, $\rho_{UV}: \varphi(V) \to \varphi(U)$, defined for all pairs of open sets for which $U \subseteq V$, such that
  (1) $\rho_{UU} = 1$ and
  (2) $\rho_{UW} = \rho_{UV} \circ \rho_{VW} \qquad (U \subseteq V \subseteq W)$.

When the overlapping patchwork of Proposition 1.4 holds in a presheaf, it becomes equivalent to a sheaf. We write it as `sheaf` to distinguish it from the earlier form of sheaf given in Definition 1.2.

1.7. DEFINITION. A presheaf over $\boldsymbol{X}$ with data φ and ρ is a `sheaf` whenever it has overlapping patchwork: for an open cover $\langle U_i \mid i \in I \rangle$ of an open subset U of $\boldsymbol{X}$ with elements σ_i in $\varphi(U_i)$, if

$$\rho_{U_i \cap U_j, U_i}(\sigma_i) = \rho_{U_i \cap U_j, U_j}(\sigma_j) \qquad (i, j \in I),$$

then there is a unique σ defined over U such that

$$\rho_{U_i, U}(\sigma) = \sigma_i \qquad (i \in I).$$

The transition from a sheaf $\langle \mathcal{A}, \pi, \boldsymbol{X} \rangle$ to a `sheaf` over $\boldsymbol{X}$ with data φ and ρ enlarges the language spelled out for sheaf spaces. For an open set U of $\boldsymbol{X}$, let $\Gamma(U, \mathcal{A})$ be the set of all sections σ from U to $\mathcal{A}$. and $\boldsymbol{\Gamma}(U, \mathcal{A})$ be its algebra, with the operations defined pointwise on stalks. Define φ and ρ by

$$\begin{aligned} \varphi(U) &= \boldsymbol{\Gamma}(U, \mathcal{A}) && (U \text{ open in } \boldsymbol{X}), \\ \rho_{UV}(\sigma) &= \sigma | U && (U, V \text{ open in } \boldsymbol{X} \text{ and } \sigma \in \Gamma(V, \mathcal{A})). \end{aligned}$$

The data φ and ρ on $\boldsymbol{X}$ is then a `sheaf`.

The other direction showing equivalence is not as trivial. Let a `sheaf` over $\boldsymbol{X}$ be given with data φ and σ. For each point x of $\boldsymbol{X}$ consider the collection, $\mathcal{U}_x = \{U \in \mathcal{T} \mid x \in U\}$, of open sets surrounding x. This is a neighborhood system of x directed by inclusion. The corresponding collection of algebras $\varphi(U)$ is directed by the homomorphisms, $\rho_{UV}\colon \varphi(V) \to \varphi(U)$. It has a direct limit $\boldsymbol{A}_x$ since the category of algebras of a given type is complete [**Berg98**, Prop. 8.1.6]. The disjoint union of these stalks $\boldsymbol{A}_x$ is given the equalizer topology of Sect. IV.2, and so we have sheaf. See [**Tenn75**, chaps. 1 and 2] for details.

Presheaves arise where sections are defined naturally on some open sets. This is the case in complex variables for an analytic function of one argument defined by a power series that converges only on a proper open subset of the plane. By way of contrast, where stalks come from quotients of an algebra, a sheaf space is most natural to define first. As this is typical of the sheaves in this book, we say nothing more about `sheaf`.

Representing algebras by sheaves has many merits. One is this. Each operation ω passes pointwise from the stalks to the sheaf. Because of this, identities are preserved:

$$\bigcap_{x \in X} \operatorname{Id} \boldsymbol{A}_x \subseteq \operatorname{Id} \boldsymbol{\Gamma}(\mathcal{A}).$$

Here, $\operatorname{Id} \boldsymbol{A}$ denotes the set of all identities satisfied by the algebra $\boldsymbol{A}$. If the stalks are in a particular equational class, then so is $\boldsymbol{\Gamma}(\mathcal{A})$. This is true since $\boldsymbol{\Gamma}(\mathcal{A})$ is a subalgebra of a product of the stalks $\boldsymbol{A}_x$.

Usually, $\boldsymbol{\Gamma}(\mathcal{A})$ is a subdirect product of the stalks $\boldsymbol{A}_x$, that is each stalk is exhausted — more precisely, through each element of each stalk there passes a global section. In the next chapter, we will see that this is always the case when $\boldsymbol{X}$ is a Boolean space (Proposition V.2.4(b)). Then sentences of the more general form $\forall\exists(\Psi = \Omega)$, where Ψ and Ω are terms of the given type, transfer back and forth between the stalks $\boldsymbol{A}_x$ and $\boldsymbol{\Gamma}(\mathcal{A})$

(Proposition V.2.7). This is so since sheaf topologies give a gluing more subtle than subdirect products alone can provide.

As an example, let X be any set with the discrete topology and let $\mathcal{A}$ be the disjoint union $\biguplus_{x\in X} A$ of the repeated algebra $A = A_x$. In this case, $\mathcal{A}$ is $X \times A$. For this sheaf $\mathcal{A}$, also with the discrete topology, we get the direct power $\Gamma(\mathcal{A}) = A^X$. In a related vein, if X is a Boolean space and if, for each member of the repeated stalk, there is a global section taking only this value, then we have essentially the **Boolean extension** of Foster [**Fost53**], although he did not invoke topologies. When the stalks may be different but still over a Boolean space and with the equalizers always clopen, Burris and Werner [**BurWe79**] call this sheaf a **Boolean product**, of which the **Boolean power** is a special case. These will be formally discussed in Sect. V.3.

With Proposition 1.3 in hand, a simpler but equivalent definition of global section arises by identifying a section with its range. A global section then becomes any subset of $\mathcal{A}$ such that

(i) Each element of σ belongs to a distinct stalk
(ii) Every stalk has an element in σ and
(iii) σ is open in $\mathcal{A}$.

In the sequel, however, we follow the original definition.

Here is an example showing that the topology on the base space X does not determine that of the sheaf $\mathcal{A}$. Write $\mathcal{Q} = \{0, 1\}$, which as an algebra has no operations; set $X = \mathcal{Q}$ and $A = \mathcal{Q} \uplus \mathcal{Q}$, (using again the original definition of disjoint union.) Assign to X the indiscrete topology; $\mathcal{T} = \{\emptyset, \mathcal{Q}\}$. On $\mathcal{A}$ there are two possible topologies:

$$\{\emptyset,\ \{\langle 0,0\rangle,\ \langle 1,0\rangle\},\ \{\langle 0,1\rangle,\ \langle 1,1\rangle\},\ \mathcal{A}\}, \text{ or}$$
$$\{\emptyset,\ \{\langle 0,0\rangle,\ \langle 1,1\rangle\},\ \{\langle 0,1\rangle,\ \langle 1,0\rangle\},\ \mathcal{A}\},$$

which give different sheaves with the same base space X. Contrast this with the fact that the topology of $\mathcal{A}$ always determines that of X as a quotient space.

We allow the empty sheaf of a given type, meaning that $X = \varnothing$, which makes $\mathcal{A}$ empty, and corresponds to an algebra $\Gamma(\mathcal{A})$ with one global section $\varnothing$. To keep this in perspective, one might review the Stone representation of Boolean algebras, where the empty topological space is dual to the one-element Boolean algebra.

2. Constructions

This section constructs sheaves from complexes. Of central importance is the Gel'fand morphism representing the original algebra in the algebra of all global sections of its constructed sheaf. Close to complexes are systems of congruences, which relate complexes to subdirect products. There follows a preview of how the construction of sheaves leads to the next chapter. We close this section with a brief discussion of Hausdorff sheaves.

Any sheaf $\langle \mathcal{A}, \pi, \boldsymbol{X} \rangle$ naturally yields a complex $\langle \boldsymbol{A}, :, \boldsymbol{X} \rangle$ — take $\boldsymbol{A}$ to be $\boldsymbol{\Gamma}(\mathcal{A})$, the algebra of global sections of the last section, and define a binary operation,

$$\sigma : \tau = \{x \in X \mid \sigma(x) = \tau(x)\},$$

which is the natural equalizer of the global sections σ and τ. Going the other way from complexes to sheaves is much more interesting and useful.

From a precomplex $\langle \boldsymbol{A}, :, \boldsymbol{X} \rangle$ we may construct a sheaf $\langle \mathcal{A}, \pi, \boldsymbol{X} \rangle$ of algebras as follows. For each x in X define a congruence θ_x of $\boldsymbol{A}$ by

$$a \ \theta_x \ b \text{ iff } x \in a : b.$$

Gather the quotient algebras, $A_x = \boldsymbol{A}/\theta_x$, into their disjoint union, $\mathcal{A} = \biguplus_{x \in X} A_x$. Stated otherwise, $\mathcal{A} = \{[a]_x \mid a \in A,\ x \in X\}$, where the germ $[a]_x$ is $\langle x, a/\theta_x \rangle$.[2] The projection π is defined as expected: $\pi([a]_x) = x$.

The operations of $\mathcal{A}$ are now partial and defined just when all arguments are in the same stalk. To introduce a topology, define for each element a of A and each open set U of X a subset $[a]_U$ of $\mathcal{A}$:

$$[a]_U = \{[a]_x \mid x \in U\}.$$

Note that $[a]_U \cap [a]_V = [a]_{U \cap V}$. The family of all such subsets of A serves as a basis for the topology of $\mathcal{A}$.

That the $[a]_U$ have been declared open secures continuity of the projection, $\pi : \mathcal{A} \to \boldsymbol{X}$, verifying that $\mathcal{A}$ is a sheaf. In fact, this is the smallest topology that could be put on $\mathcal{A}$ that would make π continuous. That each germ $[a]_x$ comes from an element of A ensures that π is a local homeomorphism. And finally, that each operation ω of the given type is continuous in $\mathcal{A}$ is explained as follows. If

$$\omega(a_1, \ldots, a_n) = b$$

in $\boldsymbol{A}$, then

$$\omega([a_1]_x, \ldots, [a_n]_x) = [b]_x$$

in the xth stalk of $\mathcal{A}$. Since $\omega^{\mathcal{A}}$ exists only when all its arguments are in the same stalk,

$$\omega([a_1]_U, \ldots, [a_n]_U) = [b]_U,$$

for any open set U of $\boldsymbol{X}$. Hence,

$$\omega^{-1}([b]_U) = \bigcup_{\omega(a_1,\ldots,a_n)=b} [a_1]_U \times \cdots \times [a_n]_U.$$

[2]This definition of the mapping $[a]_x$ of an element a of A into the xth component of the disjoint union appears convoluted, but care has to be taken not to let $[a]_x$ be just the coset a/θ_x since it may be that $a/\theta_x = a/\theta_y$ but $x \neq y$. However, the casual reader may read $[a]_x$ carelessly as a/θ_x, probably without confusion.

Therefore, the inverse image under ω of an open set in $\boldsymbol{\mathcal{A}}$ is a join of open sets in the product topology. Thus, ω is continuous in $\boldsymbol{\mathcal{A}}$.

We say that $\boldsymbol{\mathcal{A}}$ is the sheaf **associated** with the complex $\langle \boldsymbol{A},\ :\ , \boldsymbol{X}\rangle$. Its basis was chosen to guarantee a representation of the original algebra within the algebra of all global sections of the sheaf. To see this, consider the function γ from A to $\Gamma(A)$ defined by

$$\gamma(a)(x) = [a]_x \qquad (a \in A,\ x \in X). \tag{2.1}$$

This is often called the **Gel'fand morphism**. Two good things may happen:

(i) this function is injective;
(ii) this function is surjective.

The first is essential in applications, and the second is desirable. Note that, once the topology on X is given, the topology chosen for the sheaf $\boldsymbol{\mathcal{A}}$ is the largest compatible with π being a projection and with each image $\gamma(a)$ of an element a of A being continuous.

Before stating the theorem that relates complexes to sheaves and insures injectivity, we make a parenthetical comment. We are using the notation " : " for equalizers between global sections of sheaves and also between elements in complexes. There is no conflict or disparity, at least when the sheaf comes from a complex, because

$$\gamma(a){:}\gamma(b) = \{x \mid [a]_x = [b]_x\} = \{x \mid a\ \theta_x\ b\} = a{:}b.$$

The following result is theorem 1 of Swamy [**Swam74**] put into the language of complexes, and Swamy's theorem in turn is based on theorem 2.7 of Keimel [**Keim71**].

2.1. THEOREM. *Let $\langle \boldsymbol{A},\ :\ , \boldsymbol{X}\rangle$ be a precomplex of algebras and $\boldsymbol{\mathcal{A}}$ its associated sheaf. Construct from it the algebra $\boldsymbol{\Gamma}(\boldsymbol{\mathcal{A}})$ of global sections.*

(a) *The Gel'fand morphism is a homomorphism: $\gamma\colon \boldsymbol{A} \to \boldsymbol{\Gamma}(\boldsymbol{\mathcal{A}})$.*
(b) *The precomplex is actually a complex if, and only if, γ is an injection.*

PROOF. Before anything else we must show that the range of γ is within $\Gamma(\boldsymbol{\mathcal{A}})$. That is, we must prove that $\gamma(a)$ is a global section for each a in A. This reduces to showing that $\gamma(a)\colon \boldsymbol{X} \to \boldsymbol{\mathcal{A}}$ is continuous. It suffices to demonstrate that the inverse image of any set in the basis of $\boldsymbol{\mathcal{A}}$ is again open. So consider a typical set in the basis: $[a]_U = \{[a]_x \mid x \in U\}$, for some open set U of $\boldsymbol{X}$. Clearly

$$\gamma(a)^{-1}([a]_U) = U,$$

by the definition of $\gamma(a)$. Hence, $\gamma(a)$ is continuous and in $\Gamma(\boldsymbol{\mathcal{A}})$.

(a) By the pointwise definition of the operations, γ is a homomorphism.

(b) $\Rightarrow$. In order to prove that γ is an injection when $\langle \boldsymbol{A},\ :\ , \boldsymbol{X}\rangle$ is a complex, it suffices to prove that $\bigcap_{x\in X} \theta_x = 0_{\mathrm{Con}\,\boldsymbol{A}}$. To that end assume that a and b are related by the left side, that is, $a\ \theta_x\ b$ for all x in X. From the definition of the θ_x, we have $a{:}b = X$. Since we started with a complex, it follows that $a = b$.

$\Leftarrow$. Assume that γ is injective and that $a:b = X$; we wish to show that $a = b$. The latter means that $x \in a : b$ for all x in X. Therefore, by the definitions of θ_x and γ,

$$\gamma(a)(x) = [a]_x = [b]_x = \gamma(b)(x) \quad (x \in X).$$

Hence $\gamma(a) = \gamma(b)$. Since γ is injective, $a = b$. □

With this theorem, we relate the identities of $\boldsymbol{A}$ to those of $\boldsymbol{\Gamma}(\boldsymbol{\mathcal{A}})$.

2.2. COROLLARY. *Let $\langle \boldsymbol{A}, :, \boldsymbol{X} \rangle$ be a precomplex and $\boldsymbol{\mathcal{A}}$ its sheaf.*

(a) $\operatorname{Id} \boldsymbol{A} \subseteq \operatorname{Id} \boldsymbol{\Gamma}(\boldsymbol{\mathcal{A}})$.

(b) *If we actually start with a complex, then* $\operatorname{Id} \boldsymbol{A} = \operatorname{Id} \boldsymbol{\Gamma}(\boldsymbol{\mathcal{A}})$.

In other words when we have a precomplex, if $\boldsymbol{A}$ belongs to a particular equational class, then so does $\boldsymbol{\Gamma}(\boldsymbol{\mathcal{A}})$, and when we have a complex, the converse is also true.

PROOF. (a) This follows from the fact that $\boldsymbol{\Gamma}(\boldsymbol{\mathcal{A}})$ is a subalgebra of a direct product of homomorphic images of $\boldsymbol{A}$.

(b) Since γ is injective, $\boldsymbol{A}$ is isomorphic to a subalgebra of $\boldsymbol{\Gamma}(\boldsymbol{\mathcal{A}})$. □

Global sections have other characterizations. The first is due to Krauss and Clark [**KraCl79**, lemma 2.27]. It should be clear in (b) below that a finite open cover suffices when $\boldsymbol{X}$ is compact.

2.3. COROLLARY. *For a precomplex $\langle \boldsymbol{A}, :, \boldsymbol{X} \rangle$ over a topological space, $\boldsymbol{X} = \langle X, \mathcal{T} \rangle$, with the associated sheaf $\boldsymbol{\mathcal{A}}$:*

(a) $\Gamma(\boldsymbol{\mathcal{A}}) = \{\sigma : X \to \mathcal{A} \mid \forall a \in A\ (\gamma(a):\sigma \in \mathcal{T})\}$;

(b) *$\sigma \in \Gamma(\boldsymbol{\mathcal{A}})$ if, and only if, there is an open cover $\langle U_i \mid i \in I \rangle$ of X and there are b_i in A such that $\sigma|U_i = \gamma(b_i)|U_i$ for i in I.*

PROOF. (a) There are two inclusions to verify.

$\subseteq$. Obvious.

$\supseteq$. It suffices to verify that $\sigma^{-1}([a]_U) = U \cap (\gamma(a):\sigma)$ for a in A and U in $\mathcal{T}$.

(b) There are two implications to verify.

$\Rightarrow$. Clear from (a).

$\Leftarrow$. This follows from part (a) since $\gamma(a):\sigma = \bigcup_{i\in I} \big(U_i \cap (a:b_i)\big)$. □

As regards the condition (ii) that the Gel'fand map, $\gamma : \boldsymbol{A} \to \boldsymbol{\Gamma}(\boldsymbol{\mathcal{A}})$, be surjective, this is much harder to come by. In fact there is no known useful criterion that is both necessary and sufficient. A sufficient hypothesis for surjectivity is that we have a Boolean algebra of factor congruences, and this will be used extensively from Chap. VI onward,.

The simplest example with γ not being surjective arises from the complex consisting of the three-element distributive lattice, $\boldsymbol{C}_3 = \langle \{a, b, c\}; \vee, \wedge \rangle$ with $a < b < c$, over the two-element discrete space, $\Omega = \{0, 1\}$. The equalizer in this complex is specified completely by its effect on three pairs: $a:b = \{0\}$, $b:c = \{1\}$, and $a:c = \emptyset$. The associated sheaf $\boldsymbol{\mathcal{A}}$ will have the discrete topology and hence $\boldsymbol{\Gamma}(\mathcal{A})$ is the direct power $(\boldsymbol{C}_2)^2$. Later chapters have more examples in which the Gel'fand morphism is not surjective. This completes the essential points we wanted to make in this section, and now we round it out with peripheral matters.

Complexes closely resemble subdirect products; in fact, so much so that one might question the need for a new notion. We briefly examine the relationship between the two. Call the pair $\langle \boldsymbol{A}, \Theta \rangle$ a **system of congruences** if $\boldsymbol{A}$ is an algebra where each θ of Θ is a congruence of $\boldsymbol{A}$. If $\bigcap \Theta = 0_{\mathbf{Con}\,\boldsymbol{A}}$, then $\boldsymbol{A}$ is isomorphic to a subdirect product of the quotients $\boldsymbol{A}/\theta$ as θ runs over all of the congruences in the family Θ. In any case, $\boldsymbol{A}/\bigcap\Theta$ is isomorphic to a subdirect product of all the $\boldsymbol{A}/\theta$ for θ in Θ.

Each precomplex $\langle \boldsymbol{A}, :, \boldsymbol{X} \rangle$ gives rise to a system $\langle \boldsymbol{A}, \Theta \rangle$ of congruences:

$$\Theta = \{\theta_x \mid x \in X\}, \text{ where } \theta_x = \{\langle a, b \rangle \mid x \in a:b\}.$$

Notice that the topology of $\boldsymbol{X}$ is lost. This, of course, plays a role in the sheaf $\mathcal{A}$ constructed from the complex. In fact, it is easy to show directly from the topology on $\mathcal{A}$ that $\boldsymbol{A}/\bigcap\Theta$ is isomorphic to a subalgebra of $\boldsymbol{\Gamma}(\mathcal{A})$. As an aside, some of the points of X may coalesce in the sense of generating equal congruences: $\theta_x = \theta_y$; but this does not happen if the base space is T_0 and the equalizers form a subbasis of it. We will elaborate on this, which is related to the notion of irredundancy of Krauss and Clark [**KraCl79**] [1979, p. 10 and p. 41, theorem 3.30].

Going from a system $\langle \boldsymbol{A}, \Theta \rangle$ of congruences to a precomplex $\langle \boldsymbol{A}, :, \boldsymbol{\Theta} \rangle$ is slightly more subtle. Take the space to be Θ, and to construct a topology on Θ define the equalizer as before, $a:b = \{\theta \in \Theta \mid a \,\theta\, b\}$, and let these equalizers be a subbasis for a topology $\mathcal{T}$ of open sets on Θ.[3] Again, one shows that $\boldsymbol{A}/\bigcap\Theta$ is isomorphic to a subalgebra of $\boldsymbol{\Gamma}(\mathcal{A})$, where $\mathcal{A}$ is the sheaf coming from the complex $\langle \boldsymbol{A}, :, \boldsymbol{\Theta} \rangle$.

Krauss and Clark call this topology the **equalizer** topology and $\boldsymbol{\Gamma}(\mathcal{A})$ the **global closure** of $\langle \boldsymbol{A}, \Theta \rangle$, from which they develop their monograph [**KraCl79**]. When the Gel'fand morphism is surjective, they call this a **global subdirect product**. Their chaps. 2–5 give many criteria for a topology to exist on the index set so that a subdirect product is global. Here is their theorem 2.28, rephrased in our language.

2.4. THEOREM. *Let $\langle \boldsymbol{A}, \Theta \rangle$ be a system of congruences with the equalizer topology $\mathcal{T}$ given on Θ, and let $\boldsymbol{B}$ be its subdirect product. Then the Gel'fand morphism is surjective if, and only if, for every ρ in $\prod_{\theta\in\Theta} A/\theta$, if $\rho:\sigma \in \mathcal{T}$ for all σ in $\boldsymbol{B}$, then $\rho \in \boldsymbol{B}$.*

---

[3]Davey [**Davey73**, sect. 2] considers this construction in more generality where a topology is already specified for Θ, but it does not always yield a sheaf. He gives several conditions under which a sheaf is guaranteed, as in his lemma 2.1.

What if we go full circle? Starting with a system $\langle \boldsymbol{A}, \Theta \rangle$ of congruences, going to a precomplex $\langle \boldsymbol{A}, :, \boldsymbol{\Theta} \rangle$ and then returning to a system will give back the same system that we started with. What is less obvious is that starting with an arbitrary precomplex $\langle \boldsymbol{A}, :, \boldsymbol{X} \rangle$ and then going to a system of congruences will not return us to the original precomplex. The original base space X may not agree with the new base space $\boldsymbol{\Theta}$. There may be collapsing of points, as noted above. But even if not, the old topology may be finer than the new: we conclude that there might be global sections in $\boldsymbol{\Gamma}(\mathcal{A}_X)$ that are not in $\boldsymbol{\Gamma}(\mathcal{A}_{\boldsymbol{\Theta}})$, where $\boldsymbol{\Gamma}(\mathcal{A}_X)$ is the sheaf coming from the complex $\langle \boldsymbol{A}, :, \boldsymbol{X} \rangle$ and $\boldsymbol{\Gamma}(\mathcal{A}_\Theta)$ comes from $\langle \boldsymbol{A}, :, \boldsymbol{\Theta} \rangle$. Thus, for a precomplex there are the embeddings:

$$\frac{\boldsymbol{A}}{\bigcap \Theta} \underset{inj.}{\rightarrow} \boldsymbol{\Gamma}(\mathcal{A}_{\boldsymbol{\Theta}}) \underset{inj.}{\rightarrow} \boldsymbol{\Gamma}(\mathcal{A}_X).$$

The following theorem sums up this discussion. It states that, under the hypotheses discovered earlier, one can go full circle without losing anything essential. We leave its proof to the reader.

2.5. THEOREM. (a) *If $\langle \boldsymbol{A}, \Theta \rangle$ is a system of congruences, with $\langle \boldsymbol{A}, :, \boldsymbol{X} \rangle$ being the resulting precomplex, and $\langle \boldsymbol{A}, \Psi \rangle$ is the system of congruences obtained from this precomplex, then $\Theta = \Psi$. The intermediate space $\boldsymbol{X}$ is T_0 and the equalizers form a subbasis.*

(b) *Suppose that $\langle \boldsymbol{A}, :, \boldsymbol{X} \rangle$ is a precomplex, with $\langle \boldsymbol{A}, \Theta \rangle$ the resulting system of congruences, and $\langle \boldsymbol{A}, \overset{\star}{\underset{\star}{}}, \boldsymbol{Y} \rangle$ the resulting precomplex. Then there is a continuous map $\varphi: \boldsymbol{X} \to \boldsymbol{Y}$ such that $\varphi(a:b) = a \overset{\star}{\underset{\star}{}} b$. If $\boldsymbol{X}$ is T_0 and the equalizers form a subbasis for $\boldsymbol{X}$, then φ is a homeomorphism.*

Two remarks are in order about part (b). First, lest anyone think that dealing with just precomplexes is the reason one cannot go exactly full circle without some hypotheses, note that this theorem remains true even if one replaces 'precomplex' by 'complex' and 'system of congruences' by 'system of congruences whose intersection is zero'. Second, although the hypothesis that the equalizers form a subbasis appears rather weak, we will not be able to guarantee its fulfillment until §V.2.

We single out an important instance of this construction.

2.6. DEFINITION. For any algebra $\boldsymbol{A}$ its **subdirect sheaf representation** is the sheaf as constructed above when Θ is the set of all subdirectly irreducible congruences of $\boldsymbol{A}$.

2.7. THEOREM. *The subdirect sheaf representation of an algebra is injective. When the algebra is nontrivial and not subdirectly irreducible, then its subdirect sheaf representation is nontrivial, that is, it has at least two nontrivial stalks.*

PROOF. By Birkhoff's Theorem II.2.40, $\bigcap \Theta = 0_{\mathbf{Con}\,\boldsymbol{A}}$. Hence we have a complex. By Theorem 2.1, the representation is injective. □

One may analyze also the relationship between complexes and sheaves and to what extent they resemble each other. We have already investigated how to construct a sheaf $\mathcal{A}$ from any complex $\langle \boldsymbol{A}, :, \boldsymbol{X} \rangle$ and examined in detail how the obvious complex $\langle \boldsymbol{\Gamma}(\mathcal{A}), :, \boldsymbol{X} \rangle$ coming from $\mathcal{A}$ relates to the original complex.

Might we conceivably need some day a sheaf not constructible from a complex? In the next section, we show that this never happens, that is, each sheaf satisfying a modest algebraic assumption always comes from a complex. This so because in general, in going from any sheaf to a complex to a sheaf again, there is a simple relationship between the old and new sheaves; they are in a nice one-to-one correspondence, at least up to isomorphism.

Complexes are not essential. We could bypass them in the applications we are interested in, and many times work directly from Boolean subsemilattices, to be defined in the next chapter. However, complexes are a convenient and intuitive halfway house on the road to sheaf representations. Each is a repository for all the information needed to bring about a representation, including the original $\boldsymbol{A}$ and its extension $\boldsymbol{\Gamma}(\mathcal{A})$, as well as both of the topologies, either explicitly or implicitly.

As some definitions given in this section and later may appear to be backwards or upside down to some readers, we discuss the choices that had to be made. These choices make no difference in the truth of the mathematics, but they do make a difference in how we intuitively view the construction of sheaves. Whether to use the equalizer ":" or its complement ";", the inequalizer, was a toss-up. Should open or closed sets be the fundamental notion in topology? Equalizers and open sets fit well together and opting for them simplifies some proofs. Since this is nontraditional in sheaves of classical algebras, we explain our reasons.

Stating the axioms of a complex in terms of equalizers brings out their essence as just the conditions defining a relation to be congruence. Also, equalizers being open makes their appearance in applications simpler.

On the other hand, the inequalizer measures the extent to which two elements are separated, and to get a faithful representation – one which is isomorphic to the original algebra – all pairs of elements of it must be separated by at least one congruence. The axioms for a precomplex in terms of inequalizers, which must now be closed subsets of $\boldsymbol{X}$, become:

$$
\begin{aligned}
a;a &= \emptyset, \\
a;b &= b;a, \\
a;c &\subseteq a;b \cup b;c, \\
\omega(a_1, \ldots, a_n);\omega(b_1, \ldots, b_n) &\subseteq a_1;b_1 \cup \cdots \cup a_n;b_n.
\end{aligned}
$$

If, in addition,

$$a;b = \emptyset \quad \text{implies} \quad a = b,$$

then all elements are separated and it is a complex. This rewriting makes the concept of complex look like some kind of generalized metric space. Thus, it is easy to 'think metric'.

The deciding factor against using inequalizers as the fundamental notion is that in applications a negation, such as 'not $a\theta b$', crops up at the very beginning, and more negations appear later on to undo it, all adding up to some unappealing double negatives. By sticking with open sets, most formulas are positive, making it easier to devise and follow arguments. We have opted for choices that make transparent the algebraic arguments, but unfortunately at the expense of classical parallels with analysis and algebraic geometry. Many who are not algebraists will find it easier to follow some of the succeeding arguments if they rewrite them in their set-theoretical duals, using the inequalizer instead.

The discussion should now be closed, but there is one last subtlety that initially confused the author and that may be bothering some readers. In analogy with a continuous real-valued function, where the range space is Hausdorff and hence the set of zero values is always closed, it would appear that an abstract equalizer also ought to be closed rather than open. How can the literature have it open in some cases and closed in others; won't it make a difference between right and wrong theorems? The nub of the matter is that the equalizer of continuous functions in general need be neither open nor closed.

At the other extreme are equalizers that are always both open and closed, and hence so are their inequalizers. The definition of a sheaf insures that the equalizer of two global sections will be open. Although peripheral to our purposes, conditions guaranteeing that the equalizers are also closed will be sprinkled throughout this book. Clopen equalizers make the sheaf structure more transparent and are one of the defining properties of Boolean products, which will be discussed briefly in the next chapter.

One sufficient condition for the equalizers in a sheaf $\langle \boldsymbol{\mathcal{A}}, \pi, \boldsymbol{X} \rangle$ to be clopen is that its sheaf space $\boldsymbol{\mathcal{A}}$ be Hausdorff. Such sheaves are called **Hausdorff sheaves** and are more general than Boolean products. It will help to fill out the theory later to have necessary and sufficient conditions assuring this. When the sheaf comes from a complex these appear in [**Swam74**, theorem 1] and [**KraCl79**, lemma 2.32], although their language is different from ours.

This next proposition has a hypothesis that needs an explanation. Clearly, $\boldsymbol{\Gamma}(\boldsymbol{\mathcal{A}})$ is a subalgebra of the product $\prod_{x \in X} \boldsymbol{A}_x$. The only question is whether we have a subdirect product, that is, whether each projection,

$$\Gamma(\boldsymbol{\mathcal{A}}) \to A_x : \sigma \mapsto \sigma(x),$$

is surjective. Conceivably, there might be unused elements in some of the stalks. That is, for each element a in each stalk A_x, is there a global section σ such that $a = \sigma(x)$? If the sheaf comes from a complex, then this is so.

If the sheaf is over a Boolean space, we will see in Proposition V.2.4 that this is again guaranteed. If neither, it must be postulated, as expressed next.

2.8. DEFINITION.
(SS) Through each point of a sheaf there passes a global section.

Sometimes this is called a 'subdirect' sheaf. It is equivalent to saying that, for each element a of the sheaf space, there is a subset of it containing a that is homeomorphic by the projection π to the base space $\boldsymbol{X}$. Most of the sheaves of subsequent chapters satisfy this proviso.

2.9. PROPOSITION. *Let $\mathcal{A}$ be a sheaf of algebras that satisfies* SS. *Then $\mathcal{A}$ is a Hausdorff sheaf if, and only if,*
(a) *its base space $\boldsymbol{X}$ is a Hausdorff space, and*
(b) *$\sigma : \tau$ is clopen* $\quad (\sigma, \tau \in \Gamma(\mathcal{A}))$.

PROOF. $\Rightarrow$. (a) Let x and y be distinct points of X, and assume $\sigma \in \Gamma(A)$. Since the stalks are disjoint, $\sigma(x) \neq \sigma(y)$; thus, there exist disjoint neighborhoods $\mathcal{U}$ and $\mathcal{V}$ respectively of $\sigma(x)$ and $\sigma(y)$ in $\mathcal{A}$. Hence $\sigma^{-1}(\mathcal{U})$ and $\sigma^{-1}(\mathcal{V})$ are disjoint neighborhoods of x and y in X.
For (b), since inequalizers of global sections in a sheaf are always closed, it suffices to show that any inequalizer $\sigma;\tau$ is open. Assume $\sigma, \tau \in \Gamma(\mathcal{A})$ and $x \subset \sigma;\tau$. Since $\mathcal{A}$ is Hausdorff, there are disjoint neighborhoods $\mathcal{U}$ and $\mathcal{V}$ of $\sigma(x)$ and $\tau(x)$, respectively. Because the projection, $\pi: \mathcal{A} \to \boldsymbol{X}$, is a local homeomorphism, we may, without loss of generality, shrink $\mathcal{U}$ to $\mathcal{U} \cap \operatorname{rng} \sigma$ so that it is homeomorphic to a neighborhood U of x in X, and similarly reduce $\mathcal{V}$ to be homeomorphic to a neighborhood V of x. What we have proven is that any x in $\sigma;\tau$ has a neighborhood $U \cap V$ entirely within $\sigma;\tau$. This is true since $\mathcal{U} \cap \mathcal{V} = \emptyset$.
$\Leftarrow$. Let a_x and b_y be distinct elements of $\mathcal{A}$ in stalks A_x and A_y. By hypothesis, there are global sections σ and τ such that $\sigma(x) = a_x$ and $\tau(y) = b_y$. There are two cases.
Case 1. $x \neq y$. Since $\boldsymbol{X}$ is a Hausdorff space, there are disjoint neighborhoods U and V of x and y, respectively. And so their inverse images, $\pi^{-1}(U)$ and $\pi^{-1}(V)$, must be disjoint neighborhoods of a_x and b_y.
Case 2. $x = y$. As already implicitly shown by the other direction of implication, there are three neighborhoods – $\mathcal{U}$ of $\sigma(x)$ in $\mathcal{A}$, $\mathcal{V}$ of $\tau(x)$ in $\mathcal{A}$, and W of x in $\boldsymbol{X}$ – that are all homeomorphic. Since $\sigma;\tau$ is clopen, the disjoint sets,

$$\mathcal{U} \cap \sigma(\sigma;\tau) \text{ and } \mathcal{V} \cap \tau(\sigma;\tau),$$

are neighborhoods of $\sigma(x)$ and $\tau(x)$, respectively. Hence $\mathcal{A}$ is Hausdorff. $\square$

As an aside illustrating this proposition and the limitations of our use of sheaves, consider, without details, the sheaf of germs of holomorphic functions over $\mathbb{C}$, the complex numbers; this is one of the classical examples motivating the definition of sheaves [**SeeSS70**]. Then $\sigma : \tau$ is open for all global sections σ and τ, by the definition of sheaf. These global sections in this context are entire: they are holomorphic functions on the whole complex plane, $\boldsymbol{X} = \mathbb{C}$. Their equalizer is also closed since any two entire functions, σ and τ, that agree on an open set, that is, their germs are equal at a point, agree everywhere on the complex plane. Hence, $\sigma : \tau$ is clopen for all global sections, and so $\mathcal{A}$ is Hausdorff by the proposition. Therefore, if σ and τ are distinct, they must be disjoint, that is, $\sigma : \tau = \emptyset$.

From this observation, one might conclude that sheaves over the base space $\mathbb{C}$ are uninteresting and of no importance. On the contrary, if we do not insist that $\boldsymbol{\Gamma}(\mathcal{A})$ be a subdirect product of the stalks, then, in the sheaf of germs of holomorphic functions, the exponential function represents a global section, since it is entire, but the logarithmic function, on the one hand, is less than a global section since it is not definable at the origin and, on the other hand, it is 'more' since its domain of holomorphy is multivalued.

3. Categorical Reformulation

An adjunction of their categories relates complexes to sheaves more fully than was done in the previous sections. We spend time now setting the stage for each of the final sections of the next several chapters, which will summarize their principal constructions categorically. Each succeeding construction becomes more specialized, and eventually categorical adjunction becomes categorical equivalence; see Table 1 at the end of Sect. VI.4. These constructions can be thought of as generalizing the duality between Boolean algebras and Boolean spaces.

In this section, we define the categories of complexes and sheaves as well as a pair of functors going from one category to the other. Much that needs to be proven is routine but sometimes complicated. The main difficulty lies in keeping track of the various levels of abstraction going all the way from elements of algebras through topologies up to the natural transformations defining adjunctions. Otherwise, once the definitions are set up and the theorems correctly formulated, many of the proofs are inevitable. Details are gone into only when we want to get a taste of the argument, or when some special technique or trick is required.

All algebras under discussion are assumed to have the same fixed type. Recall that categories were introduced nonobjectively in Sect. III.2 as partial algebras of morphisms with the operations of composition, domain, and codomain, where unities play the role of objects. This style will be followed when defining $\mathfrak{Complex}$, $\mathfrak{Sheaf}$, and other categories; but occasionally it is clearer to write objectively.

3.1. DEFINITION. The category $\mathfrak{Complex}$ consists of all morphisms between all complexes $\langle \boldsymbol{A}, :, \boldsymbol{X} \rangle$ of a given algebraic type. A **morphism** from one complex, $\mathcal{A} = \langle \boldsymbol{A}, :, \boldsymbol{X} \rangle$, to another, $\mathcal{B} = \langle \boldsymbol{B}, \overset{\star}{\star}, \boldsymbol{Y} \rangle$, is a pair $\langle \varphi, \chi \rangle$ of functions,

$$\varphi: \boldsymbol{A} \to \boldsymbol{B} \text{ and } \chi: \boldsymbol{Y} \to \boldsymbol{X},^{4}$$

of which φ is a homomorphism and χ is a continuous function, such that

$$\chi^{-1}(a_1 : a_2) \subseteq \varphi(a_1) \overset{\star}{\star} \varphi(a_2) \qquad (a_1, a_2 \in A).$$

Express this as $\langle \varphi, \chi \rangle: \mathcal{A} \to \mathcal{B}$, or in a pinch abbreviate it as $\langle \varphi, \chi \rangle$. The **domain** and **codomain** of this morphism are the respective unities:

$$1_{\mathcal{A}} = \langle 1_{\boldsymbol{A}}, 1_{\boldsymbol{X}} \rangle: \mathcal{A} \to \mathcal{A} \text{ and } 1_{\mathcal{B}} = \langle 1_{\boldsymbol{B}}, 1_{\boldsymbol{Y}} \rangle: \mathcal{B} \to \mathcal{B},$$

or objectively just $\mathcal{A}$ and $\mathcal{B}$. The **composition** of two such morphisms,

$$\langle \varphi_1, \chi_1 \rangle: \langle \boldsymbol{A}, :, \boldsymbol{X} \rangle \to \langle \boldsymbol{B}, \overset{\star}{\star}, \boldsymbol{Y} \rangle \text{ and } \langle \varphi_2, \chi_2 \rangle: \langle \boldsymbol{B}, \overset{\star}{\star}, \boldsymbol{Y} \rangle \to \langle \boldsymbol{C}, \overset{\bigstar}{\bigstar}, \boldsymbol{Z} \rangle,$$

defined whenever the middle complexes are the same, is the composition of their component functions:

$$\langle \varphi_2, \chi_2 \rangle \circ \langle \varphi_1, \chi_1 \rangle = \langle \varphi_2 \circ \varphi_1, \chi_1 \circ \chi_2 \rangle: \langle \boldsymbol{A}, :, \boldsymbol{X} \rangle \to \langle \boldsymbol{C}, \overset{\bigstar}{\bigstar}, \boldsymbol{Z} \rangle.$$

3.2. PROPOSITION. *$\mathfrak{Complex}$ is a category.*

PROOF. Check that the composition of morphisms is again a morphism of complexes and that $\mathfrak{Complex}$ satisfies the axioms for a category. □

To define the category of sheaves, several preliminary definitions are needed, including a special kind of product. As explained in Sect. IV.1, one may specify a sheaf by a triple $\langle \mathcal{A}, \pi, \boldsymbol{X} \rangle$ or just by the sheaf space $\mathcal{A}$ alone.

3.3. DEFINITION. Let $\mathcal{A}$ be a sheaf with base space $\boldsymbol{X}$ and projection $\pi: \mathcal{A} \to \boldsymbol{X}$, and similarly let $\mathcal{B}$ be a sheaf with base space $\boldsymbol{Y}$ and projection $\rho: \mathcal{B} \to \boldsymbol{Y}$. Whenever we have a continuous function $\chi: \boldsymbol{Y} \to \boldsymbol{X}$, we may form the **pullback**:

$$\mathcal{A} \times_\chi Y = \{\langle a, y \rangle \mid \pi(a) = \chi(y)\}.$$

To give $\mathcal{A} \times_\chi Y$ a topology, restrict the product topology on $\boldsymbol{A} \times \boldsymbol{Y}$. Recall that sheaves are disjoint unions of algebras:

$$\mathcal{A} = \biguplus_{x \in X} \boldsymbol{A}_x \text{ and } \mathcal{B} = \biguplus_{y \in Y} \boldsymbol{B}_y.$$

Likewise, $\mathcal{A} \times_\chi Y$ can also be turned into a partial algebra; it is the disjoint union of algebras $A_x \times \{y\}$ with $\chi(y) = x$:

$$\mathcal{A} \times_\chi \boldsymbol{Y} = \biguplus_{y \in Y} (\boldsymbol{A}_{\chi(y)} \times \{y\}),$$

[4]Note that the functions φ and χ go in opposite directions. This is typical of many of the compound morphisms in this book.

a third morphism $\langle \psi_3, \chi_3 \rangle : \mathcal{C} \to \mathcal{D}$, where $\mathcal{D} = \langle \boldsymbol{D}, \upsilon, \boldsymbol{W} \rangle$. Then the two-fold composition of these three morphisms, with either association, evaluates as

$$\big(\langle \psi_1, \chi_1 \rangle \circ \langle \psi_2, \chi_2 \rangle \circ \langle \psi_3, \chi_3 \rangle\big)\ (a, w)$$
$$= \psi_3\big(\psi_2\big(\psi_1(a, (\chi_2 \circ \chi_3)(w)), \chi_3(w)\big), w\big)\big)$$
$$(a \in \mathcal{A} \ \text{ and } \ w \in W).$$

The remaining axioms (ii) and (iii) are straightforward to verify. □

3.8. EXERCISE. Recall Definition III.2.9 that in a category a morphism is defined to be an isomorphism if it is invertible. Prove that two sheaves, $\langle \mathcal{A}, \pi, \boldsymbol{X} \rangle$ and $\langle \mathcal{B}, \rho, \boldsymbol{Y} \rangle$, are isomorphic in $\mathfrak{Sheaf}$ if, and only if, there are a homeomorphism $\psi : \mathcal{A} \to \mathcal{B}$ (of one argument) that is also an isomorphism of partial algebras, and another homeomorphism $\chi : \boldsymbol{X} \to \boldsymbol{Y}$ such that $\rho \circ \psi = \chi \circ \pi$.

As noted in Sect. 2, each complex leads to a sheaf, and vice versa. This correspondence can be formalized as two functors going in opposite directions and adjoint to each other.s First, we will define the functor, $\Phi_1 : \mathfrak{Complex} \to \mathfrak{Sheaf}$. Recall the Gel'fand morphism γ, given by $\gamma(a)(x) = [a]_x$, which embeds the algebra $\boldsymbol{A}$ of a complex $\langle \boldsymbol{A}, :, \boldsymbol{X} \rangle$ into the algebra $\boldsymbol{\Gamma}(\mathcal{A})$ of all global sections of the corresponding sheaf, $\langle \mathcal{A}, \pi, \boldsymbol{X} \rangle = \Phi_1(\langle \boldsymbol{A}, :, \boldsymbol{X} \rangle)$. Here, $\mathcal{A} = \{[a]_x \mid a \in A, x \in X\}$ and $\pi([a]_x) = x$.

3.9. DEFINITION. It suffices to say how Φ_1 acts on morphisms. Let $\langle \varphi, \chi \rangle$ be a morphism of complexes from $\langle \boldsymbol{A}, :, \boldsymbol{X} \rangle$ to $\langle \boldsymbol{B}, \overset{\star}{\underset{\star}{}}, \boldsymbol{Y} \rangle$ with $\mathcal{A}$ and $\mathcal{B}$ the associated sheaves created by Φ_1. The sheaf $\mathcal{A}$ will have $\boldsymbol{X}$ as a base space and $\mathcal{B}$ will have $\boldsymbol{Y}$. Define $\Phi_1(\langle \varphi, \chi \rangle)$ to be $\langle \psi, \chi \rangle$ where $\psi : \mathcal{A} \times_\chi Y \to \mathcal{B}$ acts as

$$\psi([a]_x, y) = [\varphi(a)]_y \tag{3.1}$$

whenever $a \in A$, $x \in X$, $y \in Y$, and $\chi(y) = x$.

3.10. PROPOSITION. *Φ_1 a functor from $\mathfrak{Complex}$ to $\mathfrak{Sheaf}$.*

PROOF. Definition III.2.13 tells us that a functor from $\mathfrak{Complex}$ to $\mathfrak{Sheaf}$ would be a function $\Phi_1 : \mathfrak{Complex} \to \mathfrak{Sheaf}$ such that, for all morphisms $\langle \varphi_1, \chi_1 \rangle$, $\langle \varphi_2, \chi_2 \rangle$ and $\langle \varphi, \chi \rangle$ in $\mathfrak{Complex}$,

(i) $\Phi_1(\langle \varphi_1, \chi_1 \rangle \circ \langle \varphi_2, \chi_2 \rangle) = \Phi_1(\langle \varphi_1, \chi_1 \rangle) \circ \Phi_1(\langle \varphi_2, \chi_2 \rangle)$ whenever the composition $\langle \varphi_1, \chi_1 \rangle \circ \langle \varphi_2, \chi_2 \rangle$ exists;
(ii) $\Phi_1(\mathrm{dom}\langle \varphi, \chi \rangle) = \mathrm{dom}\, \Phi_1(\langle \varphi, \chi \rangle)$;
(iii) $\Phi_1(\mathrm{cod}\langle \varphi, \chi \rangle) = \mathrm{cod}\, \Phi_1(\langle \varphi, \chi \rangle)$.

To prove that Φ_1 is a function to $\mathfrak{Sheaf}$, that the pair, $\langle \psi, \chi \rangle = \Phi_1\big(\langle \varphi, \chi \rangle\big)$, is truly a morphism of sheaves whenever $\langle \varphi, \chi \rangle$ is a morphism of complexes, means checking through the clauses defining a sheaf morphism, given in

Definition 3.4. Unchanged, χ is continuous. That $\rho\big(\psi([a]_x, y)\big) = y$ whenever $\chi(y) = x$ follows from the definitions of ψ and ρ, which is the projection from $\mathcal{B}$ to $\boldsymbol{Y}$. Since each ψ_{xy} is defined from stalk to stalk as a composition of two homomorphisms,

$$\psi_{xy}([a]_x) = [\varphi(a)]_y \qquad (a \in A \text{ and } x = \chi(y)),$$

then the whole function ψ is a homomorphism of partial algebras from $\mathcal{A} \times_\chi \boldsymbol{Y}$ to $\mathcal{B}$.

To exhibit the continuity of ψ, it suffices to show that the inverse image of any of the sets $[b]_V$ is again open. Here, $b \in B$ and V is open in $\boldsymbol{Y}$ so that $[b]_V$ is in the basis for the topology of $\mathcal{B}$. Then

$$\begin{aligned}\psi^{-1}([b]_V) &= \{\langle [a]_x, y\rangle \mid \varphi(a) = b, x = \chi(y), y \in V\}\\ &= \bigcup_{\varphi(a)=b} \{\langle [a]_x, y\rangle \mid x = \chi(y), y \in V\}\\ &= \bigcup_{\varphi(a)=b} [a]_{\chi(V)} \times_\chi V.\end{aligned}$$

Thus, the inverse image of $[b]_V$ is a union of open sets of the basis for the topology on $\mathcal{A} \times_\chi \boldsymbol{Y}$.

Now we can move on to prove clauses (i)–(iii) above. To prove (i), that Φ_1 preserves the composition of morphisms, assume that there are two morphisms of complexes with a common middle complex:

$$\langle \varphi_1, \chi_1\rangle : \langle \boldsymbol{A}, :, \boldsymbol{X}\rangle \to \langle \boldsymbol{B}, \overset{\star}{\underset{\star}{:}}, \boldsymbol{Y}\rangle \text{ and } \langle \varphi_2, \chi_2\rangle : \langle \boldsymbol{B}, \overset{\star}{\underset{\star}{:}}, \boldsymbol{Y}\rangle \to \langle \boldsymbol{C}, \overset{\bigstar}{\underset{\bigstar}{:}}, \boldsymbol{Z}\rangle.$$

Within (i), set $\langle \varphi, \chi\rangle = \langle \varphi_2, \chi_2\rangle \circ \langle \varphi_1, \chi_1\rangle$. The left side of (i), $\langle \psi_l, \chi_l\rangle = \Phi_1(\langle \varphi, \chi\rangle)$ with $\varphi = \varphi_2 \circ \varphi_1$, is then easy to compute. For $\psi_l([a]_x, z) = [\varphi(a)]_z$ when $\chi_l(z) = x$. Also $\chi_l = \chi_1 \circ \chi_2$.

For the right side of (i), set $\langle \psi_r, \chi_r\rangle = \Phi_1(\langle \varphi_1, \chi_1\rangle) \circ \Phi_1(\langle \varphi_2, \chi_2\rangle)$. Then $\chi_r = \chi_1 \circ \chi_2$. If one sets $\langle \psi_1, \chi_1\rangle = \Phi_1(\langle \varphi_1, \chi_1\rangle)$, then one knows by (3.1) that $\psi_1([a]_x, y) = [\varphi_1(a)]_y$ when $a \in A$ and $\chi_1 y = x$. Similarly, setting $\langle \psi_2, \chi_2\rangle = \Phi_2(\langle \varphi_2, \chi_2\rangle)$, one knows that $\psi_2([b]_y, z) = [\varphi_2(b)]_z$ when $b \in B$ and $\chi_2 z = y$. One computes, when $\chi z = x$, that

$$\begin{aligned}\psi_r([a]_x, z) &= \psi_2\big(\psi_1([a]_x, \chi_2(z)), z\big) = \psi_2\big([\varphi_1(a)]_y, z\big)\\ &= [\varphi_2(\varphi_1(a)]_z = [\varphi(a)]_z = \psi_l([a]_x, z).\end{aligned}$$

Proving (ii) and (iii), that Φ_1 preserves identity morphisms, is easy. □

In preparation for defining the second functor Γ_1, recall that $\boldsymbol{\Gamma}(\mathcal{A})$ is the algebra of all global sections of a sheaf $\mathcal{A}$. Recall also that for any sheaf $\mathcal{A}$ over a space $\boldsymbol{X}$ there is the complex $\langle \boldsymbol{\Gamma}(\mathcal{A}), :, \boldsymbol{X}\rangle$ where $\sigma : \tau = \{x \mid \sigma(x) = \tau(x)\}$.

3.11. Definition. Our second functor, Γ_1: 𝔖𝔥𝔢𝔞𝔣 → ℭ𝔬𝔪𝔭𝔩𝔢𝔵, goes in the opposite direction from Φ_1, producing from any sheaf a complex as just described. Given a morphism of sheaves, $\langle \psi, \chi \rangle : \mathcal{A} \to \mathcal{B}$, define a morphism of their complexes,

$$\langle \varphi, \chi \rangle : \langle \boldsymbol{\Gamma}(\mathcal{A}), :, \boldsymbol{X} \rangle \to \langle \boldsymbol{\Gamma}(\mathcal{B}), \overset{\star}{\underset{\star}{,}}, \boldsymbol{Y} \rangle,$$

where the first component $\varphi : \boldsymbol{\Gamma}(\mathcal{A}) \to \boldsymbol{\Gamma}(\mathcal{B})$ evaluates as

$$\varphi(\sigma)(y) = \psi\big(\sigma(\chi(y)), y\big) \qquad (\sigma \in \Gamma(\mathcal{A}), y \in Y).$$

Set $\Gamma_1(\langle \psi, \chi \rangle) = \langle \varphi, \chi \rangle$.

3.12. PROPOSITION. *Γ_1 is a functor from* $\mathfrak{Sheaf}$ *to* $\mathfrak{Complex}$.

PROOF. To prove that Γ_1 is a functor, one must show that morphisms in $\mathfrak{Sheaf}$ go to morphisms in $\mathfrak{Complex}$, that Γ_1 preserves composition of morphisms, and that it preserves unities. To prove that sheaf morphisms go to complex morphisms, one must show four things about $\langle \varphi, \chi \rangle = \Gamma_1(\langle \psi, \chi \rangle)$:

(1) φ takes global sections of $\Gamma(\mathcal{A})$ to global sections of $\Gamma(\mathcal{B})$;
(2) φ is a homomorphism;
(3) χ is continuous;
(4) $\chi^{-1}(\sigma_1 : \sigma_2) \subseteq \varphi(\sigma_1) \overset{\star}{\underset{\star}{:}} \varphi(\sigma_2) \qquad (\sigma_1, \sigma_2 \in \Gamma(\mathcal{A}))$.

For (1), if $\sigma \in \Gamma(\mathcal{A})$ then $\varphi(\sigma)$ is continuous, since it is defined as a composition of continuous functions; also by its definition $\varphi(\sigma)$ is a right inverse of the projection ρ of $\mathcal{B}$.

For (2), one proves that φ preserves any operation ω by invoking the definition of φ again, noting that ψ is a homomorphism, and using the pointwise definition of operations ω in $\Gamma(\mathcal{B})$.

In (3), the same old χ is still continuous.

For (4) one has on the left side that

$$\begin{aligned} y \in \chi^{-1}(\sigma_1 : \sigma_2) \;\text{ iff }\; & \chi(y) \in \sigma_1 : \sigma_2 \\ \text{iff }\; & \sigma_1\big(\chi(y)\big) = \sigma_2\big(\chi(y)\big). \end{aligned}$$

On the right side, since $\varphi(\sigma)(y) = \psi\big(\sigma(\chi(y)), y\big)$ for all σ in $\Gamma(\mathcal{A})$ and y in Y, it follows that

$$\begin{aligned} y \in \varphi(\sigma_1) \overset{\star}{\underset{\star}{:}} \varphi(\sigma_2) \;\text{ iff }\; & \varphi(\sigma_1)(y) = \varphi(\sigma_2)(y) \\ \text{iff }\; & \psi\big(\sigma_1(\chi(y)), y\big) = \psi\big(\sigma_2(\chi(y)), y\big). \end{aligned}$$

This settles the inclusion.

To show that Γ_1 preserves the composition of sheaf morphisms, assume that $\mathcal{A} \xrightarrow{\langle \psi_1, \chi_1 \rangle} \mathcal{B} \xrightarrow{\langle \psi_2, \chi_2 \rangle} \mathcal{C}$. Let $\langle \varphi_1, \chi_1 \rangle$ and $\langle \varphi_2, \chi_2 \rangle$ be the respective morphisms of complexes obtained by applying Γ_1, and compose them as $\langle \varphi_l, \chi_l \rangle$. Let $\langle \psi, \chi \rangle : \mathcal{A} \to \mathcal{C}$ be the composite of the sheaf morphisms, and write $\langle \varphi_r, \chi_r \rangle$ for its image under Γ_1. In symbols,

$$\langle \varphi_l, \chi_l \rangle = \langle \varphi_2 \circ \varphi_1, \chi_1 \circ \chi_2 \rangle \quad \text{and} \quad \langle \varphi_r, \chi_r \rangle = \Gamma_1\big(\langle \psi_2, \chi_2 \rangle \circ \langle \psi_1, \chi_1 \rangle\big).$$

One must demonstrate that $\langle \varphi_l, \chi_l \rangle = \langle \varphi_r, \chi_r \rangle$. For the homomorphisms, the definition of Γ_1 yields

$$\begin{aligned}\varphi_l(\sigma)(z) &= (\varphi_2 \circ \varphi_1)(\sigma)(z) = \big(\varphi_2(\varphi_1(\sigma))\big)(z) = \psi_2\big(\varphi_1(\sigma)(y), z\big) \\ &= \psi_2\big(\psi_1(\sigma(x), y), z\big) = \psi\big(\sigma(x), z\big) = \varphi_r(\sigma)(z),\end{aligned}$$

whenever $\sigma \in \Gamma(\mathcal{A})$, $z \in Z$, $y = \chi_2(z)$, and $x = \chi_1(y)$.

Lastly, one easily shows that identity morphisms pass through Γ_1. □

We are now ready to formulate an adjoint situation, Definition III.2.21. Before anything else, we define the natural transformations:

$$\eta_1 : 1_{\mathfrak{Complex}} \to \Gamma_1 \circ \Phi_1 \quad \text{and} \quad \varepsilon_1 : \Phi_1 \circ \Gamma_1 \to 1_{\mathfrak{Sheaf}}.$$

The first will yield essentially the Gel'fand morphisms already introduced. To simplify the notation for natural transformations, we break partly with the nonobjective approach to categories and identify a complex with its identity morphism, and similarly for sheaves.

3.13. Definition. For the first natural transformation, η_1 of a complex, $\mathcal{A} = \langle \boldsymbol{A}, :, \boldsymbol{X} \rangle$, is a morphism of complexes,

$$\eta_1(\mathcal{A}) = \langle \gamma, 1_{\boldsymbol{X}} \rangle : \mathcal{A} \to \mathcal{B},$$

where $\mathcal{B} = \Gamma_1(\Phi_1(\mathcal{A})) = \langle \Gamma(\mathcal{A}), \overset{\star}{\underset{\star}{:}}, \boldsymbol{X} \rangle$. For the sheaf in the middle, $\mathcal{A} = \biguplus_{x \in X}(A/\theta_x)$, where $a\ \theta_x\ b$ iff $x \in a{:}b$. Also $\sigma \overset{\star}{\underset{\star}{:}} \tau = \{x \in X \mid \sigma(x) = \tau(x)\}$. For the two functions in the morphism, $1_{\boldsymbol{X}}$ is the identity function on $\boldsymbol{X}$, and γ is the Gel'fand morphism, $\gamma : \boldsymbol{A} \to \Gamma(\mathcal{A})$, given already by

$$\gamma(a)(x) = [a]_x \qquad (a \in A \text{ and } x \in X).$$

For the second natural transformation, ε_1 of a sheaf $\mathcal{A}$ over $\boldsymbol{X}$ is a morphism of sheaves,

$$\varepsilon_1(\mathcal{A}) = \langle \psi, 1_{\boldsymbol{X}} \rangle : \mathcal{B} \to \mathcal{A},$$

where $\mathcal{B} = \Phi_1(\Gamma_1(\mathcal{A})) = \biguplus_{x \in X}\big(\boldsymbol{\Gamma}(\mathcal{A})/\theta_x\big)$, and the sheaf $\mathcal{B}$ is over the same base space $\boldsymbol{X}$. The function $\psi : \mathcal{B} \times_{1_X} \boldsymbol{X} \to \mathcal{A}$ is given by

$$\psi([\sigma]_x, x) = \sigma(x) \qquad (\sigma \in \Gamma(\mathcal{A}) \text{ and } x \in X). \tag{3.2}$$

It is necessary, and unfortunately tedious, to check that η_1 and ε_1 are natural transformations (and also that ε_1 is well defined).

3.14. Proposition. (a) *η_1 is a natural transformation from the functor $1_{\mathfrak{Complex}}$ to the functor $\Gamma_1 \circ \Phi_1$.*
(b) *ε_1 is a natural transformation from $\Phi_1 \circ \Gamma_1$ to $1_{\mathfrak{Sheaf}}$.*

Proof. First, one verifies that η_1 and ε_1 give morphisms of the required kind. That $\eta_1(\mathcal{A})$ is a morphism of complexes for each complex $\mathcal{A}$ is readily seen from its definition, $\eta_1(\mathcal{A}) = \langle \gamma, 1_{\boldsymbol{X}} \rangle$, and the fact that $\gamma(a) \overset{\star}{\underset{\star}{:}} \gamma(b) = a{:}b$.

Similarly, one must show for each sheaf $\mathcal{A}$ that $\varepsilon_1(\mathcal{A})$ is a sheaf morphism. There are several clauses to check in the definition of a sheaf morphism; all but one are fairly trivial. The nontrivial argument concerns the continuity of $\psi: \mathcal{B} \times_{1_X} X \to \mathcal{A}$, defined in (3.2). For any open subset $\mathcal{U}$ of $\mathcal{A}$, one needs to prove that its inverse image,

$$\psi^{-1}(\mathcal{U}) = \{\langle [\sigma]_x, x\rangle \mid \sigma(x) \in \mathcal{U}, \sigma \in \Gamma(\mathcal{A}) \text{ and } x \in X\},$$

is open in $\mathcal{B} \times_{1_X} X$. Since the range of any global section σ is open in $\mathcal{A}$, the set $\operatorname{rng}\sigma \cap \mathcal{U}$ is also open. Its projection, $U = \pi(\operatorname{rng}\sigma \cap \mathcal{U})$, is open in X since π is an open function. So an open set $[\sigma]_U$ is obtained in the basis of the topology of $\mathcal{B}$ (see Sect. 2), and then an open set $[\sigma]_U \times_{1_X} X$ in the basis of $\mathcal{B} \times_{1_X} X$. But

$$[\sigma]_U \times_{1_X} X = \psi^{-1}(\operatorname{rng}\sigma \cap \mathcal{U}).$$

Thus,

$$\begin{aligned}\psi^{-1}(\mathcal{U}) &= \bigcup_{\sigma \in \Gamma(\mathcal{A})} \psi^{-1}(\operatorname{rng}\sigma \cap \mathcal{U}) \\ &= \bigcup_{\sigma \in \Gamma(\mathcal{A})} [\sigma]_U \times_{1_X} X,\end{aligned}$$

and hence it is a union of open subsets. Therefore, ψ is continuous.

Second, to complete the proof for both parts (a) and (b), one must confirm that these two diagrams commute, with $\langle \varphi, \chi\rangle : \mathfrak{A} \to \mathfrak{B}$, a morphism of complexes, and $\langle \psi, \chi\rangle : \mathcal{A} \to \mathcal{B}$, a morphism of sheaves, respectively:

$$\begin{array}{ccc} \mathfrak{A} & \xrightarrow{\eta_1(\mathfrak{A})} & (\Gamma_1 \circ \Phi_1)(\mathfrak{A}) \\ {\scriptstyle \langle\varphi,\chi\rangle}\downarrow & & \downarrow{\scriptstyle (\Gamma_1\circ\Phi_1)(\langle\varphi,\chi\rangle)} \\ \mathfrak{B} & \xrightarrow[\eta_1(\mathfrak{B})]{} & (\Gamma_1 \circ \Phi_1)(\mathfrak{B}) \end{array}$$

$$\begin{array}{ccc} (\Phi_1 \circ \Gamma_1)(\mathcal{A}) & \xrightarrow{\varepsilon_1(\mathcal{A})} & \mathcal{A} \\ {\scriptstyle (\Phi_1\circ\Gamma_1)(\langle\psi,\chi\rangle)}\downarrow & & \downarrow{\scriptstyle \langle\psi,\chi\rangle} \\ (\Phi_1 \circ \Gamma_1)\mathcal{B} & \xrightarrow[\varepsilon_1(\mathcal{B})]{} & \mathcal{B} \end{array}$$

The definitions of η_1, ε_1, Φ_1, Γ_1 and the Gel'fand morphism will prove commutativity. Here is the proof for the first diagram; the second is left to the reader. Now $\eta_1(\mathfrak{A}) = \langle \gamma_{\mathfrak{A}}, 1_X\rangle$ and $\eta_1(\mathfrak{B}) = \langle \gamma_{\mathfrak{B}}, 1_Y\rangle$ where $\mathfrak{A} = \langle A, :, X\rangle$ and $\mathfrak{B} = \langle B, \star_\star, X\rangle$. Tracing the first component of complex morphisms around the lower left corner yields the homomorphism $\gamma_{\mathfrak{B}} \circ \varphi$. To go around the upper right corner, set $\Phi_1(\langle \varphi, \chi\rangle = \langle \psi, \chi\rangle$ where

$\psi([a]_x, y) = [\varphi(a)]_y$ when $y \in Y$ and $x = \chi(y)$; and set $\Gamma_1(\langle\psi, \chi\rangle = \langle\widetilde{\varphi}, \chi\rangle$ where $\widetilde{\varphi}(\sigma)(y) = \psi\big(\sigma(x), y\big)$ when $x = \chi(y)$. It follows that

$$\begin{aligned}\widetilde{\varphi}\big(\gamma_{\mathcal{A}}(a)\big)(y) &= \psi\big(\gamma_{\mathcal{A}}(a)(x), y\big) \qquad (a \in A, y \in Y, x = \chi(y))\\ &= \psi\big([a]_x, y\big)\\ &= [\varphi(a)]_y\\ &= \gamma_{\mathcal{B}}\big(\varphi(a)\big)(y).\end{aligned}$$

Therefore, $\widetilde{\varphi} \circ \gamma_{\mathcal{A}} = \gamma_{\mathcal{B}} \circ \varphi$, thus demonstrating equality of the first components around both corners. The second component χ stays the same. □

3.15. THEOREM. *The functors,*

$$\Phi_1 \colon \mathfrak{Complex} \rightleftarrows \mathfrak{Sheaf} \colon \Gamma_1,$$

together with the natural transformations,

$$\eta_1 \colon 1_{\mathfrak{Complex}} \to \Gamma_1 \circ \Phi_1 \quad \text{and} \quad \varepsilon_1 \colon \Phi_1 \circ \Gamma_1 \to 1_{\mathfrak{Sheaf}},$$

form an adjunction:

$$\langle \eta_1, \varepsilon_1 \rangle \colon \Phi_1 \dashv \Gamma_1 \colon \langle \mathfrak{Sheaf}, \mathfrak{Complex} \rangle.$$

That is, Φ_1 is a left adjoint of Γ_1, and Γ_1 is a right adjoint of Φ_1.

PROOF. Since previous propositions of this section have shown that Φ_1 and Γ_1 are functors, and likewise that η_1 and ε_1 are natural transformations, it suffices to verify clause (iii) in Definition III.2.21 of adjunction. To do this, prove commutativity of the corresponding diagrams by tracing elements through them:

$$\begin{array}{ccc} \Gamma_1(\boldsymbol{\mathcal{A}}) & \xrightarrow{\eta_1(\Gamma_1(\boldsymbol{\mathcal{A}}))} & \Gamma_1(\Phi_1(\Gamma_1(\boldsymbol{\mathcal{A}}))) \\ & {\scriptstyle 1_{\Gamma_1(\boldsymbol{\mathcal{A}})}} \searrow & \downarrow {\scriptstyle \Gamma_1(\varepsilon_1(\boldsymbol{\mathcal{A}}))} \\ & & \Gamma_1(\boldsymbol{\mathcal{A}}) \end{array} \qquad \begin{array}{ccc} \Phi_1(\mathcal{A}) & \xrightarrow{\Phi_1(\eta_1(\mathcal{A}))} & \Phi_1(\Gamma_1(\Phi_1(\mathcal{A}))) \\ & {\scriptstyle 1_{\Phi_1(\mathcal{A})}} \searrow & \downarrow {\scriptstyle \varepsilon_1(\Phi_1(\mathcal{A}))} \\ & & \Phi_1(\mathcal{A}) \end{array}$$

where $\boldsymbol{\mathcal{A}} \in \mathfrak{Sheaf}$ and $\mathcal{A} \in \mathfrak{Complex}$.

Here is a check of the commutativity of the second diagram. It suffices to trace just the first component of the sheaf morphisms since the second component is 1_X. Given a complex, $\mathcal{A} = \langle \boldsymbol{A}, :, \boldsymbol{X}\rangle$, write $\Phi_1(\mathcal{A})$ as the sheaf $\langle \boldsymbol{\mathcal{A}}, \pi, \boldsymbol{X}\rangle$; a typical element of $\boldsymbol{\mathcal{A}}$ is $[a]_x$ where $a \in A$ and $x \in X$. The morphism, $\eta_1(\mathcal{A}) = \langle\varphi, 1\rangle$ from $\mathcal{A}$ to $\Gamma_1\Phi_1\mathcal{A}$, evaluates by Definition 3.13 as $\varphi(a) = \langle [a]_x \mid x \in X\rangle$ for a in A. Using Definition 3.9 to apply Φ_1 to $\langle\varphi, 1\rangle$ yields a sheaf morphism $\langle\psi_1, 1\rangle$:

$$\psi_1([a]_x, x) = [\varphi(a)]_x = [\langle [a]_x \mid x \in X\rangle]_x.$$

Now follow this by the downward morphism $\varepsilon_1(\Phi_1(\mathcal{A}))$. The action of its first component is defined to be $\psi_2([\sigma]_x, x) = \sigma(x)$ for σ in $\boldsymbol{\mathcal{A}}$. So the composite action is

$$(\psi_2 \circ \psi_1)([a]_x, x) = \psi_2([\langle [a]_x \mid x \in X\rangle]_x, x) = \langle [a]_x \mid x \in X\rangle(x) = [a]_x.$$

Therefore, the composite morphism around the corner is $1_{\Phi_1(\mathcal{A})}$ on the diagonal. □

It is worthwhile to limit our sheaves to those whose global sections form a subdirect product, as in the next definition.

3.16. DEFINITION. Let $\mathfrak{SheafAlg}$ be the full subcategory of $\mathfrak{Sheaf}$ of sheaves of a given algebraic type that satisfy axiom SS of Definition 2.8, that is, through each point of the sheaf space there passes a global section.

When so restricted, ε_1 becomes a natural isomorphism. Since any sheaf $\Phi_1(\mathcal{A})$ coming from a complex $\mathcal{A}$ has to satisfy axiom SS, the category $\mathfrak{SheafAlg}$ will not be mentioned after this chapter. The next proposition and theorem were reported in the abstract [**Knoe92b**].

3.17. PROPOSITION. *With the functor Γ_1 restricted to $\mathfrak{SheafAlg}$, the natural transformation ε_1 becomes a natural isomorphism from $\Phi_1 \circ \Gamma_1$ to $1_{\mathfrak{SheafAlg}}$.*

PROOF. To prove that ε_1 is a natural isomorphism, one needs to show – using the language of the preceding theorem – that $\varepsilon_1(\boldsymbol{\mathcal{A}})$ is an isomorphism of the sheaves $\boldsymbol{\mathcal{B}}$ and $\boldsymbol{\mathcal{A}}$. Here, $\boldsymbol{\mathcal{B}} = \Phi_1(\Gamma_1(\boldsymbol{\mathcal{A}}))$ and $\langle \psi, 1_{\boldsymbol{X}} \rangle : \boldsymbol{\mathcal{B}} \to \boldsymbol{\mathcal{A}}$, as given in the definition of ε_1. But what does it means for two sheaves to be isomorphic? In categorical terms this would mean that there exists a sheaf morphism $\langle \overline{\psi}, \overline{\chi} \rangle$ inverse to $\langle \psi, \chi \rangle$:

$$\langle \overline{\psi}, \overline{\chi} \rangle : \boldsymbol{\mathcal{A}} \rightleftharpoons \boldsymbol{\mathcal{B}} : \langle \psi, \chi \rangle,$$

that is, composition in either order should give identity morphisms. Tracing this through Definitions 3.4 and 3.5 of sheaf morphisms and their composition, and breaking it down by components, one arrives at four equations to be proven:

$$\begin{aligned}
\psi\big(\overline{\psi}(a, \chi(x)), x\big) &= a \qquad (a \in \boldsymbol{\mathcal{A}},\ x \in X \text{ with } \pi(a) = \chi(x)), \\
\overline{\psi}\big(\psi(b, \overline{\chi}(y)), y\big) &= b \qquad (b \in \boldsymbol{\mathcal{B}},\ y \in Y \text{ with } \rho(b) = \overline{\chi}(y)), \\
\overline{\chi} \circ \chi &= 1_{\boldsymbol{X}}, \\
\chi \circ \overline{\chi} &= 1_{\boldsymbol{Y}}.
\end{aligned}$$

By the definition of ε_1, $\boldsymbol{Y} = \boldsymbol{X}$ and $\chi = 1_{\boldsymbol{X}}$. For the last two equations to be satisfied, it must also be that $\overline{\chi} = 1_{\boldsymbol{X}}$. To define $\overline{\psi}$ use axiom SS, which distinguishes $\mathfrak{SheafAlg}$ from $\mathfrak{Sheaf}$, namely, that for each element a of $\boldsymbol{\mathcal{A}}$ we have $a = \sigma(x)$ for some global section σ and x in X. So, define

$$\overline{\psi}(a, x) = [\sigma]_x.$$

It is easy to check that $\overline{\psi}$ is well defined and also that $\langle \overline{\psi}, 1_{\boldsymbol{X}} \rangle$ is a sheaf morphism. Then the first two equations are verified:

$$\begin{aligned}
\psi\big(\overline{\psi}(a, x), x\big) &= \psi([\sigma]_x, x) = \sigma(x) = a; \\
\overline{\psi}\big(\psi(b, x), x\big) &= \overline{\psi}(\tau(x), x) = [\tau]_x = b,
\end{aligned}$$

since $b = [\tau]_x$ for some τ in $\Gamma(\boldsymbol{\mathcal{A}})$. □

3.18. THEOREM. *The functors,* $\Phi_1 : \mathfrak{Complex} \rightleftarrows \mathfrak{SheafAlg} : \Gamma_1$, *appropriately restricted, together with the natural transformations,*

$$\eta_1 : 1_{\mathfrak{Complex}} \to \Gamma_1 \circ \Phi_1 \quad \text{and} \quad \varepsilon_1 : \Phi_1 \circ \Gamma_1 \to 1_{\mathfrak{SheafAlg}},$$

form an adjunction:

$$\langle \eta_1, \varepsilon_1 \rangle : \Phi_1 \dashv \Gamma_1 : \langle \mathfrak{SheafAlg}, \mathfrak{Complex} \rangle$$

in which ε_1 *is a natural isomorphism.*

PROOF. Use the preceeding proposition and theorem. □

Thus, the algebra in a complex is embedded into a potentially larger algebra, which might be called its 'completion'. With this in mind, recall Definition III.2.19 that, for a complex $\mathcal{A}$ with a sheaf $\boldsymbol{\mathcal{A}} = \Phi_1(\mathcal{A})$, by a Γ_1-universal map $\langle \eta_1(\mathcal{A}), \boldsymbol{\mathcal{A}} \rangle$ one means that, for each sheaf $\boldsymbol{\mathcal{B}}$ and each complex morphism, $\langle \varphi, \chi \rangle : \mathcal{A} \to \Gamma_1(\boldsymbol{\mathcal{B}})$, there exists a unique sheaf morphism $\langle \psi, \chi \rangle : \boldsymbol{\mathcal{A}} \to \boldsymbol{\mathcal{B}}$ such that this triangle commutes:

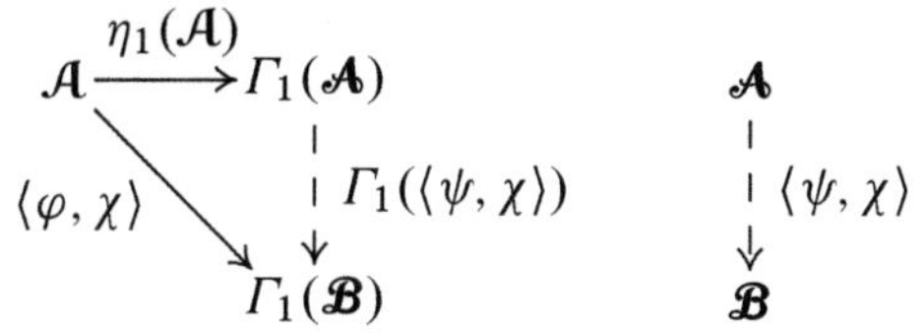

where ψ may be defined by

$$\psi(\langle [a]_x, y \rangle) = \varphi(a)(y) \qquad (a \in A,\ x = \chi(y),\ y \in Y).$$

To be precise, one should assert the existence of a sheaf morphism $\langle \psi, \widehat{\chi} \rangle$ with a new continuous function $\widehat{\chi}$; but, since one knows a priori that the χ's will be the same, the hat is omitted to simplify the picture.

3.19. PROBLEM. For a complex $\mathcal{A}$, what properties has this completion, $\eta_1(\mathcal{A}) : \mathcal{A} \to \Gamma_1(\boldsymbol{\mathcal{A}})$?

We close this chapter by pointing out that, for most of the sheaves to be used in this book, their base spaces will be Boolean spaces, or those spaces weakened to be locally compact. Sheaves used elsewhere have a variety of base spaces; for example, see [**Berg73**] and [**SeeSS70**].

For those readers who have delved into the proofs, the motto for this section might well be:

"The devil is in the details."

V

Boolean Subsemilattices

Ever since Stone represented Boolean algebras topologically [**Stone36**], many have extended his theorem to diverse algebraic systems. Structurally, it suffices to discover a fragment of the congruence lattice that is Boolean. The value of such a fragment, to be called a Boolean subsemilattice in the first section, is flexibility. By varying the Boolean subsemilattice to suit the context, different representation theorems follow automatically. In a later chapter, for example, by looking at all the factor ideals of a unital ring in which the annihilator of any element is a principal ideal generated by an idempotent, we obtain stalks with no zero divisors. With sheaves this theorem can be extended well beyond ring theory.

The main result in the second section is that each Boolean subsemilattice of congruences of an algebra $\boldsymbol{A}$ has a dual Boolean space of prime ideals of congruences, and in turn this leads to a representation of $\boldsymbol{A}$ as a sheaf over the space.

The third section looks at special sheaves that occur frequently in practice: Boolean products, powers and extensions, and Hausdorff sheaves.

In the final section of this chapter, we introduce for each algebraic type the new category $\mathfrak{BooleBraceRed}$ whose objects are algebras with selected Boolean subsemilattices of congruences. The categories defined in the previous chapter are now restricted to those over Boolean spaces. Some axioms must be added to the definitions of all these categories in order to rule out trivia. With appropriately defined functors and natural transformations, $\mathfrak{BooleBraceRed}$ and the category $\mathfrak{CompBooleRed}$ of complexes

A. Knoebel, *Sheaves of Algebras over Boolean Spaces*,
DOI 10.1007/978-0-8176-4642-4_V,

over Boolean spaces form an adjoint situation. This adjoint situation is inherited from that between complexes and sheaves of the previous chapter when they are suitably restricted.

1. Identifying the Congruences

The first part of this section defines a Boolean semilattice [**Knoe72**], and establishes that it is really a Boolean algebra. The intended application is to a Boolean algebra of congruences whose join may diverge from the traditional join of congruences. The middle part discusses its spectrum, the space of its prime ideals. The last part proves a lemma asserting the existence of prime ideals extending a given ideal under restrictive side conditions. See Sects. II.1 and III.4 for algebras not defined here.

1.1. DEFINITION. *A* **semilattice with a nullity**, $\boldsymbol{B} = \langle B; \wedge, 0\rangle$, *is a semilattice with a constant 0 satisfying*

$$0 \wedge b = 0 \qquad b \in B.$$

It is a **Boolean semilattice** *if for each b of $\boldsymbol{B}$ there is a b' in $\boldsymbol{B}$ satisfying:*

$$\text{(1.1)} \qquad c \wedge b' = 0 \ \textit{ if, and only if, } \ c \leq b \qquad (c \in B).$$

It is a **Boolean subsemilattice** *of another semilattice with a nullity if one is a subalgebra of the other. Recall that the partial order $\leq$ of a semilattice comes from its binary operation: $b \leq c$ if $b \wedge c = b$. Thus $0 \leq c$ for c in $\boldsymbol{B}$.*

Note the subtle difference between condition (1.1) and **pseudocomplementation**: for each b of $\boldsymbol{B}$ there is a b' in $\boldsymbol{B}$ satisfying:

$$\text{(1.2)} \qquad c \wedge b = 0 \ \text{ if, and only if, } \ c \leq b' \qquad (c \in B).$$

Pseudocomplementation is not enough to make $\boldsymbol{B}$ a Boolean algebra [**Birk67**]. However, the set of pseudocomplements by itself is always a Boolean algebra [**Frink62**].

In a bounded lattice $\langle A; \vee, \wedge, 0, 1\rangle$ in this paragraph, the chain $\{0, 1\}$ is always a Boolean subsemilattice. But so is any other two-element subset $\{0, a\}$, where $a \in A$. In $(\boldsymbol{C}_2)^3$, which is the three-dimensional hypercube, the subset $\{\langle 0,0,0\rangle, \langle 0,0,1\rangle, \langle 0,1,0\rangle, \langle 1,1,1\rangle\}$ is a Boolean subsemilattice. More generally in a bounded lattice, its center, defined in Sect. II.1, is a Boolean sublattice.

The complement of an element b depends very much in which subsemilattice it lives. In Fig. 1a, for example, the complement of b in the solid two-element sublattice $\{0, b\}$ is 0, whereas in the whole lattice it is b'.

In general, there is not a unique maximal Boolean subsemilattice. Consider in Fig. 1b the five-element modular lattice $\boldsymbol{M}_3$ that is not distributive. There are a number of Boolean subsemilattices, of which three are

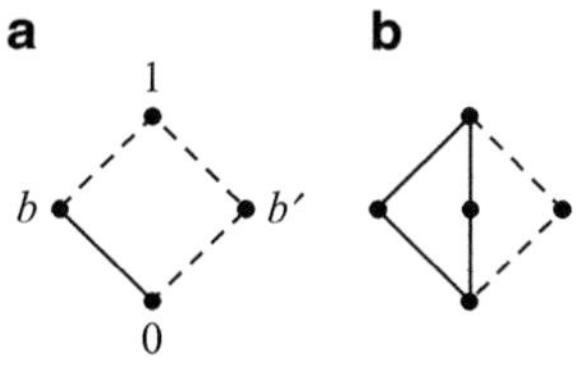

FIGURE 1. Boolean subsemilattices

maximal, each with four elements; one is shown in solid lines. Notice that the lower four elements of $\boldsymbol{M}_3$ form a meet subsemilattice but this is not Boolean. The lattice $\boldsymbol{M}_3$ is the lattice of congruences of the vier-group, $\boldsymbol{V}_4 = \mathbb{Z}_2 \times \mathbb{Z}_2$, and each maximal Boolean subsemilattice corresponds to a nontrivial product decomposition of $\boldsymbol{V}_4$. By way of contrast, there may not exist any maximal Boolean subsemilattices of a bounded lattice, much less a unique one, as shown in Example 1.7 at the end of this section.

The next proposition justifies the adjective in the phrase 'Boolean subsemilattice', and incidentally shows that complements are unique.

1.2. PROPOSITION. *Let* $\boldsymbol{B}$ *be a Boolean subsemilattice of a semilattice* $\boldsymbol{S}$ *with a nullity* 0. *Complements in* $\boldsymbol{B}$ *are unique. Define a constant 1 and a binary operation* $\sqcup$ *by*

$$1 = 0', \tag{1.3}$$

$$b \sqcup c = (b' \wedge c')' \qquad (b, c \in B). \tag{1.4}$$

Then, $\langle B; \sqcup, \wedge,', 0, 1\rangle$ *is a Boolean algebra.*

PROOF. We head for E. V. Huntington's [**Hunt33**] defining set of three axioms for Boolean algebras, as summarized by [**Birk67**, p. 44], and stated below as (h1)–(h3); they are also in Sect. III.4. In the statement and the proof of the preliminary assertions (a)–(g) below let us assume that $b \in B$ and b' is related to b as in (1.1) above; and likewise b'' is related to b', as well as b''' to b''. Until assertion (c) is proven, we do not know that complements are uniquely determined, that complementation is a function. We use fully the properties of semilattices.

(a) $b \wedge b' = 0$. This is true since $b \le b$.
(b) $b'' = b$. First, $b'' \le b$ since $b'' \wedge b' = 0$. Second, it follows from this that $b'' \wedge b = b''$, and hence $b''' = b' \wedge b'''$. Therefore,

$$b \wedge b''' = b \wedge (b' \wedge b''') = (b \wedge b') \wedge b''' = 0 \wedge b''' = 0.$$

Thus, $b \le b''$ and consequently double complements are unique.
(c) Complements themselves are unique, that is, if b' and $b^{\backprime}$ are both related to b as in (1.1), then $b' = b^{\backprime}$. To establish this, let b', b'', b''' be a

sequence of successive complementations, and let $b^{\backprime}, b^{\backprime\backprime}, b^{\backprime\backprime\backprime}$ be another sequence. Then, $b'' = b = b^{\backprime\backprime}$ by (b). So now our sequences look like:

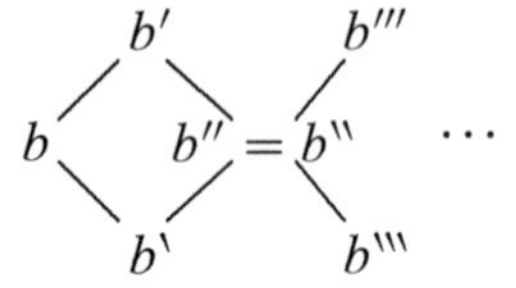

Hence, b''' is not only a double complement of b' but also of $b^{\backprime}$. So $b''' = b^{\backprime\backprime\backprime}$. Thus, $b' = b''' = b^{\backprime\backprime\backprime} = b^{\backprime}$, again by (b).

(d) $b \leq c$ iff $c' \leq b'$. A series of logical equivalences yields

$$b \leq c \text{ iff } b \wedge c' = 0 \text{ iff } c' \wedge b'' = 0 \text{ iff } c' \leq b'.$$

(e) $b \leq c$ iff $b \sqcup c = c$. Using (d), the definition of $\leq$, complementation, and the definition of $\sqcup$, we obtain in that order,

$$b \leq c \text{ iff } c' \leq b' \text{ iff } b' \wedge c' = c' \text{ iff } (b' \wedge c')' = c \text{ iff } b \sqcup c = c.$$

(f) $b \wedge c = (b' \sqcup c')'$. Complement (1.4).

(g) $b \leq d$ and $c \leq d$ iff $b \sqcup c \leq d$. The preceding property yields the corresponding dual property for $\wedge$, which holds in any semilattice.

(h1) $b \sqcup c = c \sqcup b$. This commutative law, as well as the associative law to come next, follow directly from the corresponding laws for $\wedge$ and the definition of $\sqcup$ in terms of $\wedge$ and $'$.

(h2) $b \sqcup (c \sqcup d) = (b \sqcup c) \sqcup d$.

(h3) $(b \wedge c) \sqcup (b \wedge c') = b$.

Designating the left side by L, we see that $L \leq b$ is implied by an application of (g). The reverse inequality is harder. Working backwards while writing the meet $\wedge$ as juxtaposition, and noting that $L' = (bc)'(bc')'$, we find by repeatedly using (1.4) and properties of $\wedge$ that

$$\begin{aligned} b \leq L \text{ iff } & bL' = 0 \\ \text{iff } & b(bc)'(bc')' = 0 \\ \text{iff } & b(bc)' \leq bc' \\ \text{iff } & b(bc)'bc' = b(bc)' \\ \text{iff } & b(bc)'c' = b(bc)' \\ \text{iff } & b(bc)' \leq c' \\ \text{iff } & bc(bc)' = 0. \end{aligned}$$ □

As an aside, we note in a corollary that this proposition yields a simple set of axioms for Boolean algebras.

1.3. COROLLARY. *Let $\boldsymbol{B}$ be an algebra $\langle B; \wedge, ', 0\rangle$ of type $\langle 2, 1, 0\rangle$. Assume the following axioms are satisfied for all b, c, d in $\boldsymbol{B}$:*

$$b \wedge b = b,$$
$$b \wedge c = c \wedge b,$$
$$b \wedge (c \wedge d) = (b \wedge c) \wedge d,$$
$$b \wedge 0 = 0,$$
$$c \wedge b' = 0 \ \textit{ iff } \ c \wedge b = c.$$

Then $\langle B; \sqcup, \wedge, ', 0, 1\rangle$ is a Boolean algebra, where the additional operations are defined as before.

When $\boldsymbol{B}$ is a set of congruences, the next proposition is handy to know.

1.4. PROPOSITION. *If $\boldsymbol{L}$ is a lattice with nullity, $\langle L; \vee, \wedge, 0\rangle$, and $\boldsymbol{B}$ is a Boolean subsemilattice of $\langle L; \wedge, 0\rangle$, then*

$$b \vee c \leq b \sqcup c \quad (b, c \in B),$$

where $\sqcup$ is the join (1.4) of the Boolean algebra in the previous proposition.

PROOF. Since $b \geq b \wedge c$, then $b \leq b \sqcup c$ and also $c \leq b \sqcup c$. Because $b \vee c$ is the least upper bound of b and c in $\boldsymbol{B}$, the conclusion follows. □

The divergence of these two joins does occur and is easily illustrated in Fig. 2. In (a), the solid dots are elements of B and the open one is in $L{\sim}B$. Thus, $b \vee c = a$ but $b \sqcup c = 1$. In (b), there are more divergent joins, and these do not involve the top element.

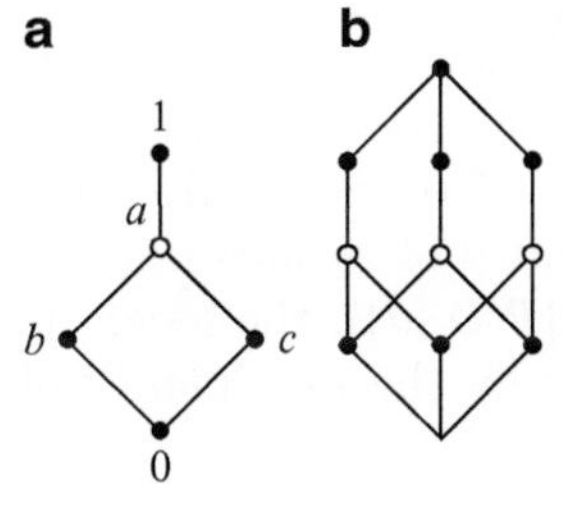

FIGURE 2. Divergence of joins

Let $\boldsymbol{B}$ be a Boolean subsemilattice of $\mathbf{Con}\,\boldsymbol{A}$, where $\boldsymbol{A}$ is an algebra. To get a single congruence from each ideal of congruences we take the join, that is, the supremum $\bigvee P$, in $\mathbf{Con}\,\boldsymbol{A}$ of each prime ideal P in $\boldsymbol{B}$; in other words, $\bigvee P$ is the smallest congruence in $\mathbf{Con}\,\boldsymbol{A}$ containing all the congruences that are in P. These suprema exist since the lattice $\mathbf{Con}\,\boldsymbol{A}$ of all congruences is complete. It would not do to take this supremum in $\boldsymbol{B}$ since, after all, $\boldsymbol{B}$ itself may not be complete. In this context, suprema are unions, as the next proposition spells out.

1.5. PROPOSITION. *If $\boldsymbol{B}$ is a Boolean subsemilattice of the meet semilattice of* $\mathbf{Con}\,\boldsymbol{A}$ *where $\boldsymbol{A}$ is an algebra, then for any ideal I of $\boldsymbol{B}$:*

$$\bigvee I = \bigcup I.$$

PROOF. Since I is upward directed by inclusion, $\bigcup I$ is a congruence. Hence it is the least upper bound of I. □

In order to prove in the next section that every Boolean subsemilattice $\boldsymbol{B}$ of the lattice of congruences of an algebra gives rise to a complex, we need a typical lemma on extending proper ideals of $\boldsymbol{B}$ to prime ideals.

1.6. LEMMA. *Let $\boldsymbol{B}$ be a Boolean subsemilattice of* $\mathbf{Con}\,\boldsymbol{A}$. *Suppose that I is a proper ideal of $\boldsymbol{B}$, and elements a and b in A are such that*

$$\langle a,b\rangle \notin \bigvee I.$$

Then there exists a prime ideal P of $\boldsymbol{B}$ such that

$$\langle a,b\rangle \notin \bigvee P \quad \text{and} \quad I \subseteq P. \tag{1.5}$$

PROOF. Recall our notation that $1_{\boldsymbol{B}}$ is the unity of $\boldsymbol{B}$, or what is the same, the largest congruence of $\boldsymbol{B}$, which need not be $1_{\mathbf{Con}\,\boldsymbol{A}}$, the universal congruence of $\boldsymbol{A}$. The proof splits into two cases according to whether a and b are related by $1_{\boldsymbol{B}}$ or not.

Case 1: $a \; 1_{\boldsymbol{B}} \; b$. A standard algebraic application of Zorn's Lemma III.1.7 yields an ideal P of $\boldsymbol{B}$ that is maximal with respect to (1.5). First, note that P is a proper ideal of $\boldsymbol{B}$ since otherwise $1_{\boldsymbol{B}} \in P$, which contradicts the assumptions that $a \; 1_{\boldsymbol{B}} \; b$ and $\langle a,b\rangle \notin \bigvee P$. The hard part is to show that P is maximal among all proper ideals of $\boldsymbol{B}$. In a Boolean algebra this is equivalent to P being prime. By way of contradiction, suppose not. From the characterization of prime ideals (property (4.1) of Sect. III.4), there must exist an element θ of $\boldsymbol{B}$ such that θ and its complement θ' are both not in P. Set $J_\theta = [P \cup \{\theta\}]$, the ideal of $\boldsymbol{B}$ generated by P and θ. Similarly, define $J_{\theta'}$. Now by the maximality of P with respect to (1.5), it must follow that both $\langle a,b\rangle \in \bigvee J_\theta$ and $\langle a,b\rangle \in \bigvee J_{\theta'}$. In other words, there are ζ and η in P such that $\langle a,b\rangle \in \zeta \sqcup \theta$ and $\langle a,b\rangle \in \eta \sqcup \theta'$, where $\sqcup$ is the join derived in $\boldsymbol{B}$. Hence, by distributivity,

$$\begin{aligned}
\langle a,b\rangle &\in (\zeta \sqcup \theta) \cap (\eta \sqcup \theta')\\
&= (\zeta \cap \eta) \sqcup (\zeta \cap \theta') \sqcup (\theta \cap \eta) \sqcup (\theta \cap \theta')\\
&\subseteq \zeta \sqcup \zeta \sqcup \eta \sqcup 0\\
&= \zeta \sqcup \eta\\
&\in P.
\end{aligned}$$

But this contradicts the assumption that $\langle a,b\rangle \notin \bigvee P$.

Case 2: not $a \; 1_{\boldsymbol{B}} \; b$. A standard result in Boolean algebras and ring theory states, via Zorn's lemma, that any proper ideal I extends to a prime ideal

P [**Hung74**, theorem VIII.2.1]. Since $\bigvee P = \bigcup P \subseteq 1_{\boldsymbol{B}}$, it follows that $\langle a, b\rangle \notin \bigvee P$. □

There are many examples of Boolean subsemilattices of the lattice of congruences of an algebra. Some of these are maximal; see the discussion concerning Fig. 1b. But contrast this with what comes next.

1.7. EXAMPLE. Maximal Boolean subsemilattices of $\mathbf{Con}\, \boldsymbol{A}$ need not exist in a particular algebra $\boldsymbol{A}$. To exhibit such an algebra, it suffices to use the representation theorem of Grätzer and Schmidt asserting that every algebraic lattice is isomorphic to the lattice of all congruences of some algebra (see [**GrätSc63**] and [**Grät79**, Appendix 7]). Let X be any infinite set and X^{fin} the collection of its finite subsets. The required lattice $\boldsymbol{L}$ is $X^{\text{fin}} \cup \{X\}$ ordered by set-inclusion. Its compact elements are those in X^{fin}. Since $X = \bigcup X^{\text{fin}}$, every element of L is a union of compact elements. Of course, this lattice is complete since arbitrary intersections exist. Therefore, $\boldsymbol{L}$ is algebraic, and thus $\boldsymbol{L} \cong \mathbf{Con}\, \boldsymbol{A}$ for some algebra $\boldsymbol{A}$.

Now let $\boldsymbol{B}$ be any Boolean subsemilattice of $\mathbf{Con}\, \boldsymbol{A}$; it must be finite. To see this, assume that b and b' are nonzero complements in the sense of (1.1). By Proposition 1.2, $\boldsymbol{B}$ a Boolean algebra. Hence, for all c in $\boldsymbol{B}$ (like in $\boldsymbol{L}$),

$$c = (c \cap b) \sqcup (c \cap b').$$

Since b and b' are finite, their intersections with all c can range only over a finite number of finite subsets. Hence, $\boldsymbol{B}$ is finite.

If $1_{\boldsymbol{B}} \in X^{\text{fin}}$, then double $\boldsymbol{B}$ by joining to each of its elements a fixed, nonempty finite set F that is disjoint from $1_{\boldsymbol{B}}$ (while keeping all its original subsets). Thus $\boldsymbol{B}$ could not have been maximal. If $1_{\boldsymbol{B}} = X$, then a more elaborate process also doubles $\boldsymbol{B}$, making it again impossible for $\boldsymbol{B}$ to have been maximal.

A notion similar to a Boolean subsemilattice is defined and studied by Davey [**Davey73**, sect. 4]. This appears to be less general than ours but this is only superficially so. The only significant difference concerns the unity of the Boolean subsemilattice. In Davey's presentation it must be $1_{\mathbf{Con}\,\boldsymbol{A}}$. In ours it is floating. But our floating unity, which must be greater than any other element in B, can always be redefined as $1_{\mathbf{Con}\,\boldsymbol{A}}$, still leaving the rest of B as is. This in no wise affects the sheaf to be constructed presently, since we will only look at the prime ideals of $\boldsymbol{B}$, and by definition these cannot contain the unity, wherever it may be.

Nevertheless, there are some differences in the two approaches. The first is that our definition of Boolean subsemilattice is quite brief, and this is often an advantage in verifying it in applications. Secondly, our use of complexes make the development of a sheaf rather different.

Hofmann [**Hofm72**, p. 323], has a third approach, a very general formulation within category theory, which he calls the 'Boolean decomposition principle', and which appears to be more restricted, since each congruence of $\boldsymbol{B}$ must be the kernel of some endomorphism of $\boldsymbol{A}$. However, on reflection, one sees that his Boolean decomposition principle can handle our Boolean algebras of sesquimorphisms of the next chapter simply by specifying our sesquimorphisms as his endomorphisms of a category, although it apparently does not encompass the more general Boolean subsemilattices of this chapter. Hofmann's principle could be applied to some structures that are not algebras. Another version is found in [**Keim70**].

2. Constructing the Complex

It is time to construct a sheaf out of any Boolean subsemilattice $\boldsymbol{B}$ of the congruence lattice of an algebra $\boldsymbol{A}$. We construct this sheaf over a Boolean space by showing that we have a complex, and then it follows, by the previous chapter, that we also have a sheaf. There is the converse that any sheaf over a Boolean space must come in this fashion from a Boolean subsemilattice of congruences. Along the way we discuss the preservation of sentences between a sheaf and its stalks. From now on, let us assume that $\boldsymbol{B}$ has at least two elements whenever $\boldsymbol{A}$ has at least two.

As a prelude before proving anything, this sheaf is briefly described [**Knoe72**]. The base space of the sheaf will be **Spec** $\boldsymbol{B}$, the topological space of prime ideals of $\boldsymbol{B}$, as set forth in Sect. III.4. The sheaf itself will be the disjoint union of the quotient algebras $\boldsymbol{A}/\theta$, where θ runs over suprema of prime ideals of **Spec** $\boldsymbol{B}$. To that end introduce the notation,

(2.1) $$\boldsymbol{A}/\!\!/\boldsymbol{B} = \biguplus \left\{ \frac{\boldsymbol{A}}{\bigvee P} \;\middle|\; P \in \mathrm{Spec}\,\boldsymbol{B} \right\},$$

for the sheaf induced by a Boolean subsemilattice $\boldsymbol{B}$ of **Con** $\boldsymbol{A}$. An element of the Pth stalk of $\boldsymbol{A}/\!\!/\boldsymbol{B}$ is written as $[a]_P = \langle P, a/\theta \rangle$ where $\theta = \bigvee P$. A basis for the topology of $\boldsymbol{A}/\!\!/\boldsymbol{B}$ is the collection of sets, $[a]_U = \{[a]_P \mid P \in U\}$, running over all a in A and all sets U open in **Spec** $\boldsymbol{B}$. The algebra of all global sections will be the algebra of all continuous functions, $\sigma\colon \mathbf{Spec}\,\boldsymbol{B} \to \boldsymbol{A}/\!\!/\boldsymbol{B}$, for which $\pi \circ \sigma = 1_{\mathbf{Spec}\,\boldsymbol{B}}$, and which is notated $\boldsymbol{\Gamma}(\boldsymbol{A}/\!\!/\boldsymbol{B})$.

An application, to appear in detail in Chap. VII, illustrates this construction as summarized in the next theorem. In a unital ring $\boldsymbol{R}$, the central idempotents form a Boolean algebra: the central idempotents are those ring elements e satisfying $ee = e$ and $ea = ae$ for all a in R, with Boolean operations $e_1 \vee e_2 = e_1 + e_2 - e_1e_2$, $e_1 \wedge e_2 = e_1e_2$, and $e' = 1 - e$. The principal ideals generated by these central idempotents then form a Boolean subsemilattice $\boldsymbol{B}$ of **Ideal** $\boldsymbol{R}$, the lattice of all ideals of $\boldsymbol{R}$. The idea here is to replace congruences by two-sided ideals, which can always be done in rings. In fact, Pierce [**Pier67**] equivalently uses the space dual to the Boolean algebra of all central idempotents as his base space. In this situation the sheaf $\boldsymbol{R}/\!\!/\boldsymbol{B}$ is often called a 'ringed space'.

2.1. THEOREM. *If* $\boldsymbol{B}$ *is a Boolean subsemilattice of the lattice* $\mathbf{Con}\,\boldsymbol{A}$ *of congruences of an algebra* $\boldsymbol{A}$*, then* $\langle \boldsymbol{A}, :, \mathbf{Spec}\,\boldsymbol{B} \rangle$ *is a complex with*

$$a:b = \left\{ P \in \operatorname{Spec} \boldsymbol{B} \mid a \bigvee P\, b \right\}.$$

Hence, $\boldsymbol{A} /\!\!/ \boldsymbol{B}$ *is the corresponding sheaf of algebras with the projection,* $\pi: \boldsymbol{A} /\!\!/ \boldsymbol{B} \to \mathbf{Spec}\,\boldsymbol{B}$*, being given by* $\pi([a]_P) = P$*. Thus,*

$$\gamma: \boldsymbol{A} \to \boldsymbol{\Gamma}(\boldsymbol{A} /\!\!/ \boldsymbol{B})$$

by the injective Gel'fand morphism γ *defined in equation (IV.2.1):*

$$\gamma(a)(P) = [a]_P \qquad (a \in A \ \text{ and } \ P \in \operatorname{Spec} \boldsymbol{B}).$$

PROOF. We first show that $\langle \boldsymbol{A}, :, \mathbf{Spec}\,\boldsymbol{B} \rangle$ is a precomplex. The first thing to demonstrate is that $a:b$ is open in $\mathbf{Spec}\,\boldsymbol{B}$ for any a and b in A. For this it is sufficient to find, for every prime ideal P in $a:b$, a neighborhood of P contained in $a:b$. Since $a \bigvee P\, b$ and $\bigvee P = \bigcup P$, there is a congruence θ in P such that $a\ \theta\ b$. The required neighborhood, by Stone duality, is the open set U_θ of $\mathbf{Spec}\,\boldsymbol{B}$ associated with θ:

$$U_\theta = \{ Q \in \operatorname{Spec} \boldsymbol{B} \mid \theta \in Q \}.$$

Now P belongs to U_θ. Moreover, $U_\theta \subseteq a:b$. This is true since, for any prime ideal Q in U_θ, it follows that $\theta \in Q$. Since $a\theta b$, then $a \bigvee Q\, b$ and $Q \in a:b$. Thus, each P in $a:b$ has a neighborhood U_θ contained in $a:b$ — hence $a:b$ is open.

Since the suprema $\bigvee P$ are congruences, the four axioms of Sect. IV.1 for a precomplex are easily checked. Nevertheless, as a sample of the kind of argument required here, we prove transitivity:

$$a:b \cap b:c \subseteq a:c \qquad (a, b, c \subset A).$$

Suppose a prime ideal P of $\boldsymbol{B}$ is in the left side. Then $a \bigvee P\, b$ and $b \bigvee P\, c$. Therefore, $a \bigvee P\, c$ since $\bigvee P$ is a congruence, and thus P is in the right side.

To demonstrate that we have a complex, rather than just a precomplex, we must show for any a and b in A that $a:b = X$ implies $a = b$. To prove the contrapositive, assume that $a \neq b$. We have to find a prime ideal P in $\operatorname{Spec} \boldsymbol{B}$ such that $\langle a, b \rangle \notin \bigvee P$, that is, $a:b \neq X$. But this is just the content of Lemma 1.6; for this purpose, take the initial ideal I to be $\{0_{\mathbf{Con}\,\boldsymbol{A}}\}$.

That we also obtain the sheaf $\boldsymbol{A} /\!\!/ \boldsymbol{B}$ from the complex $\langle \boldsymbol{A}, :, \mathbf{Spec}\,\boldsymbol{B} \rangle$ follows from the construction in Sect. IV.2. □

2.2. PROBLEM. This explores vector spaces as represented by sheaves. Let $\boldsymbol{V}$ be a vector space over a field $\boldsymbol{F}$. There are many maximal Boolean subsemilattices of $\mathbf{Con}\,\boldsymbol{V}$, but do they all give essentially the same sheaf representation? Narrow the possibilities with the axiom of choice by finding a basis B of $\boldsymbol{V}$. For each subset S of B define a sesquimorphism μ_S of $\boldsymbol{V}$:

$$\mu_S\Big(\sum_{b \in B} f_b b\Big) = \sum_{b \in S} f_b b \qquad (f_b \in F).$$

The corresponding congruences θ_S form a Boolean subsemilattice, $\Theta = \{\theta_S \mid S \subseteq B\}$, of **Con** $\boldsymbol{V}$. In fact, $\boldsymbol{\Theta}$ is anti-isomorphic to $\mathscr{P}B$. Is it maximal? Describe the topology of the space **Spec** $\boldsymbol{\Theta}$. Finally, with the sheaf created therefrom, $\mathcal{A} = \boldsymbol{A} /\!\!/ \boldsymbol{\Theta}$, is there an isomorphism, $\boldsymbol{\Gamma}(\mathcal{A}) \cong \boldsymbol{V}$?

Sheaves of algebras over Boolean spaces are significant for a number of reasons. First, many common algebras have such representations, see for example the memoir of Krauss and Clark [**KraCl79**]. Second, the stalks may have special properties; examples are the Baer–Stone shells of Chap. VIII, where the stalks have no divisors of zero. But note that the factors may be infinite, and so may their number, even up to isomorphism; this contrasts sharply with the Boolean products of part II of the memoir of Burris and McKenzie [**BurMc81**], where, in their Boolean representable varieties, there are a finite number of finite stalks, up to isomorphism. Third, the sections of $\boldsymbol{\Gamma}(\mathcal{A})$ patch nicely in a number of ways; and there are other useful properties now to be stated, such as interpolation.

The next result shows how to shrink a finite clopen covering in a Boolean space so it becomes disjoint. This lemma will help to prove Proposition 2.4.

2.3. LEMMA. *Let $x_1, x_2, \ldots, x_n$ be a finite number of distinct points of a Boolean space $\boldsymbol{X}$ with clopen neighborhoods $U_1, U_2, \ldots, U_n$ of them that cover the space. There is a finite disjoint clopen covering $V_1, V_2, \ldots, V_n$ of $\boldsymbol{X}$ such that $x_i \in V_i$ and $V_i \subseteq U_i$ for $i = 1, 2, \ldots, n$.*

PROOF. The U_i generate a finite Boolean subalgebra $\boldsymbol{B}$ of **Clop** $\boldsymbol{X}$. Each x_i is in an atom of $\boldsymbol{B}$, say A_i. If an A_i has only one x_i, then set V_i to A_i. If several of the x_i are in the same atom, then split it into disjoint clopen subsets V_i, a unique one for each x_i in it. Any remaining atoms not used can be adjoined to the V_i so that each U_i includes an enlarged V_i. □

The next proposition describes how global sections satisfy the properties of patchwork, partitioning and interpolation over Boolean spaces. Being Boolean is essential for its conclusions. For part (d), remember from (2.1) of Sect. IV.2 that $\gamma(a_i)(x) = [a_i]_x$, which is the component of a_i in $\mathcal{A}$ at the stalk $\boldsymbol{A}/\theta_x$ over the point x. The Foster extensions in Sect. 3 use (d).

2.4. PROPOSITION. *Let $\mathcal{A}$ be a sheaf of algebras over a Boolean space $\boldsymbol{X}$ with projection $\pi \colon \mathcal{A} \to \boldsymbol{X}$, and let $\boldsymbol{\Gamma}(\mathcal{A})$ be its algebra of global sections.*

(a) *(Interpolation) For a finite number of points, $a_1, \ldots, a_n$ in $\mathcal{A}$, with projections onto $\boldsymbol{X}$, $\pi(a_i) = x_i$, there is a global section σ agreeing with these, that is,*

$$\sigma(x_i) = a_i \qquad (i = 1, \ldots, n).$$

(b) *The algebra $\boldsymbol{\Gamma}(\mathcal{A})$ is a subdirect product of the stalks $\boldsymbol{A}_x$, that is, $\mathcal{A}$ satisfies axiom* SS.

(c) *(Disjoint patchwork) If* $\langle U_i \mid i \in I \rangle$ *is a finite family of disjoint clopen subsets of* $\boldsymbol{X}$ *and* σ_i *are global sections for* i *in* I, *then there is a global section* σ *such that*

$$\sigma(x) = \sigma_i(x) \qquad (i \in I \ \ \textit{and} \ \ x \in U_i).$$

(d) *If the sheaf* $\mathcal{A}$ *arises from a complex* $\langle \boldsymbol{A}, :, \boldsymbol{X} \rangle$ *over a Boolean space* $\boldsymbol{X}$, *then for each global section* σ *there are a finite number of elements* $a_1, \ldots, a_n$ *of* A *and an equal number of disjoint clopen subsets* $U_i, \ldots, U_n$ *covering* X *such that*

$$\sigma(x) = \gamma(a_i)(x) \qquad (i = 1, \ldots, n \quad \textit{and} \quad x \in U_i).$$

PROOF. (a) From the definition of sheaves, $\mathcal{A}$ is locally homeomorphic to X. Thus, for each a_i there is a neighborhood $\mathcal{U}_i$ of a_i that is homeomorphic by π to a clopen neighborhood U_i of x_i. The local inverses of π are sections $\sigma_i : U_i \to \mathcal{U}_i$. By Lemma 2.3, there is a finite disjoint clopen covering $\{V_i \mid i = 1, \ldots, n\}$ such that $x_i \in V_i$ and $V_i \subseteq U_i$. The desired global section is the piecemeal union of the restricted sections:

$$\sigma = \bigcup_{i=1}^{n} (\sigma_i | V_i).$$

It passes through the original points, and the patchwork of Proposition IV.1.4 shows that σ is a global section.

(b) Applying (a) to single points of $\mathcal{A}$ proves that it is a subdirect sheaf.

(c) What is left over, $X \sim \bigcup_{i \in I} U_i$, is also open, so the overlapping patchwork property applies (without any overlapping).

(d) Since $\mathcal{A}$ comes from a complex, for each x in X there is an element a_x of A such that the global section $\gamma(a_x)$ agrees with σ at x. Within their equalizer $\sigma : \gamma(a_x)$ and about x, there is a clopen set V_x. Clearly these V_x cover X. Since $\boldsymbol{X}$ is compact there is a finite subcover $\{V_{x_i} \mid 1 \leq i \leq n\}$ of these. Define the required U_i to be

$$U_1 = V_1, \quad U_2 = (X \sim V_1) \cap V_2, \quad U_3 = (X \sim V_1) \cap (X \sim V_2) \cap V_3, \ \ldots.$$

The collection of these intersections is a disjoint clopen cover. (Some of the these may be empty, but that is of no account.) It follows that

$$\sigma(x) = \gamma(a_i)(x) \qquad (x \in U_i \quad \text{and} \quad i = 1, \ldots, n). \quad \square$$

The difference between the overlapping patching of Proposition IV.1.4 and the disjoint patching of Proposition 2.4 is striking. Contrast the disjoint patching in sheaves of algebras over Boolean spaces with its complete opposite in the sheaf of germs of holomorphic functions of one variable. Holomorphic functions in complex analysis patch upon overlapping open sets, but disparate functions may not usually be patched together when they are disjoint, since each function has a unique analytic continuation. Because of disjoint patching, a sheaf over a Boolean space may be decomposed in many ways, as the following corollary makes clear.

2.5. COROLLARY. *For a sheaf $\mathcal{A}$ over a Boolean space $\boldsymbol{X}$ with a clopen subset U, the algebra $\boldsymbol{\Gamma}(\mathcal{A})$ of global sections decomposes as a product,*

$$\boldsymbol{\Gamma}(\mathcal{A}) \overset{\varphi}{\cong} \boldsymbol{\Gamma}(\mathcal{A}|U) \times \boldsymbol{\Gamma}(\mathcal{A}|(X \sim U)),$$

where $\varphi(\sigma) = \langle \sigma|U, \sigma|(X \sim U)\rangle$. Here $\mathcal{A}|U$ is the sheaf over U with $\mathcal{A}$ restricted to the stalks over it.

PROOF. Use Proposition 2.4 to show that φ is an isomorphism, or define a factor band, $\beta(\sigma, \tau) = \sigma|U \cup \tau|(X \sim U))$. □

The partition and patchwork properties are discussed further in [**BurWe79**, pp. 271–272 and pp. 306–307], [**KraCl79**, 1979, sect. 4], [**Pier67**, 1967, pp. 12–13], and [**Tenn75**, 1975, pp. 14 ...]. We use these properties to show what kind of sentences are preserved in passing back and forth between the stalks of a sheaf over a Boolean space and the algebra of its global sections. First, we need a lemma. In preparation for it, note how a term-operation t is evaluated pointwise in a sheaf $\mathcal{A}$,

$$\left(t^{\boldsymbol{\Gamma}(\mathcal{A})}(\sigma_1, \ldots, \sigma_n)\right)(x) = t^{\boldsymbol{A}_x}\left(\sigma_1(x), \ldots, \sigma_n(x)\right) \qquad (\sigma_i \in \Gamma(\mathcal{A}), x \in X).$$

2.6. LEMMA. *If t and w are n-ary terms in the algebraic type of a sheaf $\mathcal{A}$ over a Boolean space $\boldsymbol{X}$, and if*

$$t^{\boldsymbol{A}_x}(s_1, \ldots, s_n) = w^{\boldsymbol{A}_x}(s_1, \ldots, s_n) \qquad (s_1, \ldots, s_n \in A_x)$$

on the xth stalk $\boldsymbol{A}_x$, then there are global sections $\sigma_1, \ldots, \sigma_n$ in $\Gamma(\mathcal{A})$ and a neighborhood U of x such that

$$\sigma_1(x) = s_1, \ \ldots, \ \sigma_n(x) = s_n, \ \text{ and}$$

$$t^{\boldsymbol{A}_y}\left(\sigma_1(y), \ldots, \sigma_n(y)\right) = w^{\boldsymbol{A}_y}\left(\sigma_1(y), \ldots, \sigma_n(y)\right) \quad (y \in U).$$

PROOF. As the sheaf $\mathcal{A}$ is over a Boolean space, there are global sections $\sigma_1, \ldots, \sigma_n$ agreeing with the s_i's at x. Just as the operations of the type are continuous, so are the term-operations. Hence, $t^{\boldsymbol{\Gamma}(\mathcal{A})}(\sigma_1, \ldots, \sigma_n)$ and $w^{\boldsymbol{\Gamma}(\mathcal{A})}(\sigma_1, \ldots, \sigma_n)$ are new global sections. Their equalizer is open by Proposition IV.1.3d, and it is the required neighborhood U of x. □

An identity is satisfied in a subdirect product if, and only if, it is satisfied in all factors. For our sheaves this is true for the more general ∀∃-sentences of identities, since Pierce's proposition 3.4 of [**Pier67**] generalizes easily to our setting in universal algebra.

2.7. PROPOSITION. *Let $\mathcal{A}$ be a sheaf over a Boolean space $\boldsymbol{X}$ and let v and w be $(m+n)$-ary terms in the type of $\mathcal{A}$. Consider the sentence:*

$$\forall y_1, \ldots, y_m \, \exists z_1, \ldots, z_n \; v(y_1, \ldots, y_m, z_1, \ldots, z_n) \approx w(y_1, \ldots, y_m, z_1, \ldots, z_n). \tag{2.2}$$

This sentence is valid in the algebra $\boldsymbol{\Gamma}(\mathcal{A})$ of all global sections if, and only if, it is valid in each stalk.

PROOF. $\Rightarrow$. If the sentence is valid in $\boldsymbol{\Gamma}(\mathcal{A})$ and $s_1, \ldots, s_m$ are in a particular stalk A_x, then through these one can use the lemma to find global sections $\sigma_1, \ldots, \sigma_m$ that will play the role of the y_i's in (2.2). In $\boldsymbol{\Gamma}(\mathcal{A})$ there is a solution $\tau_1, \ldots, \tau_n$ that, when restricted to the stalk in question, gives a solution there.

$\Leftarrow$. To prove the other direction of implication, assume that the sentence (2.2) is true in each stalk. Let $\sigma_1, \ldots, \sigma_m$ be global sections in $\Gamma(\mathcal{A})$ for which we must find a solution $\tau_1, \ldots, \tau_n$.

In a particular stalk A_x there is a solution $t_1, \ldots, t_n$ to the equation

$$v(\sigma_1(x), \ldots, \sigma_m(x), t_1, \ldots, t_n) = w(\sigma_1(x), \ldots, \sigma_m(x), t_1, \ldots, t_n).$$

Since the base space is Boolean, there exist global extensions τ_i of the t_i. Consequently,

$$v(\sigma_1, \ldots, \sigma_m, \tau_1, \ldots, \tau_n)(x) = w(\sigma_1, \ldots, \sigma_m, \tau_1, \ldots, \tau_n)(x).$$

By the preceding lemma, for each x in X there is a neighborhood U_x of x over which this sentence is valid, the τ_i depending on x. By compactness and the partition property, there is a finite family $U_1, \ldots, U_k$ of disjoint clopen subsets covering $\boldsymbol{X}$ together with global sections τ_{ij} such that

$$v(\sigma_1, \ldots, \sigma_m, \tau_{1j}, \ldots, \tau_{nj})(x) = w(\sigma_1, \ldots, \sigma_m, \tau_{1j}, \ldots, \tau_{nj})(x).$$

is valid whenever $x \in U_j$ and $1 \leq j \leq k$.

With disjoint patchwork redefine the τ_i:

$$\tau_i = (\tau_{i1}|U_i) \cup \cdots \cup (\tau_{ik}|U_k) \qquad (i = 1, \ldots, n)$$

as new global sections. Clearly

$$v(\sigma_1, \ldots, \sigma_m, \tau_1, \ldots, \tau_n) = w(\sigma_1, \ldots, \sigma_m, \tau_1, \ldots, \tau_n).$$

Thus, our original sentence holds in $\boldsymbol{\Gamma}(\mathcal{A})$. □

Theorem 2.1, proven earlier in this section, has a partial converse (Knoebel [1972]). To set the stage, recall Sect. IV.2, where we showed that $\langle \boldsymbol{\Gamma}(\mathcal{A}), :, \boldsymbol{X} \rangle$ is a complex over the base space $\boldsymbol{X}$ of a sheaf $\mathcal{A}$. There we defined the equalizer to be $\sigma : \tau = \{x \in X \mid \sigma(x) = \tau(x)\}$. In the proof to come, we will need from Sect. III.4 the Boolean algebra **Clop** $\boldsymbol{X}$ of all clopen subsets of $\boldsymbol{X}$, which is dual to the Boolean space $\boldsymbol{X}$; the Boolean operations are the usual ones on sets. We also need some new notions. 'Reduced' insures that trivial stalks are rare.

2.8. DEFINITION. A sheaf $\mathcal{A}$ over a space $\boldsymbol{X}$ is **reduced** if the set of trivial stalks has an empty interior in $\boldsymbol{X}$, that is,

(RS) $$\operatorname{Int} \operatorname{Triv} \mathcal{A} = \varnothing,$$

where $\operatorname{Triv} \mathcal{A} = \{x \in X \mid |A_x| = 1\}$ and $\boldsymbol{A}_x$ is the stalk over x.

For each subset U of X, define a congruence θ_U on $\boldsymbol{\Gamma}(\mathcal{A})$ by

(2.3) $$\sigma\, \theta_U\, \tau \text{ if } U \subseteq \sigma : \tau.$$

That each θ_U is a congruence follows from the definition of equalizer. We extend this notation to a collection $\mathcal{U}$ of subsets of X: $\theta_{\mathcal{U}} = \{\theta_U \mid U \in \mathcal{U}\}$.

2.9. THEOREM. *Let* $\mathscr{A}$ *be a reduced sheaf of algebras over a Boolean space* $\boldsymbol{X}$. *Set* $\boldsymbol{A} = \boldsymbol{\Gamma}(\mathscr{A})$ *and* $\boldsymbol{B} = \theta_{\mathbf{Clop}\,\boldsymbol{X}}$.

(a) $\boldsymbol{B}$ *is a Boolean sublattice of* $\mathbf{Con}\,\boldsymbol{A}$;

(b) *All the congruences of* B *are factor congruences of* $\boldsymbol{A}$, *whose complements in* $\boldsymbol{B}$ *are complements as factor congruences;*

(c) *the sheaves* $\boldsymbol{\Gamma}(\mathscr{A})/\!/\boldsymbol{B}$ *and* $\mathscr{A}$ *are isomorphic.*

PROOF. (a) Let us first demonstrate that B is closed to the lattice operations of $\mathbf{Con}\,\boldsymbol{B}$. From the definition of θ_U, it is easily seen that

$$\theta_U \cap \theta_V = \theta_{U \cup V} \qquad (U, V \in \mathrm{Clop}\,X).$$

In order to prove the dual, $\theta_U \vee \theta_V = \theta_{U \cap V}$, it suffices to prove that $\theta_{U \cap V} \subseteq \theta_U \circ \theta_V$. To that end, suppose that $\sigma\,\theta_{U \cap V}\,\tau$ for some global sections σ and τ. Define a new global section by patching (see Sect. V.2):

$$\rho = \begin{cases} \sigma \text{ on } U, \\ \tau \text{ on } X \sim U. \end{cases} \tag{2.4}$$

Then, $U \subseteq \rho:\sigma$, and consequently $U \cap V \subseteq \rho:\sigma \cap \sigma:\tau \subseteq \rho:\tau$. Hence, $V = (U \cap V) \cup ((X \sim U) \cap V) \subseteq \rho : \tau$. Thus $\sigma\,\theta_U\,\rho$ and $\rho\,\theta_V\,\tau$, and $\sigma\,(\theta_U \circ \theta_V)\,\tau$.

Therefore, $\boldsymbol{B}$ is a sublattice of $\mathbf{Con}\,\Gamma(\mathscr{A})$. Distributivity follows quickly from the formulas just established. From them it also follows that $\theta_U \vee \theta_V = \theta_U \circ \theta_V$. Hence, complementation exists: $\theta_U \circ \theta_{X \sim U} = \theta_\emptyset = 1_{\mathbf{Con}\,A}$ and $\theta_U \cap \theta_{X \sim U} = \theta_X = 0_{\mathbf{Con}\,A}$.

(b) Clearly, this follows from the complementation just given.

(c) To verify the isomorphism of the sheaves, use Exercise IV.3.8 by defining a map,

$$\psi \colon \mathscr{A} \to \boldsymbol{\Gamma}(\mathscr{A})/\!/\boldsymbol{B},$$

with

$$\psi(\sigma(x)) = [\sigma]_{P_x} \qquad (\sigma \in \Gamma(\mathscr{A}) \text{ and } x \in X),$$

where

$$P_x = \{\theta_U \mid x \in U \in \mathrm{Clop}\,\boldsymbol{X}\}. \tag{2.5}$$

This comes from the homeomorphism established earlier in Chap. III:

$$\boldsymbol{X} \overset{\chi}{\cong} \mathbf{Spec}\,\boldsymbol{B} \colon x \mapsto P_x.$$

The function ψ is surjective and well defined because of Proposition 2.4b and the fact that $\sigma(x) = \sigma(y)$ can happen only when $x = y$. To prove that it is injective, recall that $[\sigma]_{P_x} = \langle P_x, \sigma / \bigvee P_x \rangle$. If $[\sigma]_{P_x} = [\tau]_{P_y}$, then $x = y$ since χ is a homeomorphism, and $\sigma\,\theta_U\,\tau$ for some U such that $x \in U \in \mathrm{Clop}\,\boldsymbol{X}$; hence $U \subseteq \sigma:\tau$, and thus $\sigma(x) = \tau(y)$.

It is routine that ψ preserves operations; thus ψ is a homomorphism.

To finish the proof via Exercise IV.3.8, verify its equation, $\rho \circ \psi = \chi \circ \pi$, where π and ρ are the projection maps in the two sheaves, $\langle \mathcal{A}, \pi, X \rangle$ and $\langle \boldsymbol{\Gamma}(\mathcal{A}) /\!\!/ \boldsymbol{B}, \rho, \mathrm{Spec}\, \boldsymbol{B} \rangle$. □

Theorem 2.9 anticipates Theorem 4.17 where one of the natural transformations making up an adjunction is a natural isomorphism.

3. Special Sheaves

A variety of comments and propositions about sheaves over Boolean spaces add to the previous two sections and fill them out. In particular, we discuss what it might mean for a sheaf to be indecomposable. Hausdorff sheaves, Boolean products and Boolean extensions are briefly described as special cases of sheaves over Boolean spaces. We redefine $\boldsymbol{\Gamma}(\boldsymbol{A} /\!\!/ \boldsymbol{B})$ as a direct limit. Rounding out this section, the characterization of varieties of Boolean products built from a finite number of finite algebras is mentioned.

The last section closed by reintroducing the congruences θ_x coming from a complex. This naturally leads into asking about the nature of these θ_x. If the Boolean subsemilattice $\boldsymbol{B}$ leading into a complex is finite, then the resulting sheaf gives a product with a finite number of factors $\boldsymbol{A}/\theta_x$. By way of contrast, when $\boldsymbol{B}$ is infinite and hence also X, the algebra $\boldsymbol{\Gamma}(\mathcal{A})$ will never be a direct product of the stalks; this is so since $\boldsymbol{\Gamma}(\mathcal{A})$ being a direct product implies that X must be discrete, which violates its being simultaneously compact and infinite.

One may wonder why prime ideals of congruences are considered at all. Why not use the maximal congruences of $\boldsymbol{B}$ rather than its maximal ideals to create quotient algebras? If we were only interested in Boolean algebras and closely related algebras such as semiprime commutative rings, we would do this and go directly to a sheaf representation with the base space **Spec** $\boldsymbol{B}$. But in a general algebra, the space of maximal congruences is not a Boolean space. For example, in Boolean rings without a unity, it no longer need be compact, only locally compact (Sect. XII.5). Even for non-Boolean unital rings there are choices since prime ideals may not be maximal. For instance, the hull-kernel topological spectrum of minimal prime ideals has been extensively studied by Henriksen and Jerison [**HenrJe65**], Kist [**Kist69**], and Peercy [**Peer70**]. Above all, the set of all maximal congruences may not faithfully represent an algebra $\boldsymbol{A}$, that is, their intersection may not be $0_{\boldsymbol{A}}$. Thus, we are led to look at prime ideals in Boolean subsemilattices.

Another question is: which nontrivial algebras are indecomposable by sheaves over Boolean spaces? We have to be a bit more precise here since there appear to be at least two new concepts. On the one hand, we could for the moment say a nontrivial algebra $\boldsymbol{A}$ is **Boolean irreducible** if **Con** $\boldsymbol{A}$ has no Boolean subsemilattices of more than two elements. If $\boldsymbol{A}$ is subdirectly irreducible, then $\boldsymbol{A}$ is Boolean irreducible; for on the contrary, if $\boldsymbol{A}$ is Boolean reducible, then there must be two incomparable nontrivial congruences θ and η whose intersection is 0, and hence $\boldsymbol{A}$ is subdirectly reducible. However, the ring $\mathbb{Z}$ is subdirectly reducible but not Boolean reducible. On the other hand, let us agree to call a nontrivial algebra **Boolean indecomposable** if it is not isomorphic to $\boldsymbol{\Gamma}(\mathcal{A})$ for some sheaf $\mathcal{A}$ over a Boolean space with at least two nontrivial stalks. But a moment's thought will convince one that this notion is the same as being directly indecomposable. To see this, suppose that $\mathcal{A}$ is a sheaf over a Boolean space $\boldsymbol{X}$ with at least two nontrivial stalks. Then $\boldsymbol{X}$ is the disjoint union of two nonempty clopen sets that decompose $\boldsymbol{\Gamma}(\mathcal{A})$ as a product. And conversely, being directly decomposable immediately gives a nontrivial sheaf over the two-element discrete space. In summary, we have the following implications:

Subdirect irreducibility
$\Rightarrow$ Boolean irreducibility
$\Rightarrow$ Boolean indecomposability
$\Leftrightarrow$ Direct indecomposability.

One last comment on this. One might think that the stalks of a sheaf over a sufficiently large Boolean space would have to be directly indecomposable. But Pierce [**Pier67**, lemma 4.2] has shown otherwise for noncommutative unital rings, even with $\boldsymbol{B}$ taken isomorphic to the Boolean algebra of all central idempotents. However, Burgess and Stephenson [**BurgSt78**] have iterated the sheaf construction for rings to eventually obtain indecomposable stalks. We will pursue this in Sect. XI.1 in a setting more general than rings.

A sheaf space $\mathcal{A}$ need not be a Hausdorff space. But the base space $\boldsymbol{X}$ being Hausdorff (since it is a Boolean space in most later chapters) leads to a corollary of Proposition IV.2.9 (see [**BurWe79**, p. 307]).

3.1. COROLLARY. *Let $\mathcal{A}$ be a sheaf of algebras over a Boolean space. Then, $\mathcal{A}$ is a Hausdorff space if, and only if,*

$$\sigma : \tau \quad \textit{is clopen} \qquad (\sigma, \tau \in \boldsymbol{\Gamma}(\mathcal{A})).$$

This leads to the equivalent notion of 'Boolean product'. See [**BurMc81**, part II, sec. 1] for a leisurely history of the evolution of this concept. Good examples are discriminator varieties, where each member is a Boolean product of simple members of the variety (see [**BurSa81**, Sect. IV.9]).

3.2. DEFINITION. *A* **Boolean product** *is a subdirect product of algebras indexed by a Boolean space $\boldsymbol{X}$ such that equalizers are clopen and the disjoint patching property is satisfied.*

This notion is equivalent to that of 'Hausdorff sheaf over a Boolean space' in the sense that either one may be readily constructed from the other with the techniques already employed.[1]

3.3. Example. To show that the notion of Boolean product really is a new concept we should have a counterexample of a sheaf $\mathcal{A}$ over a Boolean space that is not a Boolean product. It suffices to find two global sections σ and τ, the equalizer of which is merely open and not closed. We will construct this equalizer within a complex.[2] We start with an infinite Boolean space $\boldsymbol{X}$. It must have an open subset U that is not closed, since otherwise each point, which is closed, would have a clopen complement, and thus the point itself would be clopen; but this would violate compactness in an infinite space. For example, in the Cantor middle third set $\boldsymbol{X}$ we could let U be the open interval $(0, 1/4)$ [**Halm63**, p. 74]. Define an algebra $\boldsymbol{A}$ as a set with only two elements, a and b, and no operations. Turn it into a complex $\langle \boldsymbol{A}, :, \boldsymbol{X} \rangle$ by setting $a : b = U$. As before we know that the equalizer in the associated sheaf does not change $\gamma(a) : \gamma(b) = a : b$. Thus we have a sheaf of algebras over a Boolean space that is not a Boolean product.

It would be good to have a simpler example of a sheaf over a Boolean space that is not a Boolean product and that is closer to the shells coming in later chapters. Such is the bounded lattice in example 5.10 of Crown, Harding and Janowitz [**CroHJ96**, p. 188]. This distinction between representations by Boolean products and by sheaves over Boolean spaces is nicely illustrated by P-algebras and the more general notion of B-completely normal lattices; consult Sect. XII.2 and the theorems of Cignoli [**Cign78**, theorems 3.4 and 3.6].

3.4. Problem. One attraction of Boolean products is their almost algebraic characterization by Burris and Werner [**BurWe79**]. See [**BurSa81**, sects. IV.5, IV.8 and pp. 253–254], for details. Make it completely algebraic by suppressing the topology; the point is that only the clopen sets of the Boolean space matter. Concoct a Boolean product along the lines of a complex by forgetting the details of the subdirect product but keeping the equalizer. All that we need, really, is an algebra $\boldsymbol{A}$ together with a binary function ' : ' from $\boldsymbol{A}$ to a Boolean algebra $\boldsymbol{B}$. Find axioms for it; start with those for a complex and patching. Exercise this new notion by proving something about Boolean products without directly using topology. For more exercise, replace $\boldsymbol{B}$ by other algebras with known subdirect representations, such as distributive lattices.

---

[1] But beware that a Boolean product is also defined later and differently in [**BigBu90**] by patching and the weaker condition that equalizers are merely open. This is sometimes called a 'weak Boolean product' or 'locally Boolean product'. Its equivalence with sheaves over Boolean spaces is detailed in [**BurWe79**, p. 307].

[2] I am indebted to J. Kist for pointing out the value of complexes in constructing counterexamples.

Historically, Boolean products are the most common kind of sheaves of general algebras to have been studied. More specialized but important are **Boolean powers**,[3] which are Boolean products when all the stalks are identical. We present a Venn diagram in Fig. 3 with gradations from sheaves down to powers, showing the relationship of these constructions to those of the previous chapter. Here, Boolean products are equated with Hausdorff sheaves over Boolean spaces.

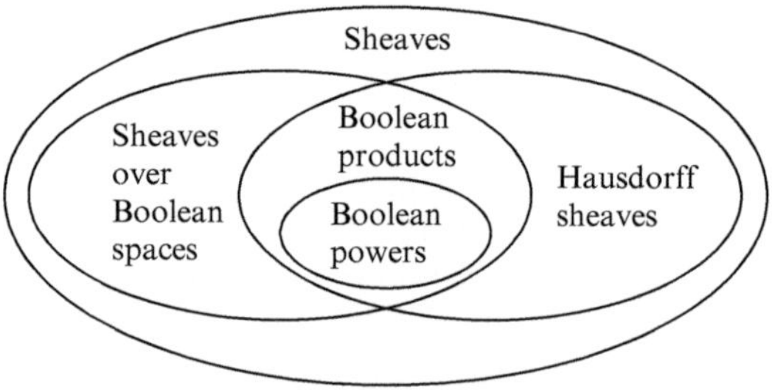

FIGURE 3. Special sheaves

We might be tempted to enclose this picture with an even vaguer region labeled 'subdirect products'. However, for sheaves in general $\boldsymbol{\Gamma}(\mathcal{A})$ need not be a subdirect product of the stalks; for not all of each stalk may be needed in $\boldsymbol{\Gamma}(\mathcal{A})$, violating the condition that all projections be surjective in a subdirect product. Ordinary products can be considered as sheaves when $\mathcal{A}$ and $\boldsymbol{X}$ both have discrete topologies. Then those, and only those, with a finite number of factors will fall within the circle of Boolean products.

The next result tells us that the distinction between Boolean products and sheaves over Boolean spaces disappears when the stalks are repeated. In it we mean by 'equal stalks' that $A_x = A_y$ $(x, y \in X)$. Then the disjoint union, $\mathcal{A} = \biguplus_{x \in X} A_x$, with elements $\langle x, a \rangle$ will separate them.

3.5. PROPOSITION. *Let $\mathcal{A}$ be a sheaf with equal stalks such that for each element a of this common stalk there is a constant global section σ_a taking only this value, in the sense that $\sigma_a(x) = \langle ax, a \rangle$ for all x in X. Then $\mathcal{A}$ is a Boolean power if, and only if, $\mathcal{A}$ is a sheaf over a Boolean space.*

PROOF. As already noted, each Boolean product is a sheaf over a Boolean space. In the other direction, consider a sheaf $\mathcal{A}$ over a Boolean space $\boldsymbol{X}$, all of whose stalks are the same algebra $\boldsymbol{A}$. We will prove that $\mathcal{A}$ is a

[3]Even though logically a power ought to be just a repeated product, Burris and Sankappanavar [**BurSa81**] use this term in a stronger sense and synonymously with that of Boolean extension. However we find it convenient to make a distinction between the two, which we will discuss more fully in a moment.

Boolean product, that is, equalizers are clopen. To that end, since equalizers are always open, it suffices to demonstrate that the equalizer of any two global sections ρ and τ is closed.

We check that their inequalizer has the form:

$$\rho;\tau = \bigcup_{\substack{a,b\in A \\ a\neq b}} (\rho:\sigma_a \cap \tau:\sigma_b).$$

If $x \in \rho;\tau$, then $\langle x,a\rangle = \rho(x) \neq \tau(x) = \langle x,b\rangle$ for some distinct elements a and b of A; thus x belongs to the union on the right side, and vice versa. Hence, $\rho;\tau$ is a union of open sets and is itself open. So its complement $\rho:\tau$ is closed, and $\mathcal{A}$ is a Boolean power by Definition 3.2. □

We give one more concept in the circle of ideas surrounding the notion of Boolean power, which is equivalent to it and of historical interest. The first construction of this kind was the Boolean extension of primal algebras of Foster [**Fost53**], who originally defined the concept algebraically, without any mention of topology, and more generally, the normal subdirect powers of Gould and Grätzer [**GouGr67**], which has in its normal transform a precursor of the quaternary discriminator term, which guarantees patching.

3.6. Definition. First define an A-**partition** of the Boolean algebra $\boldsymbol{B}$ as a function $f: A \to \boldsymbol{B}$ such that

(i) If $a \neq b$ then $f(a) \wedge f(b) = 0$ and
(ii) $\bigvee_{a\in A} f(a) = 1$.

The **Boolean extension** $\boldsymbol{A}[\boldsymbol{B}]$ of an algebra $\boldsymbol{A}$ by a Boolean algebra $\boldsymbol{B}$ has as carrier the set $A[\boldsymbol{B}]$ of all A-partitions of $\boldsymbol{B}$ and as operations $\omega^{\boldsymbol{A}[\boldsymbol{B}]}$ the convolving of the operations ω of $\boldsymbol{A}$:

$$\omega^{\boldsymbol{A}[\boldsymbol{B}]}(f_1, f_2, \ldots, f_n)(a) = \bigvee_{\omega^{\boldsymbol{A}}(a_1,a_2,\ldots,a_n)=a} f_1(a_1) \wedge f_2(a_2) \wedge \cdots \wedge f_n(a_n),$$

where $f_1, f_2, \ldots f_n \in A[\boldsymbol{B}]$ and $a \in A$. Caveat: for the join in the convolution to make sense, A must be finite or $\boldsymbol{B}$ must be complete.

These convolutions appeared in [**Fost51**] and even earlier in his papers.

3.7. Exercise. Under what hypotheses are Foster extensions equivalent to Boolean powers?

Figure 4 illustrates the interplay of these different approaches to Boolean extensions and powers. We represent an element f of the Boolean extension $\boldsymbol{A}[\boldsymbol{B}]$ by breaking it into pieces and laying them out as elements of A over the prime ideals $P_1, P_2, P_3, \ldots$ of Spec $\boldsymbol{B}$, which are laid out on the horizontal axis. Each of these pieces is to the right of an element of A on the vertical axis. A dot of a piece appears wherever a_i is a member of P_j (III.4.2). The horizontal stretches of these dots are over clopen subsets of Spec $\boldsymbol{B}$, equivalent to elements of $\boldsymbol{B}$ under Stone duality. In other words, f may be thought of either as a function from A to $\boldsymbol{B}$ in the Boolean extension, or as a function from Spec $\boldsymbol{B}$ to A, that is, as a global section in the corresponding Boolean power, thought of as a sheaf.

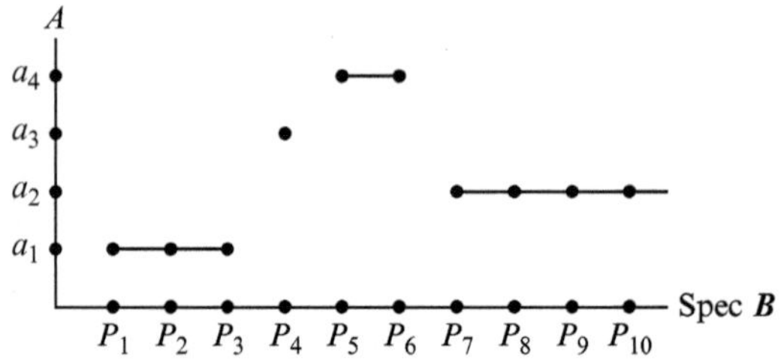

FIGURE 4. Anatomy of an element in a Boolean extension or Boolean power

For extended treatments of Boolean powers, read [**Ribe69**], [**BurWe79**], [**BanNe80**], and [**BurSa81**, Sects. IV.5, IV.8 and pp. 253–254].

Part (b) of the next proposition tells how to recognize Boolean products in complexes. Part (a) is needed for its proof.

3.8. PROPOSITION. *Let* $\langle \boldsymbol{A}, :, \boldsymbol{X} \rangle$ *be a complex* $\mathcal{A}$ *over a Boolean space* $\boldsymbol{X}$. *Assume* $\mathcal{A}$ *is the resulting sheaf.*

(a) *Then each global section* σ *in* $\boldsymbol{\Gamma}(\mathcal{A})$ *is a disjoint union of sections:*

$$\sigma = \gamma(a_1)|U_1 \,\cup\, \gamma(a_2)|U_2 \,\cup \cdots \cup\, \gamma(a_m)|U_m$$

for some finite number of $a_1, \ldots, a_m$ *in* A *and for some finite disjoint clopen covering* $U_1, \ldots, U_m$ *of* $\boldsymbol{X}$.

(b) *If all equalizers of the complex* $\mathcal{A}$ *are clopen, then we have a Boolean product, that is, all equalizers of global sections are clopen.*

PROOF. Part (a) is a restatement of property (d) in Proposition 2.4. Part (b) follows by computing the equalizer of two global sections as a finite union of clopen sets:

$$\sigma : \tau = \bigcup_{\substack{1 \le i \le m \\ 1 \le j \le n}} \big(\gamma(a_i) : \gamma(b_j)\big) \cap U_i \cap V_j,$$

where σ is as in part (a) and likewise

$$\tau = \gamma(b_1)|V_1 \,\cup\, \gamma(b_2)|V_2 \,\cup \cdots \cup\, \gamma(b_n)|V_n. \quad \square$$

A rephrasing of part (a) of this proposition in terms of direct limits without any recourse to sheaf theory is to be found in [**Davey73**, theorem 4.4]. Let the set of all finite partitions of $\boldsymbol{B}$ be designated Π. It is clearly ordered by refinement of partitions.

3.9. PROPOSITION. *If* $\boldsymbol{B}$ *is Boolean subsemilattice of* **Con** $\boldsymbol{A}$ *for an algebra* $\boldsymbol{A}$, *then* $\boldsymbol{\Gamma}(\boldsymbol{A} /\!\!/ \boldsymbol{B})$ *is isomorphic to* $\lim_{\pi \in \Pi} \prod_{\theta \in \pi} \boldsymbol{A}/\theta'$.

Significant theorems have been discovered about varieties whose members are Boolean products. Although our main interest in this book is sheaves more general than these, the results are remarkable enough to outline them here. Two theorems are stated in two areas: representation, and preservation of properties. The references to them have excellent expositions of their proofs, and give ample credit to the many whose earlier work made these possible. To present them, we need some terms.

3.10. DEFINITION. Each **discriminator** or **quasi-primal** algebra $\boldsymbol{A}$ has a quaternary term-operation t such that

$$t(a,b,c,d) = \begin{cases} c & \text{if } a = b; \\ d & \text{if } a \neq b. \end{cases} \tag{3.1}$$

A **discriminator variety** $\mathfrak{V}$ is generated by a set of discriminator algebras having a common discriminator term t.

A variety $\mathfrak{V}$ is **affine** if there is a ternary term τ such that $\tau(x,y,y) \approx x$ and $\tau(x,x,y) \approx y$ are identities of $\mathfrak{V}$, and for all n-ary operation symbols ω of $\mathfrak{V}$ there are these identities of $\mathfrak{V}$:

$$\tau\big(\omega(x_1,\dots,x_n),\omega(y_1,\dots,y_n),\omega(z_1,\dots,z_n)\big) \\ \approx \omega\big(\tau(x_1,y_1,z_1),\dots,\tau(x_n,y_n,z_n)\big).$$

Examples of discriminator varieties are Boolean algebras and their expansions, such as cylindric algebras [**Wern78**]. Affine varieties include modules and their term-reducts with $x_1 - x_2 + x_3$ a term. ('Affine' was originally 'Abelian', which now has a more general meaning [**McVa89**, p. 14].)

For a set $\mathfrak{K}$ of algebras, $\Gamma^a(\mathfrak{K})$ is the set of all Boolean products of algebras in $\mathfrak{K}$. (Do not confuse this operator Γ^a with the function Γ, which gives all global sections of a sheaf, Definition IV.1.5.) For varietal products see Definition III.1.4.

Representing finitely generated varieties by Boolean products is due to Burris and McKenzie [**BurMc81**, sect. 3 of part II] (see also [**Pinus93**, theorem 7.2]).

3.11. THEOREM. *A variety $\mathfrak{V}$ has the representation, $\mathfrak{V} = \mathsf{I}\Gamma^a(\mathfrak{K})$, for some finite set $\mathfrak{K}$ of finite algebras if, and only if, $\mathfrak{V} = \mathfrak{V}_1 \otimes \mathfrak{V}_2$ where $\mathfrak{V}_1$ is a discriminator variety with $\mathfrak{V}_1 = \mathsf{I}\Gamma^a(\mathfrak{K}_1)$ for $\mathfrak{K}_1$ a finite set of finite algebras of $\mathfrak{V}_1$, and $\mathfrak{V}_2$ is an affine variety with $\mathfrak{V}_2 = \mathsf{I}\Gamma^a(\mathfrak{K}_2)$ for $\mathfrak{K}_2$ a finite set of finite algebras of $\mathfrak{V}_2$.*

This theorem leaves open the representation of varieties that are not finitely generated. It has consequences for groups and rings.

3.12. COROLLARY. (a) *A variety $\mathfrak{V}$ of groups has the representation, $\mathfrak{V} = \mathsf{I}\Gamma^a(\mathfrak{K})$, for some finite set $\mathfrak{K}$ of finite groups if, and only if, $\mathfrak{V}$ is a variety of Abelian groups of bounded exponent.*

The full subcategory of **Complex** consisting of reduced complexes over Boolean spaces is designated **CompBooleRed**.

The first axiom RC1 insures that most stalks created in the sheaf will be nontrivial; and the second RC2 promises the injectivity of the function $U \mapsto \theta_U$. Its surjectivity is known as 'factor-transparency', which will be defined later within sheaves. In anticipation of finding a Boolean subsemilattice of congruences in any reduced complex, we next state and prove several simple consequences associated with these axioms.

4.3. LEMMA. *Assume that* $\langle \boldsymbol{A}, :, \boldsymbol{X} \rangle$ *is a complex with* U, V, W *in* $\operatorname{Clop} X$.

(a) $\theta_X = 0_{\operatorname{Con} \boldsymbol{A}}$.
(b) $\theta_U \cap \theta_V = \theta_{U \cup V}$.
If, in addition, the complex satisfies axiom RC2, *then* (c)–(g) *are true.*
(c) $U \subseteq V$ *if, and only if,* $\theta_U \supseteq \theta_V$.
(d) *In* $\theta_{\operatorname{Clop} \boldsymbol{X}}$ *the ordering* $\subseteq$ *has a join* $\sqcup$, *and it is given by*

$$\theta_U \sqcup \theta_V = \theta_{U \cap V}.$$

(e) *For all* U *there exist a* V *such that for all* W,

$$\theta_W \cap \theta_V = 0_{\operatorname{Con} \boldsymbol{A}} \quad \text{if, and only if,} \quad \theta_W \subseteq \theta_U.$$

(f) $\langle \theta_{\operatorname{Clop} \boldsymbol{X}}; \sqcup, \cap, ', 0_{\operatorname{Con} \boldsymbol{A}}, 1_{\operatorname{Con} \boldsymbol{A}} \rangle$ *is a Boolean algebra where* $(\theta_U)' = \theta_{X \sim U}$.
(g) *This Boolean algebra of* (f) *is anti-isomorphic to that on* $\operatorname{Clop} \boldsymbol{X}$.

PROOF. Throughout we apply (4.2).

(a) If $a_1 \,\theta_X\, a_2$, then $X \subseteq a_1 : a_2$ and hence $a_1 = a_2$.

(b) This follows directly from (4.2).

(c) Suppose that $\theta_U \supseteq \theta_V$. Then, by (b), $\theta_V = \theta_U \cap \theta_V = \theta_{U \cup V}$. Therefore by (RC2), $V = U \cup V$, and hence $U \subseteq V$. The other direction follows immediately from (4.2).

(d) We show that $\theta_{U \cap V}$ satisfies the definition of join, that is, $\theta_{U \cap V}$ is the smallest congruence of $\theta_{\operatorname{Clop} \boldsymbol{X}}$ containing both θ_U and θ_V. Since $U \supseteq U \cap V$, clearly $\theta_U \subseteq \theta_U \cap \theta_V$ and also $\theta_V \subseteq \theta_{U \cap V}$. Now suppose $\theta_U \subseteq \theta_W$ and $\theta_V \subseteq \theta_W$ for some clopen W. Then, $U \supseteq W$ and $V \supseteq W$ by (c), and so $U \cap V \supseteq W$. Hence, $\theta_{U \cap V} \subseteq \theta_W$. Therefore, $\theta_{U \cap V}$ is the join.

Or more conceptually, since in (c) we have an order-anti-isomorphism between $\mathbf{Clop}\, \boldsymbol{X}$ and $\boldsymbol{\theta}_{\mathbf{Clop}\, \boldsymbol{X}}$, Boolean operations in $\boldsymbol{\theta}_{\mathbf{Clop}\, \boldsymbol{X}}$, if they exist, will be the duals of those in $\mathbf{Clop}\, \boldsymbol{X}$, which do exist.

(e) We can use (a), (b), and (c), as we did in (d), to translate this statement in $\boldsymbol{\theta}_{\mathbf{Clop}\, \boldsymbol{X}}$ into an equivalent statement in the Boolean algebra of $\mathbf{Clop}\, \boldsymbol{X}$. All that is needed is to verify that

$$\forall U \in \operatorname{Clop} \boldsymbol{X} \ \exists V \in \operatorname{Clop} \boldsymbol{X} \ \forall W \in \operatorname{Clop} \boldsymbol{X} \ \big(W \cup V = X \ \text{ iff } \ W \supseteq U \big).$$

This is seen to be true by choosing the clopen set V to be $X \sim U$.

(f) Thus, by the previous parts, $\langle \theta_{\operatorname{Clop} \boldsymbol{X}}; \cap, 0_{\operatorname{Con} \boldsymbol{A}} \rangle$ is a Boolean subsemilattice of $\mathbf{Con}\, \boldsymbol{A}$. By Proposition 1.2 it becomes a Boolean algebra.

(g) The previous parts together prove this. □

For emphasis, we display these two important Boolean algebras.

4.4. COROLLARY. *In a reduced complex, the Boolean algebras,*

$$\mathbf{Clop}\,\boldsymbol{X} = \langle \mathrm{Clop}\,\boldsymbol{X}; \cup, \cap,', \varnothing, X \rangle, \quad \textit{and}$$
$$\boldsymbol{\theta}_{\mathbf{Clop}\,\boldsymbol{X}} = \langle \theta_{\mathrm{Clop}\,\boldsymbol{X}}; \sqcup, \sqcap,', 0_{\mathrm{Con}\,\boldsymbol{A}}, 1_{\mathrm{Con}\,\boldsymbol{A}} \rangle,$$

are anti-isomorphic by Lemma 4.3g. *The Boolean operations in* $\boldsymbol{\theta}_{\mathbf{Clop}\,\boldsymbol{X}}$ *are given for all U and V in* Clop $\boldsymbol{X}$ *by:*

$$\theta_U \sqcap \theta_V = \theta_{U \cup V}, \qquad \theta_U \sqcup \theta_V = \theta_{U \cap V}, \qquad (\theta_U)' = \theta_{X \sim U}.$$

One might expect that the θ_U in the next corollary would be factor congruences, but any three-element semilattice falsifies this. Recall Definition 2.8 of a reduced sheaf (RS): Int Triv $\mathcal{A} = \varnothing$ where Triv $\mathcal{A} = \{x \in X \mid |A_x| = 1\}$.

4.5. DEFINITION. The full subcategory of $\mathfrak{Sheaf}$ of reduced sheaves over Boolean spaces is designated $\mathfrak{SheafBooleRed}$.

Notice, because of patching properties, that any sheaf over a Boolean space is a member of $\mathfrak{SheafAlg}$ (Proposition 2.4b), that is to say, through any point of the sheaf passes a global section (axiom SS).

We need two more functors giving us another adjoint situation, analogous to that in Sect. IV.3: $\Phi_2\colon \mathfrak{BooleBraceRed} \rightleftharpoons \mathfrak{CompBooleRed}\colon \Gamma_2$.

4.6. DEFINITION. The first functor Φ_2 creates a complex from a Boolean brace:

$$\Phi_2(\langle \boldsymbol{A}, \boldsymbol{B} \rangle) = \langle \boldsymbol{A}, :, \boldsymbol{X} \rangle,$$

where $\boldsymbol{X} = \mathbf{Spec}\,\boldsymbol{B}$ and

$$a_1 : a_2 = \left\{ P \in \mathrm{Spec}\,\boldsymbol{B} \;\middle|\; a_1 \bigvee P \; a_2 \right\} \qquad (a_1, a_2 \in A). \tag{4.3}$$

To define it also on morphisms, start with a morphism in $\mathfrak{BooleBrace}$-$\mathfrak{Red}$, say $\langle \varphi, \upsilon \rangle \colon \langle \boldsymbol{A}, \boldsymbol{B} \rangle \to \langle \boldsymbol{A}^*, \boldsymbol{B}^* \rangle$. Write its image as $\Phi_2(\langle \varphi, \upsilon \rangle) = \langle \varphi, \chi \rangle$. This image, $\langle \varphi, \chi \rangle \colon \langle \boldsymbol{A}, :, \mathbf{Spec}\,\boldsymbol{B} \rangle \to \langle \boldsymbol{A}^*, \overset{*}{\underset{*}{:}}, \mathbf{Spec}\,\boldsymbol{B}^* \rangle$, has the same algebra homomorphism, $\varphi\colon \boldsymbol{A} \to \boldsymbol{A}^*$, but the second component, $\chi\colon \mathbf{Spec}\,\boldsymbol{B}^* \to \mathbf{Spec}\,\boldsymbol{B}$, now goes in the opposite direction:

$$\chi(Q) = \upsilon^{-1}(Q) = \{\theta \in B \mid \upsilon(\theta) \in Q\} \qquad (Q \in \mathrm{Spec}\,\boldsymbol{B}^*). \tag{4.4}$$

4.7. PROPOSITION. Φ_2 *is a functor from* $\mathfrak{BooleBraceRed}$ *to* $\mathfrak{Comp}$-$\mathfrak{BooleRed}$.

PROOF. To confirm this, we prove four things:

(i) $\Phi_2(\langle \boldsymbol{A}, \boldsymbol{B} \rangle)$ is a reduced complex whenever $\langle \boldsymbol{A}, \boldsymbol{B} \rangle$ is a reduced Boolean brace;
(ii) $\Phi_2(\langle \varphi, \upsilon \rangle)$ is a morphism of complexes whenever $\langle \varphi, \upsilon \rangle$ is a morphism of Boolean braces;

(iii) Φ_2 preserves unities, that is, identity morphisms;
(iv) Φ_2 preserves composition of morphisms.

Let $\Phi_2(\langle \boldsymbol{A}, \boldsymbol{B}\rangle) = \langle \boldsymbol{A}, :, \mathbf{Spec}\, \boldsymbol{B}\rangle$. To verify in (i) that this is a complex, we refer the reader to Theorem 2.1. To prove it is reduced, we make repeated use of Lemma 4.3. By (4.1) and (4.3), when $a_1, a_2 \in A$ and $P \in \operatorname{Spec} \boldsymbol{B}$,

$$a_1\, \theta_P\, a_2 \quad \text{iff} \quad P \in a_1 : a_2 \quad \text{iff} \quad a_1 \bigvee P\, a_2.$$

Therefore, $\theta_P = \bigvee P$, and thus (RB) implies (RC1).

Axiom RC2 follows from Lemma 1.6 asserting the existence of prime ideals of congruences in the face of adversity. For suppose $V \neq W$ for some clopen subsets V and W of $\mathbf{Spec}\, \boldsymbol{B}$; the object is to prove that $\theta_V \neq \theta_W$. By Boolean duality in Sect. III.4, there must be distinct congruences ζ and η of $\boldsymbol{B}$ such that $V = U_\zeta$ and $W = U_\eta$. Without loss of generality, assume that there are a_1 and a_2 such that $a_1\, \zeta\, a_2$ but not $a_1\, \eta\, a_2$. By the lemma already alluded to, there is a prime ideal P such that $\eta \in P$ but not $a_1 \bigvee P\, a_2$. The intention is to show from $a_1\, \zeta\, a_2$ that $a_1\, \theta_{U_\zeta}\, a_2$. From Boolean duality we know that:

$$U_\zeta = \{\, Q \in \operatorname{Spec} \boldsymbol{B} \mid \zeta \in Q \,\}; \tag{4.5}$$

If $Q \in U_\zeta$, then $\zeta \in Q$, and $a_1 \bigvee Q\, a_2$. Hence, $Q \in a_1 : a_2$ by (4.3), thus $U_\zeta \subseteq a_1 : a_2$ and so $a_1\, \theta_{U_\zeta}\, a_2$ by (4.2). Likewise, not $a_1\, \theta_{U_\eta}\, a_2$, since $P \in U_\eta$ but $P \notin a_1 : a_2$, again by (4.2). Therefore, $\theta_V = \theta_{U_\zeta} \neq \theta_{U_\eta} = \theta_W$.

To prove (ii) introduce $\langle \varphi, \chi\rangle = \Phi_2(\langle \varphi, \upsilon\rangle)$, with $\chi\colon \operatorname{Spec} \boldsymbol{B}^* \to \operatorname{Spec} \boldsymbol{B}$, defined by (4.4). First, we know that prime ideals go to prime ideals and that χ is continuous, as outlined in III.4. Second, we must show that $\langle \varphi, \chi\rangle$ is a complex morphism by verifying Definition IV.3.1:

$$\chi^{-1}(a_1 : a_2) \subseteq \varphi(a_1) \overset{\star}{\underset{\star}{:}} \varphi(a_2).$$

So assume that $Q \in \chi^{-1}(a_1 : a_2)$; this implies $\chi(Q) \in a_1 : a_2$. Set $P = \chi(Q)$. Hence, $a_1 \bigvee P\, a_2$, and so $a_1\, \theta\, a_2$ for some θ in P. Therefore, $\varphi(a_1)\, \varphi(\theta)\, \varphi(a_2)$; and thus $\varphi(a_1)\, \upsilon(\theta)\, \varphi(a_2)$ since $\varphi(\theta) \subseteq \upsilon(\theta)$ by Definition 4.1. Noting that $\upsilon(\theta) \in \upsilon(P)$ and $\upsilon(P) = \upsilon(\chi(Q)) = \upsilon(\upsilon^{-1}(Q)) = Q$, we conclude that $\varphi(a_1) \bigvee Q\, \varphi(a_2)$ and finally $Q \in \varphi(a_1) \overset{\star}{\underset{\star}{:}} \varphi(a_2)$. Thus, morphisms in $\mathfrak{BooleBraceRed}$ go to morphisms in $\mathfrak{CompBooleRed}$.

Parts (iii) and (iv) are straightforward and thus omitted. □

4.8. DEFINITION. Going in the opposite direction, the second functor,

$$\Gamma_2\colon \mathfrak{CompBooleRed} \to \mathfrak{BooleBraceRed},$$

yields a Boolean brace,

$$\Gamma_2(\langle \boldsymbol{A}, :, \boldsymbol{X}\rangle) = \langle \boldsymbol{A}, \boldsymbol{B}\rangle,$$

where the congruences of $\boldsymbol{B}$ come from clopen subsets: $\boldsymbol{B} = \boldsymbol{\theta}_{\operatorname{Clop} \boldsymbol{X}}$. To say what Γ_2 does to morphisms, let $\langle \varphi, \chi\rangle\colon \langle \boldsymbol{A}, :, \boldsymbol{X}\rangle \to \langle \boldsymbol{A}^*, \overset{\star}{\underset{\star}{:}}, \boldsymbol{X}^*\rangle$

be a morphism of complexes. Define $\langle\varphi,\upsilon\rangle\colon\langle \boldsymbol{A},\boldsymbol{B}\rangle \to \langle \boldsymbol{A}^*,\boldsymbol{B}^*\rangle$, that is, $\langle\varphi,\upsilon\rangle = \Gamma_2(\langle\varphi,\chi\rangle)$, by

$$\upsilon(\theta_U) = \theta_{\chi^{-1}(U)} \qquad (U \text{ is clopen in } \boldsymbol{X}).$$

4.9. PROPOSITION. *Γ_2 is a faithful functor going from* **CompBooleRed** *to* **BooleBraceRed**.

PROOF. For Γ_2 we should prove five things, to be labeled (i)–(v). The first four are analogous to those four parts proven for Φ_2 in the proof of Proposition 4.7. The fifth is that Γ_2 is a faithful.

(i) Set $\langle \boldsymbol{A},\boldsymbol{B}\rangle = \Gamma_2(\mathcal{A})$ for a reduced complex, $\mathcal{A} = \langle \boldsymbol{A}, :, \boldsymbol{X}\rangle$ over a Boolean space $\boldsymbol{X}$. By assembling the various parts of Lemma 4.3, we see that $\boldsymbol{B}$ is a Boolean subsemilattice of $\mathbf{Con}\,\boldsymbol{A}$. So $\langle \boldsymbol{A},\boldsymbol{B}\rangle$ is a Boolean brace.

Still left to be proven in (i) is that $\langle \boldsymbol{A},\boldsymbol{B}\rangle$ is reduced. We know, from Stone duality in Sect. III.4, Corollary 4.4 and (4.2), that any prime ideal P comes from an element x of the base space $\boldsymbol{X}$:

$$P = P_x = \{\theta_U \mid x \in U \in \operatorname{Clop}\boldsymbol{X}\}.$$

Let x be a nontrivial point of $\boldsymbol{X}$, that is, $|\theta_x| \neq 1$. We will show that $\bigvee P_x \neq 1$. By nontriviality there are elements a_1 and a_2 such that $x \notin a_1 : a_2$. Hence, $U \nsubseteq a_1 : a_2$ whenever $x \in U$. Thus it never happens that $a_1\ \theta_U\ a_2$ for such U. Hence the relationship $a_1 \bigvee P_x\ a_2$ fails. Therefore, nontrivial points go to proper primes ideals: $\operatorname{Triv}\boldsymbol{B} \subseteq \operatorname{Triv}\mathcal{A}$. Hence, $\operatorname{Int}\operatorname{Triv}\boldsymbol{B} \subseteq \operatorname{Int}\operatorname{Triv}\mathcal{A}$, and so (RC1) implies (RB).

To demonstrate (ii) that Γ_2 takes morphisms of complexes to morphisms of Boolean braces, consider the couple

$$\langle\varphi,\upsilon\rangle = \Gamma_2(\langle\varphi,\chi\rangle),$$

as spelled out in Definition 4.8. That υ preserves all the Boolean operations in going from $\boldsymbol{B}$ to $\boldsymbol{B}^*$ follows from Lemma 4.3, and the preservation of the Boolean operations on subsets of X by χ^{-1}.

Next, we want to prove that $\varphi(\theta_U) \subseteq \upsilon(\theta_U)$ for all U in $\operatorname{Clop}\boldsymbol{X}$. To that end, suppose that $\langle b_1,b_2\rangle \in \varphi(\theta_U)$. Then $b_1 = \varphi(a_1)$ and $b_2 = \varphi(a_2)$ where $a_1\ \theta\ a_2$ for some a_1,a_2 in A. So $U \subseteq a_1 : a_2$ by the definition of θ_U. Hence,

$$\chi^{-1}(U) \subseteq \chi^{-1}(a_1 : a_2) \subseteq \varphi(a_1) \overset{\star}{\underset{\star}{:}} \varphi(a_2) = b_1 \overset{\star}{\underset{\star}{:}} b_2,$$

from Definition IV.3.1 of a complex morphism. Therefore, $\langle b_1,b_2\rangle \in \theta_{\chi^{-1}(U)}$, which is equal to $\upsilon(\theta_U)$, and we conclude that $\varphi(\theta_U) \subseteq \upsilon(\theta_U)$.

Conditions (iii) and (iv) are satisfied, roughly since morphisms in both **CompBooleRed** and **BooleBraceRed** are pairs with composition by components.

To prove (v), that Γ_2 is faithful, let us assume that $\Gamma_2(\langle\varphi,\chi\rangle) = \langle\varphi,\upsilon\rangle = \Gamma_2(\langle\varphi,\psi\rangle)$. We must show that $\chi = \psi$. By way of contradiction, suppose not: that there is an x^* in X^* such that $\chi(x^*) \neq \psi(x^*)$. From the definition of Γ_2, we find for the second component that

$$\theta_{\chi^{-1}(U)} = \upsilon(\theta_U) = \theta_{\psi^{-1}(U)} \qquad (U \in \operatorname{Clop}\boldsymbol{X}).$$

Corollary 4.4 transfers this equality to its subscripts:

$$\chi^{-1}(U) = \psi^{-1}(U) \qquad (U \in \operatorname{Clop} \boldsymbol{X}). \tag{4.6}$$

Since $\boldsymbol{X}$ is Hausdorff, there is a clopen subset U and its complement V in $\boldsymbol{X}$ such that $\chi(x^*) \in U$ and $\psi(x^*) \in V$. Hence, $x^* \in \chi^{-1}(U)$, $x^* \in \psi^{-1}(V)$, and so $x^* \notin \psi^{-1}(U)$, which (4.6) can not abide. □

Before we state precisely the adjoint situation relating these functors, we need to discuss the associated natural transformations, which are

$$\eta_2 \colon 1_{\mathfrak{BooleBraceRed}} \to \Gamma_2 \circ \Phi_2 \quad \text{and} \quad \varepsilon_2 \colon \Phi_2 \circ \Gamma_2 \to 1_{\mathfrak{CompBooleRed}}.$$

4.10. DEFINITION. For the first, η_2 of a Boolean brace $\langle \boldsymbol{A}, \boldsymbol{B} \rangle$ will be a morphism in $\mathfrak{BooleBraceRed}$,

$$\eta_2(\langle \boldsymbol{A}, \boldsymbol{B} \rangle) = \langle 1_A, \widetilde{\upsilon} \rangle \colon \langle \boldsymbol{A}, \boldsymbol{B} \rangle \to \langle \boldsymbol{A}, \widetilde{\boldsymbol{B}} \rangle,$$

where $\langle \boldsymbol{A}, \widetilde{\boldsymbol{B}} \rangle = \Gamma_2\big(\Phi_2(\langle \boldsymbol{A}, \boldsymbol{B} \rangle)\big)$. In other words,

$$\widetilde{\boldsymbol{B}} = \boldsymbol{\theta}_{\operatorname{Clop} \mathbf{Spec}\, \boldsymbol{B}} \quad \text{where} \quad \widetilde{B} = \{\theta_U \mid U \in \operatorname{Clop} \mathbf{Spec}\, \boldsymbol{B}\}.$$

The function, $\widetilde{\upsilon} \colon \boldsymbol{B} \to \widetilde{\boldsymbol{B}}$, is given as

$$\widetilde{\upsilon}(\theta) = \theta_{U_\theta} \qquad (\theta \in B),$$

where $U_\theta = \{P \in \operatorname{Spec} \boldsymbol{B} \mid \theta \in P\}$ by (4.5), and by (4.1),

$$\theta_{U_\theta} = \{\langle a_1, a_2 \rangle \mid U_\theta \subseteq a_1 \mathbin{:} a_2\}.$$

For the second, ε_2 of a complex $\mathcal{A}$ over $\boldsymbol{X}$ will be a morphism of complexes,

$$\varepsilon_2(\mathcal{A}) = \langle 1_A, \widetilde{\chi} \rangle \colon \widetilde{\mathcal{A}} \to \mathcal{A},$$

where

$$\widetilde{\mathcal{A}} = \langle \boldsymbol{A}, \widetilde{:}, \widetilde{\boldsymbol{X}} \rangle = \Phi_2\big(\Gamma_2(\mathcal{A})\big).$$

That is, $\widetilde{\boldsymbol{X}} = \mathbf{Spec}\, \boldsymbol{\theta}_{\operatorname{Clop} \boldsymbol{X}}$, and

$$a_1 \mathbin{\widetilde{:}} a_2 = \Big\{P \in \widetilde{X} \;\Big|\; a_1 \bigvee\nolimits_P a_2\Big\} \qquad (a_1, a_2 \in A),$$

with $\widetilde{\chi} \colon \boldsymbol{X} \to \widetilde{\boldsymbol{X}}$ defined by

$$\widetilde{\chi}(x) = P_x = \{\theta_U \mid x \in U \in \operatorname{Clop} \boldsymbol{X}\} \qquad (x \in X). \tag{4.7}$$

4.11. PROPOSITION. (a) *η_2 is a natural transformation.*
(b) *ε_2 is a natural isomorphism.*

PROOF. Initially we prove that η_2 and ε_2 yield morphisms of the right kind, and later demonstrate that the required square diagrams commute.

(a) Let us show for any object, $\mathsf{A} = \langle \boldsymbol{A}, \boldsymbol{B} \rangle$ in $\mathfrak{BooleBraceRed}$, that

$$\eta_2(\mathsf{A}) \colon \mathsf{A} \to \Gamma_2\big(\Phi_2(\mathsf{A})\big)$$

is a morphism in $\mathfrak{BooleBraceRed}$. Now $\eta_2(\mathsf{A}) = \langle 1_{\boldsymbol{A}}, \widetilde{\upsilon}\rangle$ where $\widetilde{\upsilon}(\theta) = \theta_{U_\theta}$. To verify that $\widetilde{\upsilon}$ is a homomorphism of Boolean subsemilattices, we only need recall from Stone duality that

$$U_{\theta\cap\xi} = U_\theta \cup U_\xi, \quad \text{and} \quad U_{0_{\mathrm{Con}\,\boldsymbol{A}}} = X;$$

and remember what we proved earlier in Lemma 4.3 that

$$\theta_{U\cup V} = \theta_U \cap \theta_V, \quad \text{and} \quad \theta_X = 0.$$

Thus $\widetilde{\upsilon}$ is a homomorphism.

Proving that $\varphi(\theta) \subseteq \widetilde{\upsilon}(\theta)$ amounts to proving in this context that $\theta \subseteq \theta_{U_\theta}$, by using Definition 4.10 extensively. To this end, suppose that $a_1 \theta a_2$. For a prime ideal P of $\boldsymbol{B}$, if $\theta \in P$, then $a_1 \bigvee P\, a_2$. Hence $U_\theta \subseteq a_1 : a_2$ by (4.3). Therefore, $a_1\, \theta_{U_\theta}\, a_2$ by (4.2).

(b) Let us show for any complex, $\mathcal{A} = \langle \boldsymbol{A}, :, \boldsymbol{X}\rangle$, in $\mathfrak{CompBooleRed}$ that the function

$$\varepsilon_2(\mathcal{A})\colon \Phi_2\big(\Gamma_2(\mathcal{A})\big) \to \mathcal{A}$$

is an isomorphism in $\mathfrak{CompBooleRed}$. Write $\varepsilon_2(\mathcal{A}) = \langle 1_{\boldsymbol{A}}, \widetilde{\chi}\rangle$ where

$$\widetilde{\chi}\colon X \to \operatorname{Spec} \boldsymbol{\theta}_{\mathbf{Clop}\,\boldsymbol{X}} \colon x \mapsto P_x$$

with $P_x = \{\theta_U \mid x \in U \in \operatorname{Clop} \boldsymbol{X}\}$. Since the composition of morphisms in $\mathfrak{CompBooleRed}$ is the composition of component functions, it is clear that the inverse of the morphism, $\varepsilon_2(\mathcal{A}) = \langle 1_{\boldsymbol{A}}, \widetilde{\chi}\rangle$, will be $\langle 1_{\boldsymbol{A}}, \widetilde{\chi}^{-1}\rangle$, if it exists. So it suffices to show that $\widetilde{\chi}$ is a homeomorphism. By Corollary 4.4, we know that the Boolean algebras $\boldsymbol{\theta}_{\mathbf{Clop}\,\boldsymbol{X}}$ and $\mathbf{Clop}\,\boldsymbol{X}$ are anti-isomorphic. Stone duality assures us that the corresponding topological spaces are homeomorphic:

$$\mathbf{Spec}\,\boldsymbol{\theta}_{\mathbf{Clop}\,\boldsymbol{X}} \cong \mathbf{Spec}\,\mathbf{Clop}\,\boldsymbol{X} \cong \boldsymbol{X}.$$

With elements, this correspondence is $P_x \leftrightarrow x$. Thus, $\varepsilon_2(\mathcal{A})$ is an isomorphism in $\mathfrak{CompBooleRed}$.

Second, for naturalness one must show that these two diagrams from Definition III.2.18 commute, where we start with $\langle\varphi, \upsilon\rangle\colon \mathsf{A} \to \mathsf{A}^*$, a morphism of Boolean braces, and $\langle \varphi, \chi\rangle\colon \mathcal{A} \to \mathcal{A}^*$, a morphism of complexes:

$$\begin{array}{ccc}
\mathsf{A} & \xrightarrow{\eta_2(\mathsf{A})} & (\Gamma_2 \circ \Phi_2)(\mathsf{A}) \\
{\scriptstyle \langle\varphi,\upsilon\rangle}\big\downarrow & & \big\downarrow{\scriptstyle (\Gamma_2\circ\Phi_2)(\langle\varphi,\upsilon\rangle)} \\
\mathsf{A}^* & \xrightarrow[\eta_2(\mathsf{A}^*)]{} & (\Gamma_2 \circ \Phi_2)(\mathsf{A}^*)
\end{array}$$

$$\begin{array}{ccc}
(\Phi_2 \circ \Gamma_2)(\mathcal{A}) & \xrightarrow{\varepsilon_2(\mathcal{A})} & (\mathcal{A}) \\
{\scriptstyle (\Phi_2\circ\Gamma_2)(\langle\varphi,\chi\rangle)}\big\downarrow & & \big\downarrow{\scriptstyle \langle\varphi,\chi\rangle} \\
(\Phi_2 \circ \Gamma_2)(\mathcal{A}^*) & \xrightarrow[\varepsilon_2(\mathcal{A}^*)]{} & (\mathcal{A}^*)
\end{array}$$

To prove that the first square commutes, introduce the notations:

$$\eta_2(\mathsf{A}) = \langle 1_{\mathsf{A}}, \widetilde{\upsilon} \rangle, \qquad \eta_2(\mathsf{A}^*) = \langle 1_{\mathsf{A}^*}, \widetilde{\upsilon}^* \rangle,$$
$$\Phi_2(\langle \varphi, \upsilon \rangle) = \langle \varphi, \chi \rangle, \qquad \Gamma_2(\langle \varphi, \chi \rangle) = \langle \varphi, \widehat{\upsilon} \rangle.$$

Note that, in tracing the first component of the morphisms, all the functions are φ, 1_{A} or 1_{A^*}; so clearly the first components commute. The second components are more troublesome. Tracing the second component around through the lower left corner in the first square, we get

$$\begin{array}{ccc} \boldsymbol{B} & & \\ \upsilon \big\downarrow & & \\ \boldsymbol{B}^* & \longrightarrow & \boldsymbol{\theta}_{\mathbf{Clop\,Spec}\,\boldsymbol{B}^*} \end{array} \qquad \begin{array}{ccc} \theta & & \\ \big\downarrow & & \\ \upsilon(\theta) & \longmapsto & \theta_{U_{\upsilon(\theta)}} \end{array}$$

Going by the upper right corner, we get

$$\begin{array}{ccc} \boldsymbol{B} & \longrightarrow & \boldsymbol{\theta}_{\mathbf{Clop\,Spec}\,\boldsymbol{B}} \\ & & \big\downarrow \widehat{\upsilon} \\ & & \boldsymbol{\theta}^*_{\mathbf{Clop\,Spec}\,\boldsymbol{B}} \end{array} \qquad \begin{array}{ccc} \theta & \longmapsto & \theta_{U_\theta} \\ & & \big\downarrow \\ & & \widehat{\upsilon}(\theta_{U_\theta}) \end{array}$$

We need to demonstrate that $\theta_{U_{\upsilon(\theta)}} = \widehat{\upsilon}(\theta_{U_\theta})$ for any θ in $\boldsymbol{B}$. Now $\widehat{\upsilon}(\theta_{U_\theta}) = \theta_{\chi^{-1}(U_\theta)}$ by the definition of Γ_2. We will be finished if we can verify that $\chi^{-1}(U_\theta) = U_{\upsilon(\theta)}$. As an inverse function,

$$\chi^{-1}(U_\theta) = \{ Q \in \mathrm{Spec}\,\boldsymbol{B} \mid \chi(Q) \in U_\theta \}.$$

Reasoning with further definitions, such as for U_θ and Φ_2, we see that

$$\chi(Q) \in U_\theta \ \text{ iff } \ \theta \in \chi(Q) = \upsilon^{-1}(Q) \ \text{ iff } \ \upsilon(\theta) \in Q.$$

Therefore,

$$\begin{aligned} \chi^{-1}(U_\theta) &= \{ Q \in \mathrm{Spec}\,\boldsymbol{B} \mid \upsilon(\theta) \in Q \} \\ &= U_{\upsilon(\theta)}. \end{aligned}$$

To prove commutativity of the second square, we write it explicitly:

$$\begin{array}{ccc} \langle \boldsymbol{A}, \widetilde{:}, \widetilde{\boldsymbol{X}} \rangle & \xrightarrow{\langle 1_{\boldsymbol{A}}, \widetilde{\chi} \rangle} & \langle \boldsymbol{A}, :, \boldsymbol{X} \rangle \\ {\scriptstyle \langle \varphi, \widehat{\chi} \rangle} \big\downarrow & & {\scriptstyle \langle \varphi, \chi \rangle} \big\downarrow \\ \langle \boldsymbol{A}^*, \widetilde{\substack{\star \\ \star}}, \widetilde{\boldsymbol{X}}^* \rangle & \xrightarrow[\langle 1^*_{\boldsymbol{A}}, \widetilde{\chi}^* \rangle]{} & \langle \boldsymbol{A}^*, \substack{\star \\ \star}, \boldsymbol{X}^* \rangle \end{array}$$

Here, $\widetilde{\boldsymbol{X}} = \mathbf{Spec}\,\boldsymbol{\theta}_{\mathbf{Clop}\,\boldsymbol{X}}$ and $\widetilde{\boldsymbol{X}}^* = \mathbf{Spec}\,\boldsymbol{\theta}_{\mathbf{Clop}\,\boldsymbol{X}^*}$. Clearly the diagram commutes in the first component: $1_{\boldsymbol{A}^*} \circ \varphi = \varphi \circ 1_{\boldsymbol{A}}$. The arrows of the second components will go in the opposite direction. Let us abbreviate the composite of the second components through the upper right corner by $U.R. = \widetilde{\chi} \circ \chi$, and the composite through the lower left by $L.L. = \widehat{\chi} \circ \widetilde{\chi}^*$.

We need to prove that $U.R. = L.L.$ To that end we find expressions for these composites in terms of Definition 4.10, letting $x^* \in X^*$:

$$U.R.(x^*) = \widetilde{\chi}\big(\chi(x^*)\big) = \{\theta_U \mid \chi(x^*) \in U \in \operatorname{Clop} \boldsymbol{X}\}\,; \text{ and}$$
$$L.L.(x^*) = \widehat{\chi}\big(\widetilde{\chi}^*(x^*)\big) = \upsilon^{-1}\{\theta_{U^*} \mid x^* \in U^* \in \operatorname{Clop} \boldsymbol{X}^*\}\,,$$

where $\upsilon\colon \boldsymbol{\theta}_{\mathbf{Clop}\,\boldsymbol{X}} \to \boldsymbol{\theta}_{\mathbf{Clop}\,\boldsymbol{X}^*}$ is the homomorphism induced by χ via Γ_2. We establish their equality by demonstrating inclusion each way.

$U.R.(x^*) \subseteq L.L.(x^*)$. To prove this, assume we have θ_U in the left side where $\chi(x^*) \in U \in \operatorname{Clop} \boldsymbol{X}$. Define $U^* = \chi^{-1}(U)$. This will be clopen, and $x^* \in U^* \in \operatorname{Clop} \boldsymbol{X}^*$. Does $\theta_U \in \upsilon^{-1}(\theta_{U^*})$, anticipating membership in $L.L.(x^*)$? Well, by Definition 4.8, $\upsilon(\theta_U) = \theta_{\chi^{-1}(U)} = \theta_{U^*}$; and so, yes, it does.

$L.L.(x^*) \subseteq U.R.(x^*)$. Assume that $\theta \in L.L.(x^*)$. That is, $\upsilon(\theta) = \theta_{U^*}$ for some U^* such that $x^* \in U^* \in \operatorname{Clop} \boldsymbol{X}^*$. Now $\theta \in B$ where $B = \theta_{\mathbf{Clop}\,\boldsymbol{X}}$; so $\theta = \theta_U$ for some U in **Clop** $\boldsymbol{X}$. We need to show that U and U^* are appropriately related: $U^* = \chi^{-1}(U)$. Notice that

$$\theta_{U^*} = \upsilon(\theta) = \upsilon(\theta_U) = \theta_{\chi^{-1}(U)},$$

by the definition of υ. Since the map $U \mapsto \theta_U$ is injective by Lemma 4.3, it follows that $U^* = \chi^{-1}(U)$. Because $x^* \in U^*$, it further follows that $\chi(x^*) \in U$. Since $U \in \operatorname{Clop} \boldsymbol{X}$, then $\theta \in U.R.(x^*)$ by its expression above. We have settled the inclusions in both directions, so equality holds between $L.L.(x^*)$ and $U.R.(x^*)$, and the diagram commutes. □

In passing we note that, while $\widetilde{\upsilon}$ (from Definition 4.10) is always an isomorphism of Boolean algebras, η_2 itself need not be a natural isomorphism of categories, but it is close (see Theorem 2.9). If nothing else, the unity of a Boolean subsemilattice of congruences may float. In other words, conceivably $\langle 1_{\boldsymbol{A}}, \widetilde{\upsilon}^{-1}\rangle$ may not be a morphism of complexes. All we can be certain of is parts (a) and (b) of the next corollary. We do have an expected, but not entirely obvious characterization of isomorphism in part (c). The next two corollaries come from the preceding proof.

4.12. Corollary. *Let $\eta_2(\mathsf{A})$ be the natural transformation $\langle 1_{\boldsymbol{A}}, \widetilde{\upsilon}\rangle$ of a reduced Boolean brace* A.

(a) $\theta \subseteq \theta_{U_\theta} \quad (\theta \in B)$.
(b) $\widetilde{\upsilon}$ *is always an isomorphism:* $\boldsymbol{B} \cong \theta_{\mathbf{Clop\,Spec}\,\boldsymbol{B}}$.
(c) η_2 *is a natural isomorphism iff* $\boldsymbol{B} = \theta_{\mathbf{Clop\,Spec}\,\boldsymbol{B}}$ *and* $\widetilde{\upsilon} = 1_{\boldsymbol{B}}$.

Proof. (a) By Definition 4.10.

(b) This is true since by definition, $\widetilde{\upsilon}(\theta) = \theta_{U_\theta}$; we have shown that the functions $\theta \mapsto U_\theta$ and $U \mapsto \theta_U$ are bijections preserving the dual Boolean operations: the first by Stone duality, and the second by Lemma 4.3g.

(c) $\Rightarrow$. If we have a categorical isomorphism $\langle 1_{\boldsymbol{A}}, \widetilde{\upsilon}\rangle$, then it has an inverse $\langle 1_{\boldsymbol{A}}, \widetilde{\upsilon}^{-1}\rangle$, where $\widetilde{\upsilon}^{-1}$ is the inverse of $\widetilde{\upsilon}$. By Definition 4.1,

$$\theta = 1_{\boldsymbol{A}}(\theta) \subseteq \widetilde{\upsilon}(\theta) \qquad (\theta \in \boldsymbol{B}), \quad \text{and}$$
$$\theta = 1_{\boldsymbol{A}}(\theta) \subseteq \widetilde{\upsilon}^{-1}(\theta) \qquad (\theta \in \theta_{\mathrm{Clop\,Spec}\,\boldsymbol{B}}).$$

Hence,

$$\theta \subseteq \widetilde{\upsilon}(\theta) \subseteq \widetilde{\upsilon}^{-1}(\widetilde{\upsilon}(\theta)) = \theta.$$

Thus, $\theta = \widetilde{\upsilon}(\theta) = \theta_{U_\theta}$ whenever $\theta \in \boldsymbol{B}$, and the conclusion follows.

$\Leftarrow$. Obvious. □

A useful consequence of the proof of Proposition 4.11 comes from cranking up the duality one more step.

4.13. COROLLARY. *Let $\varepsilon_2(\mathcal{A})$ be the natural transformation $\langle 1_{\mathcal{A}}, \widetilde{\chi}\rangle$ of a reduced complex, $\mathcal{A} = \langle \boldsymbol{A}, :, \boldsymbol{X}\rangle$, and assume that $U \in \mathrm{Clop}\,\boldsymbol{X}$.*

(a) $\theta_{U_{\theta_U}} = \theta_U$.

(b) $\widetilde{\chi}^{-1}(U_{\theta_U}) = U$.

PROOF. Recall for both parts from Stone duality and (4.7) that

$$\boldsymbol{B} = \boldsymbol{\theta}_{\mathrm{Clop}\,\boldsymbol{X}},$$
$$U_{\theta_U} = \{P \in \mathrm{Spec}\,\boldsymbol{B} \mid \theta_U \in P\},$$

and each prime ideal P of $\boldsymbol{B}$ is determined by an element x of X as

$$P = P_x = \{\theta_U \mid x \in U \in \mathrm{Clop}\,\boldsymbol{X}\}.$$

(a) This amounts to the equivalence: $U_{\theta_U} \subseteq a_1 \mathbin{\tilde{:}} a_2$ iff $U \subseteq a_1 : a_2$. But this is true since for any prime ideal, $P_x \in U_{\theta_U}$ iff $\theta_U \in P_x$ iff $x \in U$.

(b) Since $\widetilde{\chi}(x) = P_x$, the assertion amounts to proving for any $x \in X$ that $P_x \in U_{\theta_U}$ iff $x \in U$, which was proven in part (a). □

We are now ready for the first adjunction of this chapter, which will shortly be composed with the adjunction of the previous chapter.

4.14. THEOREM. *The functors,*

$$\Phi_2 \colon \mathfrak{BooleBraceRed} \rightleftarrows \mathfrak{CompBooleRed} \colon \Gamma_2,$$

together with the natural transformations,

$$\eta_2 \colon 1_{\mathfrak{BooleBraceRed}} \to \Gamma_2 \circ \Phi_2 \quad \text{and} \quad \varepsilon_2 \colon \Phi_2 \circ \Gamma_2 \to 1_{\mathfrak{CompBooleRed}},$$

are an adjunction:

$$\langle \eta_2, \varepsilon_2\rangle \colon \Phi_2 \dashv \Gamma_2 \colon \langle \mathfrak{CompBooleRed}, \mathfrak{BooleBraceRed}\rangle.$$

That is, Φ_2 is a left adjoint of Γ_2, and Γ_2 is a right adjoint of Φ_2. Moreover, ε_2 is a natural isomorphism.

PROOF. By Theorem III.2.20 it suffices to show that each brace, $\mathsf{A} = \langle \boldsymbol{A}, \boldsymbol{B} \rangle$, in $\mathfrak{BooleBraceRed}$ has a Γ_2-universal map $\langle \eta_2(\mathsf{A}), \mathcal{A} \rangle$, where $\mathcal{A} = \Phi_2(\mathsf{A})$. By a Γ_2-universal map in this context we mean that for each complex $\mathcal{A}^*$ and each morphism, $\langle \varphi, \upsilon \rangle : \mathsf{A} \to \Gamma_2(\mathcal{A}^*)$ in $\mathfrak{BooleBraceRed}$, there exists a unique complex morphism, $\langle \varphi, \chi \rangle : \mathcal{A} \to \mathcal{A}^*$, such that

$$\Gamma_2(\langle \varphi, \chi \rangle) \circ \eta_2(\mathsf{A}) = \langle \varphi, \upsilon \rangle,$$

that is, this triangle commutes:

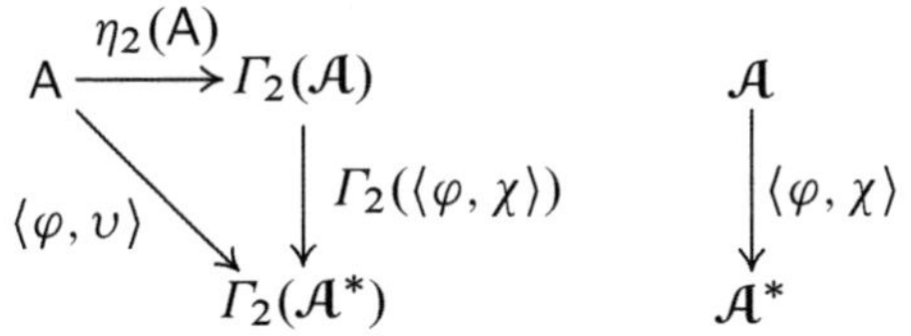

In anticipation that φ would not change, we have avoided introducing notation that would be more honest than necessary. Write out the second complex as $\mathcal{A}^* = \langle \boldsymbol{A}, \overset{\star}{\underset{\star}{,}}, \boldsymbol{X}^* \rangle$. Define the function, $\chi : \boldsymbol{X}^* \to \mathbf{Spec}\, \boldsymbol{B}$, as the dual of the function υ on congruences:

$$\chi(x^*) = \upsilon^{-1}(P_{x^*}) \qquad (x^* \in X^*). \tag{4.8}$$

This proof requires showing the following.

(a) The function $\langle \varphi, \chi \rangle$ is a morphism of complexes:
  (1) χ is continuous,
  (2) $\chi^{-1}(a_1 : a_2) \subseteq \varphi(a_1) \overset{\star}{\underset{\star}{:}} \varphi(a_2) \quad (a_1, a_2 \in \boldsymbol{A})$.
(b) The triangle above commutes.
(c) The natural transformation ε_2 fits in, that is:
  (1) $\Gamma_2\big(\varepsilon_2(\mathcal{A})\big) \circ \eta_2\big(\Gamma_2(\mathcal{A})\big) = 1_{\Gamma_2(\mathcal{A})} \qquad (\mathcal{A} \in \mathfrak{CompBooleRed})$,
  (2) $\varepsilon_2\big(\Phi_2(\mathsf{A})\big) \circ \Phi_2\big(\eta_2(\mathsf{A})\big) = 1_{\Phi_2(\mathsf{A})} \qquad (\mathsf{A} \in \mathfrak{BooleBraceRed})$.
(d) The morphism $\langle \varphi, \chi \rangle$ is the unique one making the triangle commute.

Formulas (c.1) and (c.2) amount to the commutativity of the diagrams:

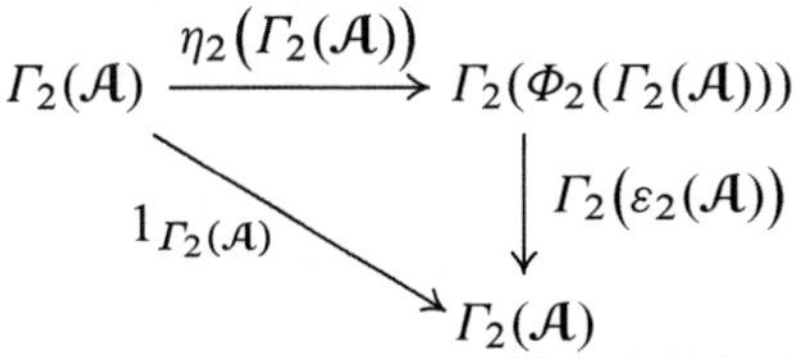

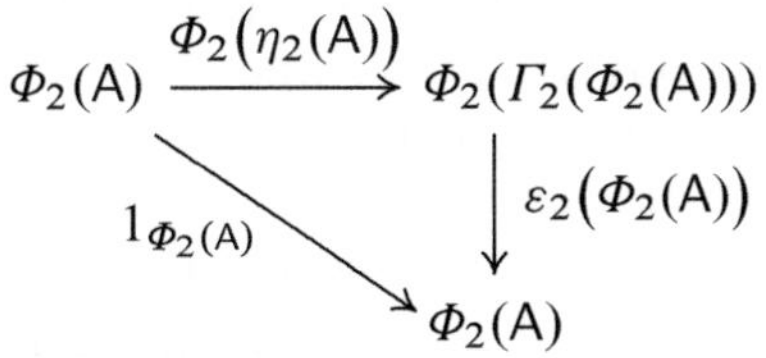

These will ensure that the natural transformation ε_2, already defined, is indeed the unique one guaranteed by Theorem III.2.20. We defer the proofs of (a) and (b), which are based upon (c.1), now to be proven. Then, after proving (a) and (b), we will finish up with (d) and (c.2).

(c.1) We have to prove the commutativity of the triangle on the left. Let $\langle \boldsymbol{A}, \boldsymbol{B} \rangle = \Gamma_2(\mathcal{A})$ for some orthodox complex $\mathcal{A}$ and write $\langle 1_{\boldsymbol{A}}, \widetilde{\upsilon} \rangle = \eta_2\big(\Gamma_2(\mathcal{A})\big)$. Clearly the first components of all the morphisms are identities.

4.15. PROPOSITION. (a) *If $\mathcal{A}$ is a reduced complex over a Boolean space $\boldsymbol{X}$, then $\Phi_1(\mathcal{A})$ is a reduced sheaf.*
(b) *If $\mathcal{A}$ is a reduced sheaf, then $\Gamma_1(\mathcal{A})$ is a reduced complex.*

PROOF. Both parts rely on patching arguments, see Sect. 2.
(a) Let us verify axiom RS using axiom RC1. By the definition of Φ_1 each stalk $\boldsymbol{A}_x$ of $\mathcal{A}$ is A/θ_x. Therefore, the trivial stalks of the sheaf are just those of the complex. Since their base topologies are the same, the interiors of their sets of trivial points are the same.
(b) Again, since the base spaces stay the same in passing from sheaves to complexes via Φ_2, one need only trace nontrivial points through this functor in order to verify axiom RC1. For any nontrivial point x in X of the sheaf $\mathcal{A}$, there are a and b in $\mathcal{A}$ such that $a \neq b$ and $\pi(a) = x = \pi(b)$. The definition of $\mathfrak{SheafAlg}$, of which $\mathfrak{SheafBooleRed}$ is a subcategory, guarantees that through both a and b there pass global sections σ and τ, respectively, differing at x of course. Thus, σ and τ differ on $\boldsymbol{A}_x$, and so x is a nontrivial point of the complex $\Gamma_1(\mathcal{A})$.
For the proof of axiom RC2, suppose $\mathcal{A}$ is a reduced sheaf over $\boldsymbol{X}$, and U, V are clopen subsets of $\boldsymbol{X}$ such that $U \neq V$; so there is an x in $V \sim U$. By (RS) there must be global sections σ, τ in $\Gamma\ (\mathcal{A})$ such that $\sigma(x) \neq \tau(x)$. With Proposition 2.4 patch together a new global section, $\rho = \sigma|(X \sim U) \cup \tau|U$. Then $\rho(x) = \sigma(x) \neq \tau(x)$ and $\rho|U = \tau|U$. Recall that $\theta_U = \{\langle\sigma_1, \sigma_2\rangle \mid U \subseteq \sigma_1 : \sigma_2\}$. Therefore, $\rho\, \theta_U\, \tau$ but not $\rho\, \theta_V\, \tau$. □

4.16. COROLLARY. *If Φ_1, Γ_1, η_1, and ε_1 are restricted to $\mathfrak{CompBooleRed}$ and $\mathfrak{SheafBooleRed}$, then there is again an adjunction:*

$$\langle \eta_1, \varepsilon_1 \rangle : \Phi_1 \dashv \Gamma_1 : \langle \mathfrak{SheafBooleRed}, \mathfrak{CompBooleRed} \rangle.$$

Composing this adjunction with the previous one gives a third adjunction of the functors:

$$\Phi_1 \Phi_2 : \mathfrak{BooleBraceRed} \to \mathfrak{SheafBooleRed},$$

$$\Gamma_2 \Gamma_1 : \mathfrak{SheafBooleRed} \to \mathfrak{BooleBraceRed}.$$

Composite notations are:

$$\Phi = \Phi_1 \circ \Phi_2, \quad \text{and} \quad \Gamma = \Gamma_2 \circ \Gamma_1.$$

A comment is called for on how this new Γ encompasses the old Γ that constructed all global sections in Sect. IV.2. Since the old Γ produces the desired algebra, with a slight abuse of notation we may imagine it simultaneously producing the associated Boolean algebra of congruences, which is isomorphic to the dual of the base space.

We calculate that

$$\Phi(\langle \boldsymbol{A}, \boldsymbol{B} \rangle) = \left\langle \biguplus_{P \in X} \frac{\boldsymbol{A}}{\bigvee P}, \pi, \boldsymbol{X} \right\rangle$$

where $\boldsymbol{X} = \mathbf{Spec}\,\boldsymbol{B}$. Notice how Φ_2 first creates $\boldsymbol{X}$ but leaves $\boldsymbol{A}$ untouched in the intermediate complex; then Φ_1 produces the sheaf over the same Boolean space. Notice a similar division of labor when we calculate

$$\Gamma(\mathcal{A}) = \langle \boldsymbol{A}, \boldsymbol{B} \rangle,$$

where $\boldsymbol{A}$ is the algebra of global sections of the sheaf $\mathcal{A}$ and $\boldsymbol{B} = \boldsymbol{\theta}_{\mathbf{Clop}\,\boldsymbol{X}}$. The functor Γ_1 first creates $\boldsymbol{A}$ but leaves the base space $\boldsymbol{X}$ untouched en passant the intermediate complex; then Γ_2 produces the dual Boolean algebra $\boldsymbol{B}$.

The new natural transformations associated with the composite adjunction are denoted as

$$\eta\colon 1_{\mathfrak{BooleBraceRed}} \to \Gamma \circ \Phi \quad \text{and} \quad \varepsilon\colon \Phi \circ \Gamma \to 1_{\mathfrak{SheafBooleRed}},$$

and operate like so:

$$\eta(\mathsf{A})\colon \mathsf{A} \to \Gamma\big(\Phi(\mathsf{A})\big) \qquad (\mathsf{A} \in \operatorname{dom} \mathfrak{BooleBraceRed});$$

$$\varepsilon(\mathcal{A})\colon \Phi\big(\Gamma(\mathcal{A})\big) \to \mathcal{A} \qquad (\mathcal{A} \in \operatorname{dom} \mathfrak{SheafBooleRed}).$$

Specifically, from (III.2.1), for any A in dom $\mathfrak{BooleBraceRed}$,

$$\eta(\mathsf{A}) = \Gamma_2\big(\eta_1(\Phi_2(\mathsf{A}))\big) \circ \eta_2(\mathsf{A}).$$

Similarly, for any $\mathcal{A}$ in dom $\mathfrak{SheafBooleRed}$,

$$\varepsilon(\mathcal{A}) = \varepsilon_1(\mathcal{A}) \circ \Phi_1\big(\varepsilon_2(\Gamma_1(\mathcal{A}))\big).$$

If we compute $\eta(\mathsf{A}) = \langle \widetilde{\varphi}, \widetilde{\upsilon} \rangle\colon \mathsf{A} \to \widetilde{\mathsf{A}}$, where $\widetilde{\mathsf{A}} = \Gamma\big(\Phi(\mathsf{A})\big)$, we find that

$$\widetilde{\varphi}(a) = \gamma(a) \ \text{ and } \ \widetilde{\upsilon}(\theta) = \theta_{U_\theta} \qquad (a \in A,\ \theta \in B).$$

If we compute $\varepsilon(\mathcal{A}) = \langle \widetilde{\psi}, \widetilde{\chi} \rangle\colon \widetilde{\mathcal{A}} \to \mathcal{A}$, where $\widetilde{\mathcal{A}} = \Phi\big(\Gamma(\mathcal{A})\big)$, we find that

$$\widetilde{\psi}(\lfloor\sigma\rfloor_{P_x}, x) = \sigma(x) \ \text{ and } \ \widetilde{\chi}(x) = P_x \qquad (\sigma \in \Gamma(\mathcal{A}),\ x \in X).$$

We formalize this adjoint situation as the principle theorem of this section, which was abstracted in [**Knoe92b**].

4.17. THEOREM. *The functors,*

$$\Phi\colon \mathfrak{BooleBraceRed} \rightleftarrows \mathfrak{SheafBooleRed}\colon \Gamma,$$

together with the natural transformations,

$$\eta\colon 1_{\mathfrak{BooleBraceRed}} \to \Gamma \circ \Phi \quad \text{and} \quad \varepsilon\colon \Phi \circ \Gamma \to 1_{\mathfrak{SheafBooleRed}},$$

form an adjunction:

$$\langle \eta, \varepsilon \rangle\colon \Phi \dashv \Gamma\colon \langle \mathfrak{SheafBooleRed}, \mathfrak{BooleBraceRed} \rangle.$$

In other words, Φ is a left adjoint of Γ, and Γ is a right adjoint of Φ. Moreover, ε is a natural isomorphism.

PROOF. Employ Theorems III.2.24, IV.3.18 and 4.14. □

The categories $\mathfrak{Complex}$ and $\mathfrak{CompBooleRed}$ have served their purpose and will no longer be seen, except implicitly when equalizers are needed.

We close this section by pointing out how second-order axioms — (RB), (RC1), (RC2) and (RS)— arose in Definitions 2.8, 4.1 and 4.2. Clearly they are aimed at ensuring the interdefinability of the different structures. Starting with a complex, to recover a Boolean subsemilattice of congruences it appears that one needs an axiom such as (RC2) in order to prove that complements exist. In turn, starting with a sheaf, to prove (RC2) in the resulting complex, one needs (RS). But to prove (RS) in a sheaf, when coming from a complex, one also needs (RC1). Similarly, (RC1) is needed to prove (RB) for Boolean braces. And, in going from Boolean braces back to complexes, (RB) proves (RC1). We see that, in order to create functors passing back and forth between the three kinds of categorical objects, the choice of axioms is restricted and interlocking.

Here is a preview of what kind of Boolean subsemilattices will be created in later chapters. In Chaps. VI, VII, and VIII, we will use suprema of factor congruences to create the base space. In Chap. IX, the base space will be created out of annihilators, and in Sect. XI.2, it will be Boolean congruences (yet to be defined).

VI

Sheaves from Factor Congruences

In studying sheaves of algebras over Boolean spaces, much more can be said when the chosen congruences are factor congruences and their join agrees with that of **Con** $\boldsymbol{A}$. In this way, factorial braces are defined in the first section. This notion appeared in the seminal paper of [**Davey73**], but not by this name. Its benefit is that internal factor objects are brought to bear on structural questions. The Gel'fand map γ is now an isomorphism.

The second section explores the many equivalent ways to define Boolean algebras of factor objects: bands, congruences and sesquimorphisms. Commutativity of these is the key.

The third section defines the algebras of [**Comer71**]. The difference between the algebras of Davey and Comer is this: in the former a factorial brace starts with some factor congruences, in the latter all factor congruences are used. Algebras with Boolean factor congruences (BFC) lead naturally to reduced and factor-transparent sheaves, and pave the way to representing shells in the next chapter.

The last section captures categorically these algebras by introducing three the new categories: $\mathfrak{FactorBraceRed}$, $\mathfrak{AlgBFC}$ and $\mathfrak{Sheaf BooleRedFt}$ – the last is the category of all reduced and factor-transparent sheaves. We will find that $\mathfrak{FactorBraceRed}$ is categorically equivalent to $\mathfrak{SheafBooleRed}$, and $\mathfrak{AlgBFC}$ to $\mathfrak{SheafBooleRedFt}$. Thus, we have two generalizations of Stone's representation theorem for Boolean algebras [**Stone37**].

A. Knoebel, *Sheaves of Algebras over Boolean Spaces*,
DOI 10.1007/978-0-8176-4642-4_VI,

1. Factorial Braces

The purpose of this section is simple. We abstract the appropriate properties of the algebra $\boldsymbol{\Gamma}(\mathcal{A})$ of all global sections of a sheaf $\mathcal{A}$ over a Boolean space. This is a special case of Chap. V and it is historically important: the Boolean subsemilattice will now actually be a sublattice of factor congruences. But more than this is true now: these conditions of [**Davey73**] are strong enough to ensure that the Gel'fand map is not only injective but also surjective: it maps the given algebra onto the algebra of all global sections. Needed in the development is a discussion about how the selected congruences may serve as moduli in the solution of various systems of equations. That pairwise solutions may imply a total solution is the Chinese remainder theorem; it requires commutativity and distributivity of the congruences.

Recall from Sect. I.2 that a congruence θ of an algebra $\boldsymbol{A}$ is a **factor congruence** if there is another congruence θ' of $\boldsymbol{A}$ such that

$$\theta \cap \theta' = 0_{\mathbf{Con}\,\boldsymbol{A}} \quad \text{and} \quad \theta \circ \theta' = 1_{\mathbf{Con}\,\boldsymbol{A}}.$$

Call such a pair **complementary**. Designate the set of all factor congruences of an algebra $\boldsymbol{A}$ by $\mathrm{Con}'\,\boldsymbol{A}$. Remember their significance: to each pair of complementary congruences corresponds a decomposition of $\boldsymbol{A}$ as a direct product of two factors.

Because $\theta \circ \theta' \subseteq \theta \vee \theta'$ for any congruences θ and θ', we realize that complimentary factor congruences are also complimentary in the lattice **Con** $\boldsymbol{A}$. But the converse is false, for example, as seen with the three-element semilattice $\boldsymbol{SL}_3$ that is not a chain:

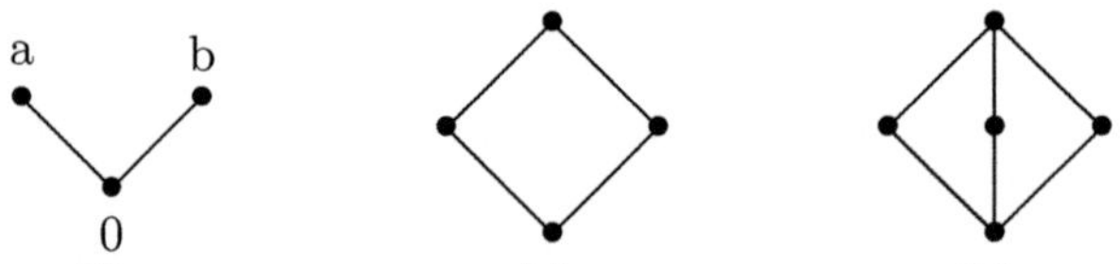

$\boldsymbol{SL}_3$ $\qquad$ $\boldsymbol{M}_2$ $\qquad$ $\boldsymbol{M}_3$

This also shows that focusing our attention on factor congruences is a definite restriction, for there are Boolean subsemilattices of congruences containing non-factor congruences. For example, in the semilattice $\boldsymbol{SL}_3$ just displayed, its lattice of congruences is the Boolean lattice $\boldsymbol{M}_2$. Nevertheless, the intermediate congruences are not a pair of complementary factor congruences.

On the other hand, the set of factor congruences is not always a Boolean subsemilattice. For example, in the vier group, $\boldsymbol{G} = \mathbb{Z}_2 \times \mathbb{Z}_2$, complements in **Con** $\boldsymbol{G}$ are not unique. For its set of factor congruences, $\mathrm{Con}'\,\boldsymbol{G} = \mathrm{Con}\,\boldsymbol{G}$, and this is the five-element modular, nondistributive lattice $\boldsymbol{M}_3$.

More broadly, vector spaces are good examples of algebras in which $\mathrm{Con}'\,\boldsymbol{A} = \mathrm{Con}\,\boldsymbol{A}$, since every linear subspace corresponds to a factor congruence. By way of contrast, the regular non-commutative unital rings $\boldsymbol{A}$ of von Neumann [**vonN36**] are historical examples in which usually $\mathrm{Con}'\,\boldsymbol{A} \neq \mathrm{Con}\,\boldsymbol{A}$.

For these reasons we study algebras in which there is a set of factor congruences satisfying special restrictions. To see what these might be we look further into the Boolean braces of Sect. V.2 that are obtained from sheaves. Conditions (a) and (b) of Theorem V.2.9 define the algebras of Davey [**Davey73**, p. 288], to which a name is now attached.

1.1. DEFINITION. A **factorial brace** $\langle \boldsymbol{A}, \boldsymbol{B} \rangle$ is an algebra $\boldsymbol{A}$ where

(i) $\boldsymbol{B}$ is a collection of factor congruences of $\boldsymbol{A}$,
(ii) it is a Boolean sublattice of $\mathbf{Con}\,\boldsymbol{A}$, written $\boldsymbol{B} = \langle B; \vee, \cap, ', 0, 1 \rangle$,
(iii) complements in $\boldsymbol{B}$ are complements as factor congruences.

Commutativity of congruences in $\boldsymbol{B}$ comes for free, as shown next.

1.2. PROPOSITION. *Let $\langle \boldsymbol{A}, \boldsymbol{B} \rangle$ satisfy the statements* (i) *and* (ii) *of Definition 1.1. Then the congruences of B commute if, and only if, they satisfy statement* (iii) *of it.*

PROOF. $\Rightarrow$. Let θ and ζ be commuting complements in the sublattice $\boldsymbol{B}$. It is always the case that $\theta \vee \zeta = (\theta \circ \zeta) \cup (\theta \circ \zeta \circ \theta) \cup (\theta \circ \zeta \circ \theta \circ \zeta) \cup \cdots$. It follows that $\theta \circ \zeta = \theta \vee \zeta = 1$. Therefore, θ and ζ are factor complements.

$\Leftarrow$. Here is a proof from [**Pier68**]. Let θ and ζ be factor congruences that are in $\boldsymbol{B}$. To prove that $\theta \circ \zeta = \zeta \circ \theta$, it suffices to show that $\theta \vee \zeta \subseteq \theta \circ \zeta$, since always $\theta \circ \zeta \subseteq \theta \vee \zeta$. To that end, suppose that $\langle a, c \rangle \in \theta \vee \zeta$. Since θ has a factor complement θ' in $\boldsymbol{B}$, there is a b such that $a \mathrel{\theta} b \mathrel{\theta'} c$. Conclude that $\langle b, c \rangle \in \theta' \cap (\theta \vee \zeta) = \theta' \cap \zeta \subseteq \zeta$, the equality courtesy of distributivity. Hence, $\langle a, c \rangle \in \theta \circ \zeta$. □

Recall that the semilattice $\boldsymbol{SL}_3$, with B as the set of all its congruences, is an example of a Boolean brace that is not factorial, for the reason that two of the congruences fail to factor. So $\boldsymbol{SL}_3$ satisfies (ii) but not (iii) of Definition 1.1.

In preparation for Theorem 1.8, we discuss the Chinese remainder theorem about the simultaneous solution of systems of equations modulo congruences. However, this theorem is about neither numbers, rings, nor algebras; it is really about equivalence relations. First comes the definition of it as a property, followed by the theorem, with its proof given by a lemma that goes to the heart of the matter. See [**Grät79**, p. 221, exercise 68] for a different version.

1.3. DEFINITION. Suppose Θ is a collection of equivalence relations on a set A. We say that Θ satisfies the **Chinese Remainder Theorem** (CRT) if, whenever a finite system (for $a_1, a_2, \ldots, a_n \in A$ and $\theta_1, \theta_2, \ldots, \theta_n \in \Theta$):

$$(1.1) \qquad x \equiv \begin{cases} a_1 \ (\theta_1), \\ a_2 \ (\theta_2), \\ \vdots \\ a_n \ (\theta_n), \end{cases}$$

is pairwise solvable, then it is totally solvable by a common x.

In the next theorem, the join $\theta \vee \eta$ of two equivalence relation θ and η is the smallest equivalence relation that includes both.

1.4. THEOREM. *Let* $\boldsymbol{\Theta}$ *be a lattice* $\langle \Theta; \vee, \cap \rangle$ *of equivalence relations on a set* A *where they commute:* $\theta \circ \eta = \eta \circ \theta \quad (\theta, \eta \in \Theta)$. *Then* Θ *satisfies the* CRT *if, and only if,* $\boldsymbol{\Theta}$ *is distributive.*

PROOF. In the presence of commutativity, the distributivity of $\vee$ over $\cap$ is equivalent to (1.2) below. The next lemma completes the proof. $\square$

1.5. LEMMA. *Let* Θ *be a set of equivalences on a set* A *closed to finite intersections. The* CRT *holds in* Θ *if, and only if,* Θ *is distributive in this special sense:*

$$(1.2) \qquad (\theta_1 \cap \theta_2) \circ \eta = (\theta_1 \circ \eta) \cap (\theta_2 \circ \eta) \qquad (\theta_1, \theta_2, \eta \in \Theta).$$

PROOF. $\Rightarrow$. One direction of inclusion $\subseteq$ always holds; we need a proof of the other inclusion $\supseteq$. To that end, assume that $a\ [(\theta_1 \circ \eta) \cap (\theta_2 \circ \eta)]\ b$. Then the system,

$$(1.3) \qquad x \equiv \begin{cases} a \ (\theta_1), \\ a \ (\theta_2), \\ b \ (\eta), \end{cases}$$

is pairwise solvable, and by the CRT it is completely solvable. Consequently, $a\ [(\theta_1 \cap \theta_2) \circ \eta]\ b$.

$\Leftarrow$. From (1.2) we will prove the CRT by induction on n when $n \geq 3$, going from $n-1$ to n. Assume that there are two sequences $a_1, a_2, \ldots, a_n$ in A and $\theta_1, \theta_2, \ldots, \theta_n$ in Θ such that $a_i \equiv a_j \ (\theta_i \circ \theta_j)$ for all i and j. When the CRT holds for $n-1$, there is an x' such that $x' \equiv a_i \ (\theta_i)$ whenever $i < n$. Since also $a_i \equiv a_n \ (\theta_i \circ \theta_n)$, we have that $x' \equiv a_n \ (\theta_i \circ \theta_n)$ if $i < n$, and hence that $x' \equiv a_n$ $(\bigcap_{i<n}(\theta_i \circ \theta_n))$. By distributivity, $x' \equiv a_n$ $((\bigcap_{i<n}\theta_i) \circ \theta_n)$. Thus, there exist an x in A such that $x \equiv x'$ $((\bigcap_{i<n}\theta_i)$ and $x \equiv a_n$ (θ_n). Therefore, $x \equiv a_i$ (θ_i) for all i. □

1.6. PROPOSITION. *Any factorial brace $\langle \boldsymbol{A}, \boldsymbol{B} \rangle$ satisfies the Chinese remainder theorem.*

1.7. EXERCISE. (a) Give counterexamples showing that both distributivity and commutativity of equivalence relations are essential for the truth of the CRT. Hint: for distributivity use the vier group $\mathbb{Z}_2^2$. For commutativity use the algebra on three elements that has no operations.

(b) Generalize pairwise solvability to n-solvability in the obvious way. Give a counterexample in $\mathbb{R}^3$ showing that, with only the commutativity of equivalence relations, 3-solvability does not imply 4-solvability.

The next and main theorem of this section, due to Davey [**Davey73**], is patterned after Pierce's construction for rings [**Pier67**]. Krauss and Clark present it in a rather different manner [**KraCl79**, pp. 55–59].

1.8. THEOREM. *Let $\langle \boldsymbol{A}, \boldsymbol{B} \rangle$ be a factorial brace, and consider its set, $X = \operatorname{Spec} \boldsymbol{B}$, of prime ideals P of $\boldsymbol{B}$. Give X the usual Stone topology as spelled out in Sect. III.4. Let $\langle \boldsymbol{A}, :, \boldsymbol{X} \rangle$ be the resulting complex, and let*

$$\mathcal{A} = \biguplus_{P \in X} \frac{A}{\bigvee P}$$

be the resulting sheaf with the usual equalizer topology. Then $\boldsymbol{A}$ is isomorphic by the Gel'fand map γ to the algebra $\boldsymbol{\Gamma}(\mathcal{A})$ of all global sections of the sheaf $\mathcal{A}$ over the Boolean space $\boldsymbol{X}$.

PROOF. Since $\boldsymbol{B}$ is a Boolean subsemilattice of **Con** $\boldsymbol{A}$, we may construct the sheaf, $\mathcal{A} = \boldsymbol{A} /\!/ \boldsymbol{B}$. By Theorem V.2.1, the Gel'fand map, $\gamma: \boldsymbol{A} \to \boldsymbol{\Gamma}(\mathcal{A})$, is an injective homomorphism. All that remains to be proven is that this injection is an isomorphism. To this end, we will use Proposition 1.6 to show that γ is surjective. Recall that this map is given by $\gamma(a)(P) = [a]_P$ when $P \in \operatorname{Spec} \boldsymbol{B}$.

To start, assume that $\sigma \in \Gamma(A)$; we must prove that there is an element s in $\boldsymbol{A}$ such that $\gamma(s) = \sigma$, that is, $\gamma(s)(P) = \sigma(P)$ for all prime ideals P in X. This last condition, translated to congruences, would mean that $\gamma(s) \bigvee P\ \sigma$. Since we are dealing with a sheaf over a Boolean space, we know by Proposition V.2.4d that there are a finite number n of elements a_i in A and an equal number of disjoint clopen subsets U_i covering X such that

$$\sigma(P) = \gamma(a_i)(P) \qquad (P \in U_i \text{ and } 1 \le i \le n).$$

But each of these clopen sets may be chosen to have the special form:

$$U_i = U_{\theta_i} = \{P \in X \mid \theta_i \in P\}$$

for some θ_i in $\boldsymbol{B}$. Recall Theorem III.4.1 establishing the isomorphism between $\boldsymbol{B}$ and the Boolean algebra **Clop Spec** $\boldsymbol{B}$ of clopen sets of prime ideals of $\boldsymbol{B}$.

We have to find one s that will simultaneously replace all of the a_i. To do this with the CRT, we need only establish pairwise solvability. Since $U_i \cap U_j = \emptyset$ when $i \neq j$, we have that $\theta_i \vee \theta_j = 1$ since $1_{\mathbf{Con}\,\boldsymbol{A}} \in B$; alternatively $\theta_i \vee \theta_j$ would otherwise belong to a prime ideal common to both U_i and U_j. By commutativity, $\theta_i \circ \theta_j = \theta_i \vee \theta_j = 1$. Hence there are s_{ij} in A such that $a_i \; \theta_i \; s_{ij} \; \theta_j \; a_j$, and we have 2-solvability. Therefore, there is a common solution to all the equations,

$$s \; \theta_i \; a_i \qquad (i \leq n).$$

It follows from the form of the U_i and the nature of the stalks that

$$s \bigvee P \; a_i \qquad (P \in U_i),$$

and hence

$$\gamma(s)(P) = \left\langle P, \frac{s}{\bigvee P} \right\rangle = \left\langle P, \frac{a_i}{\bigvee P} \right\rangle = \gamma(a_i)(P) = \sigma(P) \qquad (P \in U_i).$$

Since the U_i cover X,

$$\gamma(s)(P) = \sigma(P) \qquad (P \in X),$$

and thus $\gamma(s) = \sigma$. This proves the surjectivity of γ, and the theorem. □

As Davey did, we established the sheaf representation of factorial braces without assuming axiom RB, which stipulated that the interior of the set of trivial stalks is empty. But the corresponding axiom RS, defining reduced sheaves, will be necessary when going full circle in order to prove the equivalence of the categories to come in Sect. 4. We know of no practical application where (RB) and (RS) are violated.

1.9. PROBLEM. Find an algebra $\boldsymbol{A}$ having a proper ideal P of factor congruences whose union is an improper congruence, that is, $\bigvee P = 1_{\mathbf{Con}\,\boldsymbol{A}}$.

A slightly stronger form of axiom RB has an immediate and useful consequence. For a collection of ideals of congruences, such as Spec $\boldsymbol{B}$, write the collection of suprema of its members as

$$\bigvee \mathrm{Spec}\, \boldsymbol{B} = \left\{ \bigvee P \mid P \in \mathrm{Spec}\, \boldsymbol{B} \right\}.$$

As a quotient space, it inherits the topology of **Spec** $\boldsymbol{B}$. When all the suprema $\bigvee P$ are proper congruences, this is nothing new as the next result of [**Comer71**] shows.

1.10. PROPOSITION. *Let* $\langle \boldsymbol{A}, \boldsymbol{B} \rangle$ *be a factorial brace such that*

$$\left| \bigvee P \right| \neq 1 \qquad (P \in \mathrm{Spec}\, \boldsymbol{B}). \tag{1.4}$$

(a) *Then for prime ideals* P *of* $\boldsymbol{B}$, *the function,*

$$P \mapsto \bigvee P,$$

is bijective. Hence, the Boolean space **Spec** $\boldsymbol{B}$ *of prime ideals of* $\boldsymbol{B}$ *and the Boolean space* $\bigvee$Spec $\boldsymbol{B}$ *of their suprema are homeomorphic.*

(b) *If* θ *and* η *are different congruences of* $\bigvee$Spec $\boldsymbol{B}$, *then* $\theta \circ \eta = 1_{\mathbf{Con}\,\boldsymbol{A}}$.

(c) *If* P *is a prime ideal of* $\boldsymbol{B}$ *and* $\theta \in \boldsymbol{B}$, *then*

$$\theta \in P \textit{ if, and only if, } \theta \subseteq \bigvee P.$$

(d) *If* $\theta \in \boldsymbol{B}$, *then*

$$\theta = \bigcap\{\bigvee P \mid \theta \in P \in \operatorname{Spec} \boldsymbol{B}\}.$$

PROOF. (a) It suffices to demonstrate injectivity, that is, for any two prime ideals M and P, we must encounter a contradiction by assuming that $\bigvee M = \bigvee P$ but $M \neq P$. Suppose that there is a congruence θ in $M \sim P$. By the nature of prime ideals in a Boolean algebra, it follows that $\theta' \in P$. Remember that for any ideal N of $\boldsymbol{B}$, $\bigvee N = \bigcup N$. Therefore, $\theta \subseteq \bigvee M$ and $\theta' \subseteq \bigvee P = \bigvee M$. Hence, $1 = \theta \circ \theta' \subseteq \bigvee M$, which violates the propriety of $\bigvee M$ guaranteed by (1.4). Thus the quotient map from Spec $\boldsymbol{B}$ to $\bigvee$Spec $\boldsymbol{B}$ is bijective.

(b) Essentially, repeat the proof of (a).

(c) $\Rightarrow$. This is true since $\bigvee P = \bigcup P$ by Proposition V.1.5.

$\Leftarrow$. Assume that $\theta \subseteq \bigvee P$. If $\theta \notin P$, then $\theta' \in P$, a prime ideal. Therefore, $\theta' \subseteq \bigvee P$, and so $1 = \theta \vee \theta' \subseteq \bigvee P$, which is a contradiction to (1.4).

(d) $\subseteq$. Prove with (c).

$\supseteq$. Prove by the contrapositive. If not $a\ \theta\ b$, then use Lemma V.1.6 to find a prime ideal P of $\boldsymbol{B}$ such that $\theta \in P$ but not $a \bigvee P\ b$. Thus, a and b are not related by the right side. □

Note that when (1.4) is satisfied, then part (a) of this proposition justifies using either space, **Spec** $\boldsymbol{B}$ or $\bigvee$**Spec** $\boldsymbol{B}$, as the base space for the construction of complexes and sheaves. Comer states in [**Comer71**, proposition 2.3] a converse to part (a) when Con′ $\boldsymbol{A}$ is a Boolean sublattice of Con $\boldsymbol{A}$: such an algebra satisfies (1.4) iff the map going from regular congruences back to their ideals, $\varphi \mapsto \{\theta \in \mathbf{Con}'\,\boldsymbol{A} \mid \theta \subseteq \varphi\}$, is surjective, onto the lattice of all ideals of factor congruences. (A congruence θ is **regular** if $\theta = \bigvee \{\eta \in \operatorname{Con}'\boldsymbol{A} \mid \eta \subseteq \theta\}$.)

2. Boolean Algebras of Factor Objects

The classical example of an algebra whose factor objects form a Boolean algebra is any unital ring; the central idempotents, the factor elements serve as the starting point for constructing a sheaf over a Boolean space.

Although factor elements may not be available in an arbitrary algebra, one can still look for algebras in which other kinds of factor objects, such as factor congruences, form a Boolean algebra; in this case one says that the algebra has 'Boolean factor congruences'.

Even when all the factor congruences do not form a Boolean algebra, some subset nevertheless will. Together $0_{\mathrm{Con}\,\boldsymbol{A}}$ and $1_{\mathrm{Con}\,\boldsymbol{A}}$ always do. As already seen in the previous section, the presence of a Boolean algebra $\boldsymbol{B}$ of factor congruences of an algebra allows the algebra to be represented as all global sections of a sheaf over the prime ideal space of $\boldsymbol{B}$.

Pursued in this section is what it means for a selected set of factor objects to form a Boolean algebra. There are three parts to it: in the first not all factor objects need form a Boolean algebra – only a selected set, as in Sect. 1. In the second part they all must, as developed further in Sect. 3. In the third part, there are further characterizations, such as the strict refinement property, which is related to factorable congruences. We start off by presenting several ways to characterize Boolean algebras of factor objects: commuting bands, congruences, and sesquimorphisms, which yield isomorphic Boolean algebras.

As factor bands are most amenable to this analysis, they come first. We defined a factor band β of an algebra $\boldsymbol{A}$ in Sect. II.2 as a homomorphism from $\boldsymbol{A} \times \boldsymbol{A}$ to $\boldsymbol{A}$ that, as a binary operation, is a rectangular band. Let $\mathrm{Band}'\,\boldsymbol{A}$ denote the set of all factor bands of an algebra $\boldsymbol{A}$. In general, no structure or operations are to be found on this set. But when we start with a factorial brace $\langle \boldsymbol{A}, \boldsymbol{C} \rangle$, the associated set B of factor bands will have the form of a Boolean algebra, $\boldsymbol{B} = \langle B; \vee, \wedge, ', 0, 1 \rangle$, where the operations of $\boldsymbol{A}$ are defined for α, β in B and a, b in A:

$$(2.1) \qquad (\alpha \vee \beta)(a,b) = \alpha(\beta(a,b), b),$$

$$(2.2) \qquad (\alpha \wedge \beta)(a,b) = \alpha(a, \beta(a,b)),$$

$$(2.3) \qquad \beta'(a,b) = \beta(b,a),$$

$$(2.4) \qquad 0(a,b) = a,$$

$$(2.5) \qquad 1(a,b) = b.$$

More generally, these operations may be defined on any factor bands of any algebra, or even on any binary functions on a set A. However, unless two factor bands 'commute', their join or meet may no longer be a factor band. Two binary functions α and β on A are said to **commute** when

$$\alpha(\beta(a,b), \beta(c,d)) = \beta(\alpha(a,c), \alpha(b,d)) \qquad (a,b,c,d \in A).$$

In the matrix notation of Sect. II.1,

$$\alpha(M\beta) = (\alpha M)\beta$$

for all 2 by 2 matrices M over A. The real interest is when this is the case and these five operations – $\vee, \wedge, ', 0, 1$ – obey the identities of Boolean algebras. Note for use in future proofs that, if two factor bands commute, say α and β on $\boldsymbol{A}$, then by idempotence, for a, b, c in A:

$$(2.6) \qquad \alpha(a, \beta bc) = \beta(\alpha ab, \alpha ac) \quad \text{and} \quad \alpha(a, \beta ab) = \beta(a, \alpha ab).$$

2.1. DEFINITION. A set $\boldsymbol{B}$ of factor bands of an algebra $\boldsymbol{A}$ is called a **Boolean algebra of commuting factor bands** if

(i) $\boldsymbol{B}$ is closed to the operations $\wedge$ and $'$ (and hence closed to $\vee$, 0, 1);
(ii) These operations satisfy the identities of Boolean algebras; and
(iii) All the bands of $\boldsymbol{B}$ commute.

George Bergman calls a Boolean algebra of commuting factor bands a 'B-set' [**Berg91**]. Stephen Bloom, Zoltán Ésik and Ernest Manes show that any Boolean algebra may be represented by a Boolean algebra of commuting factor bands [**BloÉM90**].

Recall from Sect. II.2 the relationship between factor bands and pairs of complementary factor congruences:

(1) $a\ \theta\ b$ if, and only if, $\beta(a, b) = b$;
$a\ \theta' b$ if, and only if, $\beta(a, b) = a$;
(2) $\beta(a, b) = c$ if, and only if, $a\ \theta\ c$ and $c\ \theta'\ b$.

When compliments are unique, as in the next proposition, this one-to-one correspondence may be notated $\beta \mapsto \theta_\beta$ and $\theta \mapsto \beta_\theta$, dispensing with the need for pairs of congruences. To bring out the parallel nature of the concepts of this section, at the risk of redundancy with the last section, we say that a set Θ of factor congruences of an algebra $\boldsymbol{A}$ is a **Boolean algebra of commuting factor congruences** if $\langle \boldsymbol{A}, \Theta \rangle$ is a factorial brace.

2.2. PROPOSITION. *Let $\boldsymbol{A}$ be an algebra.*

(a) *For any Boolean algebra $\boldsymbol{B}$ of commuting factor bands of $\boldsymbol{A}$, the corresponding set $\theta_{\boldsymbol{B}}$ is a Boolean algebra of commuting factor congruences, and $\boldsymbol{B} \cong \boldsymbol{\theta}_{\boldsymbol{B}}$ via $\beta \mapsto \theta_\beta$.*
(b) *For any Boolean algebra $\boldsymbol{\Theta}$ of commuting factor congruences of $\boldsymbol{A}$, the corresponding set β_Θ is a Boolean algebra of commuting factor bands, and $\boldsymbol{\Theta} \cong \boldsymbol{\beta}_{\boldsymbol{\Theta}}$ via $\theta \mapsto \beta_\theta$.*
(c) *These mappings are inverses of each other: $\beta_{\theta_\beta} = \beta$ and $\theta_{\beta_\theta} = \theta$.*

PROOF. That θ_β and β_θ are factor objects follows from Theorem II.2.5. So does the bijectivity of these maps, as well as the equalities in (c).

(a) To prove the commuting, $\zeta \circ \eta = \eta \circ \zeta$, of two congruences ζ and η in $\theta_{\boldsymbol{B}}$, suppose $\zeta = \theta_\alpha$ and $\eta = \theta_\beta$ for some α and β in $\boldsymbol{B}$. Assume that $a\,(\zeta \circ \eta)\,b$, with the objective of proving that $a\,(\eta \circ \zeta)\,b$. Now there exists an x such that $a\ \zeta\ x$ and $x\ \eta\ b$; we need a y such that $a\ \eta\ y$ and $y\ \zeta\ b$. A proposed solution for y is $\alpha(b, \beta ab)$; this is also equal to $\beta(\alpha ba, b)$ for commuting factor bands. Since $a\ \zeta\ x\ \eta\ b$, $\alpha ax = x$ and $\beta xb = b$. Because ζ and η are symmetric relations, also $\alpha xa = a$ and $\beta bx = x$. To prove that $a\ \eta\ y$ and $y\ \zeta\ b$, it suffices to prove that $\beta ay = y$ and $\alpha yb = b$. From various identities (2.6) for commuting factor bands,

$$\begin{aligned}\beta ay &= \beta\big(a, \beta(\alpha ba, b)\big)\\ &= \beta(a,b)\\ &= \beta(\alpha xa, b)\\ &= \alpha(\beta xb, \beta ab)\\ &= \alpha(b, \beta ab)\\ &= y.\end{aligned}$$

Hence, $a\ \eta\ y$. More simply, also $\alpha yb = b$ and accordingly $y\ \zeta\ b$. So $a\ (\eta\circ\zeta)\ b$, and therefore, the congruences commute.

That the map $\beta \mapsto \theta_\beta$ is a homomorphism will be proven by establishing the preservation of $\wedge$ and $'$. For meet we need to prove that $\theta_{\alpha\wedge\beta} = \theta_\alpha \cap \theta_\beta$.

For one direction of inclusion, suppose $a\ \theta_{\alpha\wedge\beta}\ b$; then $b = (\alpha\wedge\beta)(a,b) = \alpha(a, \beta ab)$. First, $\alpha ab = \alpha(a, \alpha(a, \beta ab)) = \alpha(a, \beta ab) = b$; hence $a\ \theta_\alpha\ b$. Second, with the help of the last sentence and commutativity, $\beta ab = \beta(a, \alpha ab)$ $= b$; thus $a\ \theta_\beta\ b$. Therefore, $\theta_{\alpha\wedge\beta} \subseteq \theta_\alpha \cap \theta_\beta$.

For the other direction, assume $a(\theta_\alpha \cap \theta_\beta)b$. Then $\alpha ab = b$ and $\beta ab = b$. Hence, $(\alpha\wedge\beta)(a,b) = \alpha(a,\beta ab) = \alpha ab = b$. Thus $a\ \theta_{\alpha\wedge\beta}\ b$, and therefore, $\theta_\alpha \cap \theta_\beta \subseteq \theta_{\alpha\wedge\beta}$.

Similarly proven but simpler, complementation is preserved: $(\theta_\alpha)' = \theta_{\alpha'}$.

(b) The preservation of the Boolean operations comes from part (c) and their preservation in part (a).

Let α and β be factor bands in $\boldsymbol{B}$; we need to prove their commutativity:

$$\alpha(\beta ab, \beta cd) = \beta(\alpha ac, \alpha bd) \qquad (a,b,c,d \in A).$$

On the left side introduce $L = \alpha vw$ where $v = \beta ab$ and $w = \beta cd$; likewise, on the right introduce $R = \beta yz$, where $y = \alpha ac$ and $z = \alpha bd$. We must show that $L = R$. To that end, let η and η' be the complementary factor congruences associated with α, and similarly θ and θ' with β. By the correspondence between bands and congruences, the equation $L = \alpha vw$ implies that $v\ \eta\ L$ and $L\ \eta' w$; and there are ten more such relationships. From these we make the chain of relationships:

$$L\ \eta\ v\ \theta\ a\ \eta\ y\ \theta\ R.$$

Hence $L\ (\eta\vee\theta)\ R$. By parallel arguments, also $L\ (\eta\vee\theta')R$, $L\ (\eta'\vee\theta)R$, and $L\ (\eta'\vee\theta')R$. Thus, L and R are related by

$$(\eta\vee\theta)\cap(\eta\vee\theta')\cap(\eta'\vee\theta)\cap(\eta'\vee\theta'),$$

which, as a Boolean expression, is equal to $0_{\mathrm{Con}\,\boldsymbol{A}}$. Therefore, $L\ 0_{\mathrm{Con}\,\boldsymbol{A}}\ R$ and hence $L = R$. □

In contrast to Boolean subsemilattices, where there may not be a maximal one, commuting factor objects always yield a maximal Boolean algebra. There may be more than one, as in the Vier group $(\mathbb{Z}_2)^2$.

2.3. PROPOSITION. *In any algebra there is a maximal Boolean algebra of commuting factor bands.*

PROOF. Use Zorn's lemma. □

2.4. EXERCISE. Let $\boldsymbol{B}$ be a set of factor bands of an algebra $\boldsymbol{A}$, and let $\overline{\boldsymbol{B}}$ be the closure of $\boldsymbol{B}$ under $\wedge$ and $'$. The following are equivalent.

(a) All the bands of $\boldsymbol{B}$ commute.
(b) All the bands of $\overline{\boldsymbol{B}}$ commute.
(c) The closure $\overline{\boldsymbol{B}}$ is the carrier of a Boolean algebra of factor bands, $\overline{\boldsymbol{B}} = \langle \overline{\boldsymbol{B}}; \vee, \wedge, ', 0, 1\rangle$.

2.5. EXERCISE. Exercise II.2.21 showed that any factor band β of a vector space $\boldsymbol{V}$ is of the form, $\beta(v, w) = Mv + M'w$, for linear transformations M and M' such that $M^2 = M$, $M'^2 = M'$, $M + M' = 1$, and $MM' = 0 = M'M$. Prove that the join $\alpha \vee \beta$ of two factor bands, $\alpha(v, w) = Lv + L'w$ and $\beta(v, w) = Mv + M'w$, is again a factor band whenever any pair among L, L', M and M' commute. Is there is geometrical interpretation of $\alpha \vee \beta$?

Analogous results might be phrased and proved for factor congruences and sesquimorphisms, and perhaps also for factor ideals and elements. But we shall be content to correlate Boolean algebras of them with those of factor bands as already done for factor congruences. In general, by a **Boolean algebra of factor objects** of whatever kind, we mean a set F of them closed to the meet and complement appropriate for that kind of factor object such that under these two operations F forms a Boolean algebra. Keep in mind that complements over all possible factor objects need not be unique, but in a Boolean algebra they are. For congruences and ideals, meet is intersection; and for sesquimorphisms, it is composition. As always, the other three standard Boolean operations are definable with these two:

$$x \vee y = (x' \wedge y')', \quad 0 = x \wedge x', \quad 1 = 0'.$$

These definitions fit in with the already discussed Boolean algebras of factor bands and congruences. In special varieties in the subsequent chapters, many alternative and simpler forms for the Boolean operations of various factor objects will appear.

To get Boolean algebras of commuting sesquimorphisms, we need to single out one element o, called the origin, of an algebra $\boldsymbol{A}$, that is, $\langle \boldsymbol{A}, o\rangle$ is a pointed algebra. The element o need not be a constant term of $\boldsymbol{A}$.

From a Boolean algebra $\boldsymbol{B}$ of commuting factor bands of $\boldsymbol{A}$ we obtain a Boolean algebra $\boldsymbol{M}$ of commuting factor sesquimorphisms of $\boldsymbol{A}$ by the formulas of Theorem II.2.12. Designate the maps passing back and forth between factor bands and sesquimorphisms by $\beta \mapsto \mu_\beta$ and $\mu \mapsto \beta_\mu$.

Two sesquimorphisms μ and ν **commute** if $\mu \circ \nu = \nu \circ \mu$. The Boolean operations of $\boldsymbol{M}$ are

$$\begin{aligned} \mu \wedge \nu &= \mu \circ \nu, \\ \mu' &= \text{the factor complement of } \mu, \\ \mu \vee \nu &= \left(\mu' \circ \nu'\right)', \\ 0\,(a) &= o \qquad (a \in A), \\ 1\,(a) &= a \qquad (a \in A). \end{aligned}$$

For the next proposition remember that $\overset{\text{anti}}{\cong}$ means an anti-isomorphism of Boolean algebras. Anti-isomorphisms seem inevitable here, unless the Boolean operations on bands or sesquimorphisms are redefined (see Exercise 2.10). A fuller explanation of the choices made will be given at the end of Sect. VII.4.

2.6. PROPOSITION. *Let $\langle \boldsymbol{A}, o\rangle$ be a pointed algebra.*

(a) *For any Boolean algebra $\boldsymbol{B}$ of commuting factor bands of $\boldsymbol{A}$, the corresponding set $\mu_{\boldsymbol{B}}$ is a Boolean algebra of commuting factor sesquimorphisms, $\langle \mu_{\boldsymbol{B}}; \vee, \wedge, ', 0, 1\rangle$ with operations as given above. Then $\boldsymbol{B} \overset{\text{anti}}{\cong} \boldsymbol{\mu_B}$ via $\beta \mapsto \mu_\beta$ where*

$$\mu_\beta(a) = \beta(a, o) \quad (a \in A).$$

(b) *For any Boolean algebra $\boldsymbol{M}$ of commuting factor sesquimorphisms of $\boldsymbol{A}$, the corresponding set $\beta_{\boldsymbol{M}}$ is a Boolean algebra of commuting factor bands, $\langle \beta_{\boldsymbol{M}}; \vee, \wedge, ', 0, 1\rangle$ with operations as given by (2.1)–(2.5). Then $\boldsymbol{M} \overset{\text{anti}}{\cong} \boldsymbol{\beta_M}$ via $\mu \mapsto \beta_\mu$ where*

$$\beta_\mu(a,b) = c \quad \textit{iff} \quad \mu c = \mu a \quad \textit{and} \quad \mu' c = \mu' b \qquad (a, b, c \in A).$$

(c) *These mappings are inverses of each other: $\beta_{\mu_\beta} = \beta$ and $\mu_{\beta_\mu} = \mu$.*

PROOF. With the help of Sect. II.2, one establishes commutativity, bijectivity and the preservation of Boolean operations. □

2.7. PROBLEM. As an example of Boolean algebras of commuting sesquimorphisms, see George Bergman's article [**Berg72**]. Could its content really be that of universal algebra?

In contrast to factor bands, congruences, and sesquimorphisms, Boolean algebras of factor ideals and elements do not appear yet to be definable independently of the preceding factor objects (but they will be in Chap. VII). We defined ideals in Sect. II.1 in terms of congruences, starting with a pointed algebra $\langle \boldsymbol{A}, o\rangle$. From a collection Θ of congruences define a collection of ideals:

$$\frac{o}{\Theta} = \left\{ \frac{o}{\theta} \mid \theta \in \Theta \right\}.$$

2.8. PROPOSITION. *Let $\boldsymbol{\Theta}$ be a Boolean algebra of commuting factor congruences of an algebra $\boldsymbol{A}$.*

(a) *If two members of Θ have a congruence class in common, then they are equal.*

(b) *In other words, if $c/\theta = c/\eta$ for some c in A and some θ and η in Θ, then $\theta = \eta$.*

(c) *If $\boldsymbol{A}$ also has an origin o, then $\boldsymbol{\Theta}$ is isomorphic to the corresponding Boolean algebra $o/\boldsymbol{\Theta}$ of factor ideals. Its Boolean operation of meet is intersection, and the complement of o/θ is o/θ'.*

PROOF. (b). It suffices to prove that θ' and η are complementary:

$$(1)\ \ \theta' \vee \eta = 1 \ \text{ and } \quad (2)\ \ \theta' \cap \eta = 0.$$

For (1) suppose that $c/\theta = c/\eta$ and that $a, b \in A$; we show that $a\ (\theta' \vee \eta)\ b$:

$$a\ \theta'\ \mu'(a)\ \eta\ c\ \eta\ \mu'(b)\ \theta'\ b.$$

Here μ' is the factor sesquimorphism associated with θ' with respect to c as an origin. The first congruence is by Theorem II.2.12, and the second by Theorem II.2.19: $\mu'(a) \in c/\theta$ for any a, and the assumption: $c/\theta = c/\eta$.

For (2) one also obtains symmetrically from (1) that $\theta \vee \eta' = 1$. Complement both sides to finish with $\theta' \cap \eta = 0$.

(a) follows from (b).

(c) Set $c = o$ in (b) to show that factor ideals uniquely determine the factor congruences in Θ. Clearly, $o/(\theta \cap \eta) = o/\theta \cap o/\eta$. Thus the map $\theta \mapsto o/\theta$ is an isomorphism. □

Factor elements are defined in terms of factor bands or sesquimorphisms, starting with a doubly pointed algebra $\langle \boldsymbol{A}, o, t \rangle$. From a Boolean algebra $\boldsymbol{M}$ of commuting sesquimorphisms, define a collection of elements of A:

$$M(t) = \{\mu(t) \mid \mu \in M\}.$$

The Boolean operations are defined indirectly through those of $\boldsymbol{M}$:

$$\mu(t) \wedge \nu(t) = (\mu \circ \nu)(t) \quad \text{and} \quad (\mu(t))' = \mu'(t).$$

It's an open question as to how much can be said in general about these factor elements. In the next several chapters, explicit answers will be given for various types of shells.

2.9. PROBLEM. Define a **full house** to be a doubly pointed algebra $\langle A, o, t \rangle$ in which any Boolean algebra of factor elements is isomorphic to the Boolean algebra of commuting factor sesquimorphisms from which it came. Isomorphisms with the other three kind of factor objects follow from earlier results. For what algebras can an origin o and terminus t be found so that they are full houses? When can the various factor objects be independently defined? What does it mean for factor ideals to commute, for elements? When can their Boolean operations be independently defined? Chapter VII will give some answers.

2.10. EXERCISE. To better understand how natural were the choices made for the Boolean operations, consider the doubly pointed ring $\mathbb{R}^3$ where $o = \langle 0,0,0\rangle$ and $t = \langle 1,1,1\rangle$. Describe geometrically within the three-fold product $\mathbb{R}\times\mathbb{R}\times\mathbb{R}$ the different kinds of factor objects and the Boolean operations on them.

We present now a more detailed view of factoring, namely, from the vantage point of factorable congruences; then proceed to more concepts equivalent to having Boolean factor congruences, such as the strict refinement property. No doubt these concepts could also be phrased in terms of factor bands, sesquimorphisms, or ideals. The next definition captures the idea of a homomorphic image of a product inheriting the product. How it does this is part of the content of the two propositions that follow it.

2.11. DEFINITION. A congruence η of an algebra $\boldsymbol{A}$ is **factored** by a pair of complementary factor congruences θ and θ' if

$$\eta = (\eta \vee \theta) \cap (\eta \vee \theta'). \tag{2.7}$$

Under the circumstances, (2.7) equivalent to

$$\eta = (\eta \vee \theta) \sqcap (\eta \vee \theta'), \tag{2.8}$$

in the notation of Definition II.2.31. A congruence is **factorable** if it is factored by all pairs of complementary factor congruences. An algebra has **factorable congruences** if all its congruences are factorable; and a variety has **factorable congruences** if all its algebras have them. An algebra has **factorable factor-congruences** if (2.7) holds only for factor congruences η, and complementary factor congruences θ and θ'.

Figure 1 illustrates the factorable congruence η in the ring $\mathbb{Z}_{12}$. Set $\eta =$ mod 6, $\theta =$ mod 3, and $\theta' =$ mod 4. The columns are the θ-classes and the rows the θ'-classes. The η-classes are enclosed by lines.

9	1	5
3	7	11
6	10	2
0	4	8

FIGURE 1. The factorable congruence mod 6 in $\mathbb{Z}_{12}$

Grant Fraser and Alfred Horn [**FraHo70**] showed that the concept of factorable congruence is definable by a Mal'cev condition. It is called the 'Frazer-Horn-Hu property' by Bigelow and Burris [**BigBu90**], and 'directly decomposable congruence' by Chajda, Eigenthaler, and Länger [**ChaEL03**, p. 35]. It was described in Problem 40 of Grätzer's book [**Grät79**, pp. 195, 345].

Clearly, if an algebra has factorable congruences, then it has also factorable factor-congruences. This last concept will be shown in the next section to be equivalent to the better known notion of having 'Boolean factor congruences'. Here are some examples: unital nullital semigroups have factorable factor-congruences; and lattices and unital rings have also factorable congruences. Their proofs and many more examples will appear subsequently. The four-element semilattice $(\boldsymbol{C}_2)^2$ has factorable factor-congruences but not factorable congruences (Bigelow and Burris [**BigBu90**]). More general than for lattices, the following is easy to prove.

2.12. PROPOSITION. *If* $\mathbf{Con}\,\boldsymbol{A}$ *is distributive, then* $\boldsymbol{A}$ *has factorable congruences.*

PROOF. With distributivity we get (2.7):

$$(\eta \vee \theta) \cap (\eta \vee \theta') = \eta \vee (\theta \cap \theta') = \eta \vee 0 = \eta. \qquad \square$$

Examples of congruence-distributivity are Dedekind domains, lattices and median algebras. Even better, lattices with additional operations are also congruence-distributive, such as lattices with operators (Gehrke and Jónsson [**GehJó04**]) and lattice-ordered monoids (Jipsen and Tsinakis [**JipTs02**]).

2.13. PROBLEM. (a) Do the sheaves constructed from the factor congruences of nontrivial lattices ever have trivial stalks?

(b) More generally, do such sheaves of nontrivial algebras with factorable congruences ever have trivial stalks?

Here are some equivalent formulations for a congruence to be factorable that will help to prove Proposition 2.15. We need the product of congruences. If $\alpha \in \operatorname{Con}\boldsymbol{A}$ and $\beta \in \operatorname{Con}\boldsymbol{B}$, their **product** in $\boldsymbol{A} \times \boldsymbol{B}$ is defined:

$$\langle a_1, b_1 \rangle \, (\alpha \times \beta) \, \langle a_2, b_2 \rangle \quad \text{if} \quad a_1 \, \alpha \, a_2 \quad \text{and} \quad b_1 \, \beta \, b_2.$$

Not every congruence in a product has this form. However, part (c) in the next proposition amounts to saying that any factorable congruence in a product is always the product of congruences.

2.14. PROPOSITION. *Suppose that* $\eta \in \operatorname{Con}\boldsymbol{A}$ *for some algebra* $\boldsymbol{A}$ *and that* θ *and* θ' *are complementary factor congruences of* $\boldsymbol{A}$*. These statements are equivalent.*

(a) $\eta = (\eta \vee \theta) \cap (\eta \vee \theta')$, *that is,* θ *and* θ' *factor* η.

(b) $\frac{\boldsymbol{A}}{\eta} \overset{\xi}{\cong} \frac{\boldsymbol{A}}{\eta\vee\theta} \times \frac{\boldsymbol{A}}{\eta\vee\theta'}$, *where* $\xi(\frac{a}{\eta}) = \langle \frac{a}{\eta\vee\theta}, \frac{a}{\eta\vee\theta'} \rangle$.

(c) $\varphi(\eta) = \zeta \times \zeta'$ *for some* ζ *in* $\operatorname{Con}(\boldsymbol{A}/\theta)$ *and* ζ' *in* $\operatorname{Con}(\boldsymbol{A}/\theta')$, *where* φ *is the canonical isomorphism* $\boldsymbol{A} \overset{\varphi}{\cong} \frac{\boldsymbol{A}}{\theta} \times \frac{\boldsymbol{A}}{\theta'}$.

PROOF. The plan is to prove (a) $\Leftrightarrow$ (b) and (c) $\Leftrightarrow$ (a). These equivalences lean heavily on Chap. II and especially its Sect. 2.

(a) $\Rightarrow$ (b). By Proposition II.2.30, η is an inner direct product of $\eta \vee \theta$ and $\eta \vee \theta'$, and hence ξ is an isomorphism.

(b) $\Rightarrow$ (a). The inclusion, $\eta \subseteq (\eta \vee \theta) \cap (\eta \vee \theta')$, holds generally. For the other direction, assume that $a\ \big((\eta \vee \theta) \cap (\eta \vee \theta')\big)\ b$. Then

$$\xi\Big(\frac{a}{\eta}\Big) = \Big(\frac{a}{\eta \vee \theta}, \frac{a}{\eta \vee \theta'}\Big) = \Big(\frac{b}{\eta \vee \theta}, \frac{b}{\eta \vee \theta'}\Big) = \xi\Big(\frac{b}{\eta}\Big).$$

Since ξ is injective, $a/\eta = b/\eta$, and thus $a\ \eta\ b$.

(a) $\Rightarrow$ (c). Assuming (a), we need to prove that $\varphi(\eta) = \zeta \times \zeta'$ in (c). To that end, define $\zeta = (\eta \vee \theta)/\theta$ and $\zeta' = (\eta \vee \theta')/\theta'$. Recall that

$$\frac{a}{\theta}\ \frac{\eta \vee \theta}{\theta}\ \frac{d}{\theta} \quad \text{iff} \quad a\ (\eta \vee \theta)\ b, \qquad \text{and} \qquad \frac{c}{\theta'}\ \frac{\eta \vee \theta'}{\theta'}\ \frac{d}{\theta'} \quad \text{iff} \quad c\ (\eta \vee \theta')\ b.$$

Also recall that, for all a, b, c and d in A,

$$\Big\langle \frac{a}{\theta}, \frac{a}{\theta'} \Big\rangle\ \varphi(\eta)\ \Big\langle \frac{b}{\theta}, \frac{b}{\theta'} \Big\rangle \quad \text{iff} \quad a\ \eta\ b;$$

$$\Big\langle \frac{a}{\theta}, \frac{c}{\theta'} \Big\rangle\ (\zeta \times \zeta')\ \Big\langle \frac{b}{\theta}, \frac{d}{\theta'} \Big\rangle \quad \text{iff} \quad \frac{a}{\theta}\ \zeta\ \frac{b}{\theta} \quad \text{and} \quad \frac{c}{\theta'}\ \zeta'\ \frac{d}{\theta'}.$$

That $\varphi(\eta) \subseteq \zeta \times \zeta'$ is clear from the definitions just given.

To prove the other direction, $\zeta \times \zeta' \subseteq \varphi(\eta)$, assume that

$$\Big\langle \frac{a}{\theta}, \frac{c}{\theta'} \Big\rangle\ (\zeta \times \zeta')\ \Big\langle \frac{b}{\theta}, \frac{d}{\theta'} \Big\rangle.$$

Introduce x and y so that $a\ \theta\ x\ \theta'\ c$ and $b\ \theta\ y\ \theta'\ d$. Then $x\ \theta\ a\ (\eta \vee \theta)\ b\ \theta\ y$. Likewise, $x\ (\eta \vee \theta')\ y$. Hence, $x\ (\eta \vee \theta) \cap (\eta \vee \theta')\ y$, and $x\ \eta\ y$ by (a). Thus,

$$\Big\langle \frac{a}{\theta}, \frac{c}{\theta'} \Big\rangle = \Big\langle \frac{x}{\theta}, \frac{x}{\theta'} \Big\rangle\ \varphi(\eta)\ \Big\langle \frac{y}{\theta}, \frac{y}{\theta'} \Big\rangle = \Big\langle \frac{b}{\theta}, \frac{d}{\theta'} \Big\rangle.$$

(c) $\Rightarrow$ (a). Four steps are needed. First, since $\varphi(\eta) = \zeta \times \zeta'$, it follows that

$$\frac{\boldsymbol{A}}{\eta} \overset{\varphi/\eta}{\cong} \frac{\boldsymbol{A}/\theta}{\zeta} \times \frac{\boldsymbol{A}/\theta'}{\zeta'}.$$

Second, one must show that $\zeta = (\eta \vee \theta)/\theta$ and $\zeta' = (\eta \vee \theta')/\theta'$; only the first is proven.

$\zeta \subseteq (\eta \vee \theta)/\theta$. Assume that $(a/\theta)\ \zeta\ (b/\theta)$. Since $\varphi(\eta) = \zeta \times \zeta'$, then $a\ \eta\ b$, by their definitions. Hence $(a/\theta)\ \big((\eta \vee \theta)/\theta)\big)\ (b/\theta)$.

$\zeta \supseteq (\eta \vee \theta)/\theta$. Suppose that $(a/\theta)\ \big((\eta \vee \theta)/\theta)\big)\ (b/\theta)$. Then, $a\ (\eta \vee \theta)\ b$. Thus, there must exist x_i such that

$$a\ \eta\ x_1\ \theta\ x_2\ \eta\ \ldots\ \theta\ x_{n-1}\ \eta\ x_n\ \theta\ b.$$

Therefore,

$$\frac{a}{\theta}\,\zeta\,\frac{x_1}{\theta} = \frac{x_2}{\theta}\,\zeta\,\ldots = \frac{x_{n-1}}{\theta}\,\zeta\,\frac{x_n}{\theta} = \frac{b}{\theta}.$$

Third, by combining the first two steps of this proof, we find that

$$\begin{aligned}\frac{\boldsymbol{A}}{\eta} &\cong \frac{\boldsymbol{A}/\theta}{(\eta\vee\theta)/\theta} \times \frac{\boldsymbol{A}/\theta'}{(\eta\vee\theta')/\theta'}.\\ &\cong \frac{\boldsymbol{A}}{\eta\vee\theta)} \times \frac{\boldsymbol{A}}{\eta\vee\theta'}.\end{aligned}$$

Fourth, since this last is canonical, $\eta = (\eta\vee\theta)\cap(\eta\vee\theta')$, which is (a). □

Burris [**Burr86**] and Werner [**Wern74**] further characterized factorable congruences. Factorable congruences are now explained categorically by adapting a formulation due to Awad A. Iskander [**Iska96**].

2.15. PROPOSITION. *A variety $\mathfrak{V}$ has factorable congruences if, and only if, for every factorization, $\boldsymbol{A} \cong \boldsymbol{B}\times\boldsymbol{B}'$, of an algebra $\boldsymbol{A}$ in $\mathfrak{V}$, with projections π and π', and for every surjective homomorphism, $\chi\colon \boldsymbol{A}\to\boldsymbol{C}$, there is a factorization, $\boldsymbol{C}\cong\boldsymbol{D}\times\boldsymbol{D}'$, with projections ρ and ρ', and together with surjective homomorphisms, $\psi\colon\boldsymbol{B}\to\boldsymbol{D}$ and $\psi'\colon\boldsymbol{B}'\to\boldsymbol{D}'$, such that this diagram commutes:*

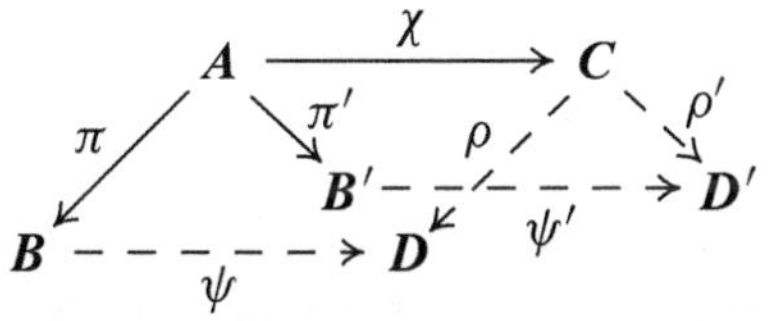

Restricting the χ in this diagram to those surjective homomorphisms for which $\ker\chi$ *is a factor congruence characterizes those varieties having factorable factor-congruences.*

PROOF. $\Rightarrow$. Given $\boldsymbol{A}\cong\boldsymbol{B}\times\boldsymbol{B}'$ with factorable congruences and with data π, π' and χ as specified, we must complete the diagram. We use (b) of Proposition 2.14. Set $\theta = \ker\pi$, $\theta' = \ker\pi'$, $\eta = \ker\chi$, $\boldsymbol{D} = \boldsymbol{A}/(\eta\vee\theta)$, and $\boldsymbol{D}' = \boldsymbol{A}/(\eta\vee\theta')$. Without loss of generality, assume that $\boldsymbol{B} = \boldsymbol{A}/\theta$, $\boldsymbol{B}' = \boldsymbol{A}/\theta'$, and $\boldsymbol{C} = \boldsymbol{A}/\eta$. Thus, $\pi(a) = a/\theta$, $\pi'(a) = a/\theta'$, and $\chi(a) = a/\eta$. Finally, define $\rho\colon\boldsymbol{C}\to\boldsymbol{D}$ by $\rho(a/\eta) = a/(\eta\vee\theta)$, and $\psi\colon\boldsymbol{B}\to\boldsymbol{D}$ by $\psi(a/\theta) = a/(\eta\vee\theta)$, etc. These easily settle commutativity: $\rho\circ\chi = \psi\circ\pi$.

$\Leftarrow$. Given the possibility of factoring any homomorphic image of a product with the commutative diagram as indicated, we will prove (c) of Proposition 2.14. To that end, let η be a congruence of $\boldsymbol{A}$ that is to be factored. We simplify the presentation by replacing isomorphisms by equalities, wherever possible, as for example φ of (c) by $1_{\boldsymbol{A}}$. That is, without loss of generality assume that $\boldsymbol{A} = \boldsymbol{B}\times\boldsymbol{B}'$. Take χ to be the canonical homomorphism from $\boldsymbol{A}$ onto $\boldsymbol{A}/\eta = \boldsymbol{C}$ The existence of a product for $\boldsymbol{C}$ may be also taken to be equality, $\boldsymbol{C} = \boldsymbol{D}\times\boldsymbol{D}'$, with χ adjusted accordingly so that $\ker\chi = \eta$. Then commutativity of the diagram above simply asserts the existence of surjective homomorphisms, $\psi\colon \boldsymbol{B} \to \boldsymbol{D}$ and $\psi'\colon \boldsymbol{B}' \to \boldsymbol{D}'$, such that $\chi = \langle \psi, \psi' \rangle$. Define $\zeta = \ker \psi$ and $\zeta' = \ker \psi'$. The equality, $\eta = \zeta \times \zeta'$, now follows easily (for $b, c \in B$ and $b', c' \in B'$):

$$\begin{aligned}
\langle b, b' \rangle \, \eta \, \langle c, c' \rangle \;&\Leftrightarrow\; \chi(\langle b, b' \rangle) = \chi(\langle c, c' \rangle) \\
&\Leftrightarrow\; \langle \psi, \psi' \rangle(\langle b, b' \rangle) = \langle \psi, \psi' \rangle(\langle c, c' \rangle) \\
&\Leftrightarrow\; \psi(b) = \psi(c) \quad \text{and} \quad \psi'(b') = \psi'(c') \\
&\Leftrightarrow\; b \, \zeta \, c \quad \text{and} \quad b' \, \zeta' c' \\
&\Leftrightarrow\; \langle b, b' \rangle (\zeta \times \zeta') \langle c, c' \rangle. \qquad \square
\end{aligned}$$

2.16. PROBLEM. Swamy and Murti [**SwaMu81**] define the Boolean center $\boldsymbol{C}$ of any algebra $\boldsymbol{A}$ to be the set of all factorable factor-congruences having a factorable complement. They show that $\boldsymbol{C}$ is a Boolean lattice of factor congruences, which gives a sheaf decomposition of $\boldsymbol{A}$ per Sect. 1. Note that $\boldsymbol{C} = \operatorname{Con}' \boldsymbol{A}$ iff $\boldsymbol{A}$ has factorable factor congruences.

(a) Is this Boolean center the intersection of all maximal Boolean lattices of factor congruences of $\boldsymbol{A}$? Contrast this with Proposition 2.3.
(b) Compare this Boolean center with $\operatorname{Elem}' \operatorname{Con} \boldsymbol{A}$, the Boolean algebra of Boolean congruences of Sect. XI.2.

In a unitary ring, which always has factorable factor-congruences, factor elements are defined as those elements that are idempotent and commute with all other elements of the ring. Thus, they need not be defined in terms of other factor objects such as bands, which has been the case so far. Notice in its 'commutative' law, $er = re$, that there are two kinds of 'variables': r which runs over all ring elements and e which is restricted to factor elements. In Chap. VII (Theorem 3.4 and Exercise 3.12, for example), we will explore more generally how to characterize factor elements in terms of such identities.

In preparation, we define the general concept of a 'factor identity' and state a criterion that makes it easy to verify such identities in algebras more general than rings. For simplicity in this discussion, let $\boldsymbol{A}$ be a doubly pointed algebra $\langle A; 0, 1, \ldots, \omega, \ldots \rangle$ in which the origin 0 sand 1 are taken as constants of $\boldsymbol{A}$. Recall from Sect. II.2 that complementary factor elements are elements e and e' of A such that $e = \beta(1, 0)$ and $e' = \beta(0, 1)$ for some factor band β of $\boldsymbol{A}$. Remember that in such a general set-up, e and e' will not always uniquely determine the factor band, from which they came.

2.17. DEFINITION. A **factor identity** is an identity

$$(2.9) \quad t_1(x_1, x_2, \ldots, x_m, z_1, z_1', z_2, z_2', \ldots, z_n, z_n') \approx t_2(x_1, x_2, \ldots, x_m, z_1, z_1', z_2, z_2', \ldots, z_n, z_n')$$

where t_1 and t_2 are terms of a given type and there are two sorts of variables: the ordinary variables $x_1, x_2, \ldots, x_m$; and the 'factor' variables, which come in pairs $z_1, z'_1, z_2, z'_2, \ldots, z_n, z'_n$. Not all variables need to occur explicitly. Such an identity is **satisfied** in a doubly pointed algebra $\boldsymbol{A}$ if

$$(2.10)\quad t_1^{\boldsymbol{A}}(a_1, a_2, \ldots, a_m, e_1, e'_1, e_2, e'_2, \ldots, e_n, e'_n) = t_2^{\boldsymbol{A}}(a_1, a_2, \ldots, a_m, e_1, e'_1, e_2, e'_2, \ldots, e_n, e'_n)$$

for all a_i in A and all pairs of complementary factor elements e_j and e'_j.

2.18. THEOREM. *Let $t_1 \approx t_2$ be a factor identity in the type of a doubly pointed algebra, $\boldsymbol{A} = \langle A; 0, 1 \ldots, \omega, \ldots \rangle$, with factorable factor-congruences. Then this identity is satisfied in $\boldsymbol{A}$ if, and only if,*

$$(2.11)\quad t_1^{\boldsymbol{A}}(a_1, a_2, \ldots, a_m, c_1, c'_1, c_2, c'_2, \ldots, c_n, c'_n) = t_2^{\boldsymbol{A}}(a_1, a_2, \ldots, a_m, c_1, c'_1, c_2, c'_2, \ldots, c_n, c'_n)$$

for all a_i in A and all pairs of complementary constants c_i and c'_i set equal to 0 and 1 respectively, or to 1 and 0 respectively, with the assignments chosen independently over all pairs.

PROOF. $\Rightarrow$. Obvious.

$\Leftarrow$. A proof with two pairs of complementary factor variables suffices to illustrate the basic idea. Let e_1 and e'_1 be the complementary factor elements coming from a factor band β_1. From the latter come the complementary factor congruences θ_1 and θ'_1. Let e_2 and e'_2 likewise be from β_2 with corresponding θ_2 and θ'_2. Then,

$$\begin{aligned} t_1^{\boldsymbol{A}}(a_1, a_2, \ldots, a_m, e_1, e'_1, e_2, e'_2) \quad (\theta_1 \vee \theta_2) \quad & t_1^{\boldsymbol{A}}(a_1, a_2, \ldots, a_m, 1, 0, 1, 0) \\ = \quad & t_2^{\boldsymbol{A}}(a_1, a_2, \ldots, a_m, 1, 0, 1, 0) \\ (\theta_1 \vee \theta_2) \quad & t_2^{\boldsymbol{A}}(a_1, a_2, \ldots, a_m, e_1, e'_1, e_2, e'_2). \end{aligned}$$

Similarly, we prove that the first and last terms are also related by $\theta_1 \vee \theta'_2$, $\theta'_1 \vee \theta_2$ and $\theta'_1 \vee \theta'_2$. These four joins meet at $0_{\mathrm{Con}'\boldsymbol{A}}$ since the factor-congruences factor. Thus, these evaluations of the terms t_1 and t_2 are equal. □

To prime the reader for the various kinds of shells and elements factoring them, to be developed in Chap. VII, we pose a leading question.

2.19. PROBLEM. To what extent might this theorem generalize to first-order formulas?

3. Algebras Having Boolean Factor Congruences

In this section, the factorial braces of Sect. 1 are restricted to those algebras $\boldsymbol{A}$ having Boolean factor congruences, which were first singled out by Comer [**Comer71**]: all factor congruences commute and form a Boolean sublattice **Con′ *A*** of Con $\boldsymbol{A}$, insuring that the resulting sheaf will be reduced and factor-transparent. This sheaf representation is unique among those that are reduced and factor-transparent. It will lead naturally into the representation of shells in the next chapter.

3.1. DEFINITION. An algebra $\boldsymbol{A}$ has **Boolean factor congruences**, BFC, if Con′ $\boldsymbol{A}$ is a Boolean sublattice of Con $\boldsymbol{A}$. By this we mean that Con′ $\boldsymbol{A}$, the set of factor congruences of $\boldsymbol{A}$, is closed to the lattice operations of Con $\boldsymbol{A}$, and factor congruences have unique complements in Con′ $\boldsymbol{A}$. Write **Con′ *A*** for this Boolean algebra $\langle \text{Con}' \boldsymbol{A}; \vee, \wedge, ', 0, 1 \rangle$. Its operations are:

$$\begin{aligned} \theta \vee \eta &= \theta \circ \eta, \\ \theta \wedge \eta &= \theta \cap \eta, \\ \theta' &= \text{the } \eta \text{ such that } \theta \circ \eta = 1 \text{ and } \theta \cap \eta = 0, \\ 0 &= 0_{\text{Con}\boldsymbol{A}}, \\ 1 &= 1_{\text{Con}\boldsymbol{A}}. \end{aligned}$$

A variety has BFC is each of its algebras does.

Examples of algebras with BFC are diverse. There are congruence -distributive algebras, such as the algebras with lattice reducts, listed after Proposition 2.12. And there are the many instances of unital half-shells listed after Definition VII.2.1. Chuan-Chong Chen [**Chen77**] showed that semilattices have BFC, but its proof falls outside the methods of this book. Ross Willard [**Will90**] showed that any centerless algebra has BFC, and Vaggione and Sanchez [**VagSá04**] that compact factor congruences imply BFC. Bigelow and Burris [**BigBu90**] proved implicitly that BFC is a Mal'cev property of varieties; see also the explicit work of Pedro Sánchez Terraf [**Sánc08**]. Conditions equivalent to BFC are given later in Theorems 3.2, 3.9, and 3.11.

The vier-group, $\boldsymbol{V}_4 = \mathbb{Z}_2 \times \mathbb{Z}_2$, is an example of an Abelian group that does not have BFC, but all of its congruences are factor congruences. Its congruence lattice is the five-element modular, nondistributive lattice $\boldsymbol{M}_3$, where complements are not unique.

This example also demonstrates that algebras with BFC are not closed to products. Therefore, the class of all algebras of a given type with BFC is not an equational class, by the HSP theorem of Birkhoff [**Birk35**] in Sect. II.1.

Having BFC is not inherited by subalgebras and homomorphic images. To see this modify the vier-group, $\boldsymbol{V}_4 = {\mathbb{Z}_2}^2$, by adding one more element 4 to obtain a new 'addition', as shown in Fig. 2. Here, in order to be brief, we have coded the ordered pairs of the original group ${\mathbb{Z}_2}^2$: $\langle a, b\rangle \mapsto 2a + b$. Then the algebra, $\boldsymbol{A} = \langle \{0, 1, 2, 3, 4\}; +\rangle$, has ${\mathbb{Z}_2}^2$ as both a subalgebra and a homomorphic image (identify 3 and 4). However, having a prime number of elements, $\boldsymbol{A}$ can have no proper factorization. Therefore, $\boldsymbol{A}$ has Boolean factor congruences whereas ${\mathbb{Z}_2}^2$ does not.

+	0	1	2	3	4
0	0	1	2	3	3
1	1	0	3	2	2
2	2	3	0	1	1
3	3	2	1	0	0
4	3	2	1	0	0

FIGURE 2. An algebra that has BFC, but contains a group that does not have BFC.

We give some other ways of viewing BFC. Sufficient conditions are of many kinds: Boolean algebras of other factor objects, refinements of direct decompositions, splitting congruences in products, and Mal'cev conditions. For the moment, assume that $\langle \boldsymbol{A}, o\rangle$ is a pointed algebra.

The next theorem follows from the previous results and proofs about commuting factor objects, and equates their Boolean algebras by means of the formulas of Sect. II.2. Curiously, each factor sesquimorphism carries just enough more information than its corresponding factor congruence to insure in part (a) that commutativity alone implies BFC. But the vier-group $\boldsymbol{V}_4$ shows that commutativity alone is not sufficient in part (c). For the algebra $\boldsymbol{A}$ in this theorem, let Band$'\boldsymbol{A}$, Con$'\boldsymbol{A}$ and Sesq$'\boldsymbol{A}$ denote its sets of factor bands, congruences and sesquimorphisms with origin o; and likewise their corresponding Boolean algebras by **Band**$'\boldsymbol{A}$, **Con**$'\boldsymbol{A}$ and **Sesq**$'\boldsymbol{A}$, when they exist.

3.2. THEOREM. *For any pointed algebra $\langle \boldsymbol{A}, o\rangle$, the following are equivalent:*

(a) *Its factor sesquimorphisms commute.*
(b) *Its factor bands commute.*
(c) *Its factor congruences commute and form a Boolean algebra. that is, the algebra has* BFC.

When these conditions hold, the factor sesquimorphisms and bands also form Boolean algebras, and these all are isomorphic or anti-isomorphic:

$$\mathbf{Con}'\boldsymbol{A} \cong \mathbf{Band}'\boldsymbol{A} \overset{\text{anti}}{\cong} \mathbf{Sesq}'\boldsymbol{A}\,.$$

PROOF. Use Propositions 2.2, 2.6, Theorem II.2.12, and variants of (2.6). □

Also for a pointed algebra with BFC, by Proposition 2.8(c) we may define the Boolean algebra **Ideal**$'\boldsymbol{A}$ on the set Ideal$'\boldsymbol{A}$ of all factor ideals; it is isomorphic to **Con**$'\boldsymbol{A}$. We will have to wait till the next chapter about shells to include factor elements in this list.

Now we name the suprema of factor congruences and ideals that are used in the next Proposition, which will be used to prove Proposition VIII.1.12.

3.3. DEFINITION. A congruence θ of a pointed algebra $\langle \boldsymbol{A}, o\rangle$ with BFC is **regular** if there is an ideal N of $\mathbf{Con}'\,\boldsymbol{A}$ such that $\theta = \bigvee N$. An ideal I of $\boldsymbol{A}$ is **regular** if there exists a regular congruence θ such that $I = o/\theta$. For any ideal N of $\mathbf{Con}'\,\boldsymbol{A}$, write for the corresponding ideal of factor ideals:

$$\frac{o}{N} = \left\{ \frac{o}{\theta} \;\middle|\; \theta \in N \right\}.$$

From the isomorphism of $\mathbf{Con}'\,\boldsymbol{A}$ and $\mathbf{Ideal}'\,\boldsymbol{A}$, we deduce that any set N of factor congruences in a pointed algebra $\boldsymbol{A}$ with BFC is a prime ideal of $\mathbf{Con}'\,\boldsymbol{A}$ iff o/N is a prime ideal of $\mathbf{Ideal}'\,\boldsymbol{A}$. We record several useful characterizations of regular ideals and maximal regular ideals. Here, 'maximal' means maximal among all regular ideals.

3.4. PROPOSITION. *Let $\langle \boldsymbol{A}, o\rangle$ be a pointed algebra with* BFC.

(a) *For any ideal N of $\mathbf{Con}'\,\boldsymbol{A}$,*

$$\bigvee \frac{o}{N} = \frac{o}{\bigvee N}.$$

(b) *For any ideal I of $\boldsymbol{A}$, I is regular if, and only if, there is an ideal $\mathcal{I}$ of $\mathbf{Ideal}'\,\boldsymbol{A}$ such that $I = \bigvee \mathcal{I}$.*

(c) *For any ideal I of $\boldsymbol{A}$, I is a maximal regular ideal if, and only if, there is a prime ideal $\mathcal{P}$ of $\mathbf{Ideal}'\,\boldsymbol{A}$ such that $I = \bigvee \mathcal{P}$.*

(d) *If $\mathcal{I}$ is an ideal of factor ideals, then*

$$\bigvee \mathcal{I} = \bigcup \mathcal{I}.$$

(e) *If J is a maximal regular ideal of $\boldsymbol{A}$, then*

$$J = \bigcup \{ I \in \mathrm{Ideal}'\,\boldsymbol{A} \mid I \subseteq J \}.$$

PROOF. Throughout we will repeatedly make use of the fact that $\bigvee N = \bigcup N$ for any ideal N of factor congruences (Proposition V.1.5).

(a) $\supseteq$. Suppose $a \in o/\bigvee N$, that is, $o \bigvee N\, a$. By the fact just stated, there is a congruence θ in N so that $o\,\theta\,a$. Hence, $a \in o/\theta \subseteq \bigcup(o/N) \subseteq \bigvee(o/N)$.

$\subseteq$. Since $o/\theta \subseteq o/\bigvee N$ for any θ in N, and since $\bigvee(o/N)$ is the smallest ideal containing all such o/θ, this direction of inclusion is clear.

(b) This follows directly from part (a).

(c) If I is a maximal regular ideal, then $I = \bigvee(o/P)$ for some prime ideal P of $\mathbf{Con}'\,\boldsymbol{A}$. We finish by part (a).

Conversely, if $I = \bigvee \mathcal{P}$ where $\mathcal{P}$ is a prime ideal of $\mathbf{Ideal}'\,\boldsymbol{A}$, then, by the isomorphism used in the proof of part (b), there is a prime ideal P of $\mathbf{Con}'\,\boldsymbol{A}$ such that $\mathcal{P} = o/P$, and we finish again by part (a).

(d) Let $\mathcal{I}$ be an ideal of factor ideals. By Proposition 2.8(c), $\mathbf{Ideal}'\,\boldsymbol{A} \cong \mathbf{Con}'\,\boldsymbol{A}$, and so there is an ideal N of factor congruences such that $\mathcal{I} = o/N$. Therefore, by Proposition V.1.5,

$$\bigvee \mathcal{I} = \bigvee \frac{o}{N} = \frac{o}{\bigvee N} = \frac{o}{\bigcup N} = \bigcup \frac{o}{N} = \bigcup \mathcal{I}.$$

(e) Using previous parts, we find a prime ideal N of $\mathbf{Con}'\,\boldsymbol{A}$ such that

$$J = \frac{o}{\bigvee N} = \bigvee \frac{o}{N} = \bigcup \{o/\theta \mid \theta \in N\}.$$

We compare this last union with the right side of (e) by replacing I by o/θ:

$$R = \bigcup \{o/\theta \mid \theta \in \mathrm{Con}'\,\boldsymbol{A} \text{ and } o/\theta \subseteq o/\bigvee N\}.$$

Clearly, $J \subseteq R$. For the other direction of inclusion, suppose $a \in R$. Then $a\ \theta\ o$ for some θ in the right side. Hence $a \bigvee N\ o$ and thus $a \in J$. □

More could be said along these lines when $\left|\bigvee P\right| \neq 1$ for all prime ideals P of factor congruences in an algebra with BFC. With that hypothesis in mind, recall Proposition 1.10 that the function, $P \mapsto \bigvee P$, is a one-to-one correspondence between prime ideals in $\mathbf{Con}'\,\boldsymbol{A}$ and maximal regular congruences.

Regular congruences are usually not factor congruences themselves, unless each prime ideal of factor congruences is principal; in this special case the generator of each principal prime ideal would be a co-atom of $\mathbf{Con}'\,\boldsymbol{A}$, and hence the factor congruences would form an atomic Boolean algebra.

3.5. PROBLEM. Let $\boldsymbol{B}$ be a Boolean lattice $\langle B; \vee, \wedge\rangle$ of commuting equivalence relations on a set S, with $\vee$ as composition and $\wedge$ as intersection. Can $\boldsymbol{B}$ be realized as the Boolean algebra $\mathbf{Con}'\,\boldsymbol{A}$ of factor congruences for some algebra $\boldsymbol{A}$ with BFC? There are two possible answers here: the abstract and the concrete: respectively,

$$\boldsymbol{B} \simeq \mathbf{Con}'\,\boldsymbol{A} \quad \text{and} \quad \boldsymbol{B} = \mathbf{Con}'\,\boldsymbol{A}.$$

The abstract is easy since $\boldsymbol{B} \cong \mathbf{Con}'\,\boldsymbol{B}$, but the concrete is open.

Our aim now is to prove that an algebra having factorable factor-congruences is equivalent to having Boolean factor congruences. We also equate this with the strict refinement property. For comparison we first state the outer refinement property; for more, see [**McMcT87**, sect. 5.6].

3.6. DEFINITION. An algebra $\boldsymbol{A}$ is said to have the **outer refinement property** if for any two factorizations, $\boldsymbol{A} \cong \prod_{i\in I} \boldsymbol{B}_i$ and $\boldsymbol{A} \cong \prod_{j\in J} \boldsymbol{C}_j$, there is a third factorization $\boldsymbol{A} \cong \prod_{i\in I,\, j\in J} \boldsymbol{D}_{ij}$ so that $\boldsymbol{B}_i \cong \prod_{j\in J} \boldsymbol{D}_{ij}$ $(i \in I)$ and $\boldsymbol{C}_j \cong \prod_{i\in I} \boldsymbol{D}_{ij}$ $(j \in J)$.

Paralleling outer and inner direct products, which are equivalent, there is also a corresponding inner refinement property, but here, surprisingly, the inner is stronger than the outer. This property holds in an algebra whenever any two inner product families of congruences have a common refinement. Since the inner is what is equivalent to BFC, we say no more about the outer. For inner direct products, we use the notation in Definition II.2.31.

3.12. PROBLEM. The collection

$$\{\zeta \vee \eta \mid \zeta \in Z \text{ and } \eta \in H\}$$

of Definition 3.7 is called the **common refinement** of Z and H. Is the following statement true or false? The common refinement of two inner products, Z and H, of an algebra is also an inner product iff their corresponding factor bands commute. These factor bands may have more than two arguments, or even an infinite number. But is finiteness needed?

We pose a problem connecting several threads running through previous sections. It tries to make a parallel between algebras whose factor congruences are factorable (BFC) and those whose congruences are all factorable, a condition stronger than BFC. For a start consult [**CroHJ96**]. Remember that statement (b) is equivalent to being a Boolean product.

3.13. PROBLEM. For an algebra $\boldsymbol{A}$ with BFC, which of these statements imply other ones? Which are equivalent? For its reduced and factor-transparent sheaf $\mathcal{A}$ over a Boolean space $\boldsymbol{X}$, we know that $\boldsymbol{\Gamma}(\mathcal{A}) \cong \boldsymbol{A}$.

(a) $\boldsymbol{A}$ has factorable congruences.
(b) $\mathcal{A}$ is Hausdorff (that is, equalizers are clopen).
(c) For each congruence η of Con $\boldsymbol{\Gamma}(\mathcal{A})$ there is a subset U of $\boldsymbol{X}$ such that for all global sections σ and τ of $\boldsymbol{\Gamma}(\mathcal{A})$, $\sigma \,\eta\, \tau$ iff $U \subseteq \sigma{:}\tau$.
(d) $\boldsymbol{A}$ has the refinement property for products (not just SRP).

Identifying sheaves that come from algebras having BFC leads to the notions of reduced and factor-transparent sheaves. The principal theorem of this section may then be paraphrased: any algebra with BFC is faithfully represented by a reduced and factor-transparent sheaf and vice versa. The first occurrence of these notions occurs in Pierce's monograph for rings in a much different form [**Pier67**, p. 15]. Adapting this to general algebras, Comer insisted on nontrivial stalks in his sheaves [**Comer71**]. However, Bigelow and Burris weakened this to the set of trivial stalks being topologically small [**BigBu90**]; then the class of algebras considered by Comer may be enlarged to all those having BFC.[1] To accommodate the more general Boolean braces of Chap. V, we have gone further and restricted the term 'reduced' to the half of their definition having to do with stalks, and now use their term, 'factor transparent', for their remaining half having to do with congruences. The role of reduced and factor-transparency will be further clarified in the next section when we speak in the language of categories; we will return to a Pierce-like definition of reduced at the end of the next chapter. For convenience we repeat the earlier Definition V.2.8.

[1] See also [**Corn77**] for another generalization of Comer's theorem.

3.14. Definition. Let $\mathcal{A}$ be a sheaf of algebras over a Boolean space $\boldsymbol{X}$. A sheaf $\mathcal{A}$ is **reduced** (RS) if

$$\mathrm{Int}(\mathrm{Triv}\,\mathcal{A}) = \varnothing.$$

The trivial stalks are identified as

$$\mathrm{Triv}\,\mathcal{A} = \{x \in X \mid |A_x| = 1\}.$$

The sheaf $\mathcal{A}$ is **factor-transparent** if

(FS) $$\mathrm{Con}'\,\boldsymbol{\Gamma}(\mathcal{A}) = \theta_{\mathrm{Clop}\,\boldsymbol{X}}.$$

Remember that $\mathrm{Clop}\,\boldsymbol{X}$ contains the clopen sets U of $\boldsymbol{X}$ and that $\theta_{\mathrm{Clop}\,\boldsymbol{X}}$ contains all θ_U for U in $\mathrm{Clop}\,\boldsymbol{X}$ where $\theta_U = \{\langle \sigma, \tau\rangle \in \boldsymbol{\Gamma}(\mathcal{A})^2 \mid U \subseteq \sigma : \tau\}$.

Recall the sheaf constructed from an algebra $\boldsymbol{A}$ with BFC. Consider its set, $X = \mathrm{Spec}\,\mathbf{Con}'\,\boldsymbol{A}$, of prime ideals of factor congruences. Give X the Stone topology as spelled out in Sect. III.4. The topology of $\boldsymbol{X}$ induces a topology on the sheaf space $\mathcal{A}$, a disjoint sum of quotient algebras by the prime ideals. As before, express $\mathcal{A}$ as $\boldsymbol{A} /\!\!/ \mathbf{Con}'\,\boldsymbol{A}$, and call it the **canonical sheaf** or **Pierce sheaf** of an algebra with BFC.

3.15. Theorem. ([**Comer71**], [**BigBu90**]) *Suppose the algebra $\boldsymbol{A}$ has Boolean factor congruences.*

(a) *Then, $\boldsymbol{A}$ is isomorphic to the algebra $\boldsymbol{\Gamma}(\mathcal{A})$ of all global sections of the sheaf, $\mathcal{A} = \boldsymbol{A} /\!\!/ (\mathbf{Con}'\,\boldsymbol{A})$, over the Boolean space* $\mathbf{Spec}\,\mathbf{Con}'\,\boldsymbol{A}$.

(b) *Moreover, the sheaf of part (a) is reduced and factor-transparent.*

(c) *This representation is unique up to isomorphism of reduced and factor-transparent sheaves over Boolean spaces. That is, if $\mathcal{B}$ is another such sheaf and $\boldsymbol{\Gamma}(\mathcal{B}) \cong \boldsymbol{A}$, then $\mathcal{B} \cong \mathcal{A}$.*

(d) *Conversely, if a sheaf $\mathcal{A}$ over a Boolean space is reduced and factor-transparent, then its algebra $\boldsymbol{\Gamma}(\mathcal{A})$ of all global sections has Boolean factor congruences.*

Proof. (a) Clearly, $\mathbf{Con}'\,\boldsymbol{A}$ is a Boolean subsemilattice of $\mathbf{Con}\,\boldsymbol{A}$ by the definition of BFC. So we construct the sheaf as described in Theorem 1.8, where the Gel'fand map is an isomorphism, $\gamma: \boldsymbol{A} \to \boldsymbol{\Gamma}(\mathcal{A})$.

(b) Consequently, $\mathbf{Con}'\,\boldsymbol{A} \cong \mathbf{Con}'\,\boldsymbol{\Gamma}(\mathcal{A})$ by the map $\theta \mapsto \theta_{U_\theta}$. Therefore, every factor congruence of $\boldsymbol{\Gamma}(\mathcal{A})$ is of the required form, and hence $\mathcal{A}$ is factor-transparent.

Duality gives a basis for $\boldsymbol{X}$ with sets: $U_\theta = \{P \in \mathrm{Spec}\,\mathbf{Con}'\,\boldsymbol{A} \mid \theta \in P\}$ when $\theta \in \mathrm{Con}'\,\boldsymbol{A}$. For each nonempty U_θ we may find a P in it such that $\bigvee P \neq 1_{\mathbf{Con}\,\boldsymbol{A}}$ (by Zorn's lemma, as used in Lemma V.1.6). Thus, $U_\theta \nsubseteq \mathrm{Int}\,\mathrm{Triv}\,\mathcal{A}$. Since this is true for every nonempty member of the basis, $\mathrm{Int}\,\mathrm{Triv}\,\mathcal{A} = \varnothing$, and $\mathcal{A}$ is reduced.

(c) We are given an isomorphism of the algebras of global sections, $\varphi: \boldsymbol{\Gamma}(\mathcal{A}) \to \boldsymbol{\Gamma}(\mathcal{B})$ coming from

$$\boldsymbol{\Gamma}(\mathcal{A}) \cong \boldsymbol{A} \cong \boldsymbol{\Gamma}(\mathcal{B}).$$

Consequently, φ pushes through the construction of their sheaves:

$$\boldsymbol{\Gamma}(\mathcal{A})/\!\!/\,\mathbf{Con}'\,\boldsymbol{\Gamma}(\mathcal{A}) \;\cong\; \boldsymbol{\Gamma}(\mathcal{B})/\!\!/\,\mathbf{Con}'\,\boldsymbol{\Gamma}(\mathcal{B}).$$

It follows from (a) that

$$\mathcal{A} \;\cong\; \boldsymbol{\Gamma}(\mathcal{A})/\!\!/\,\mathbf{Con}'\,\boldsymbol{\Gamma}(\mathcal{A}) \;\cong\; \boldsymbol{\Gamma}(\mathcal{B})/\!\!/\,\mathbf{Con}'\,\boldsymbol{\Gamma}(\mathcal{B}) \;\cong\; \mathcal{B}.$$

(d) This follows directly from factor-transparency. □

In a reduced and factor-transparent sheaf over a Boolean space, the factor bands of $\boldsymbol{\Gamma}(\mathcal{A})$ have an especially simple form. This next corollary will be used in the proof of the converse Theorem VII.5.5, where algebras with BFC will be embedded in shells.

3.16. COROLLARY. *Let* $\boldsymbol{A}$ *be an algebra with* BFC, *and* $\mathcal{A}$ *its sheaf. For any factor band* β *of* $\boldsymbol{\Gamma}(\mathcal{A})$, *with corresponding complementary factor congruences* θ *and* θ',

$$\beta\sigma\tau = \sigma|U_\theta \cup \tau|U_{\theta'} \qquad (\sigma,\tau \in \Gamma(\mathcal{A})).$$

PROOF. Set ρ to be the right side. Then $\rho\,\theta\,\sigma$ since $\rho:\sigma \supseteq U_\theta$. We know that $\beta\sigma\tau\,\theta\,\sigma$ by Theorem II.2.5. Hence $\beta\sigma\tau\,\theta\,\rho$. Similarly $\beta\sigma\tau\,\theta'\,\rho$. Therefore, $\beta\sigma\tau = \rho$. □

3.17. EXERCISE. Prove that, in an algebra $\boldsymbol{A}$ with BFC, any decomposition of it into a finite product, $\boldsymbol{A} = \prod_{i=1}^{n} \boldsymbol{A}_i$, corresponds, via the mapping $\theta \mapsto U_\theta$, to a finite disjoint clopen covering $\bigcup_{i=1}^{n} U_i$ of the base space $\boldsymbol{X}$ of its canonical sheaf. Conversely, show that any such covering of $\boldsymbol{X}$ yields, via the mapping $U \mapsto \theta_U$, a product decomposition of $\boldsymbol{A}$, and that these two processes are inverse to each other. Thus, the common refinement of two such finite products corresponds to the intersection of their corresponding partitions of $\boldsymbol{X}$. Why do these proofs not carry over to products with an infinite number of factors?

Unital rings have BFC. The stalks of the sheaf constructed for a commutative unital ring by our method are directly indecomposable. Proofs are to be found in Pierce [**Pier67**, p. 15] and Johnstone [**John82**, p. 183]. Under the assumption that the unital ring is a Baer–Stone ring, we will prove, in a more extensive setting in Chap. VIII, that the stalks have no divisors of zero.

Lattices also have BFC. Georgescu [**Geor88**, proposition 2.5] found a characterization of those bounded distributive lattices whose canonical sheaves have stalks with no divisors of zero, which we state later as Proposition XII.2.1, along with other results of a similar nature.

In general, one can say little about the stalks arising in our theorem. But look at this theorem by Bigelow and Burris in [**BigBu90**].

3.18. THEOREM. *Consider a variety with* BFC *in which the stalks of its sheaves as constructed above are all directly indecomposable. Then the class of these stalks is axiomatizable by universal sentences.*

3.19. PROBLEM. In constructing sheaves we have relied on suprema of congruences. Instead, can any of the other kinds of factor objects be used by creating limits of them?

3.20. PROBLEM. The phrase 'reduced sheaf' would seem to imply a process of reduction. Suppose we remove the set Int Triv $\mathcal{A}$ from the base space of a sheaf $\mathcal{A}$, and shrink the sheaf space accordingly. When is this new sheaf reduced? When is the new algebra of global sections isomorphic to the old?

3.21. PROBLEM. Do the canonical sheaves of Theorem 3.15 help us to understand any of the congruence-distributive algebras listed after Proposition 2.12?

This section closes by representing primal varieties, which are varieties generated by a **primal** algebra $\boldsymbol{P}$, a finite algebra for which all finitary functions on the carrier are term-operations. Examples are the p-rings (p a prime) of McCoy and Montgomery [**McCMo37**]. See Sect. I.1 for a history of this influential theorem due primarily to Foster [**Fost53**] and Hu [**Hu69**]. With all that has been proven, its proof is quite short.

3.22. THEOREM. *Any algebra $\boldsymbol{A}$ of a variety generated by a primal algebra $\boldsymbol{P}$ is isomorphic to a Boolean power of $\boldsymbol{P}$. Moreover, any primal variety is categorically equivalent to the category of Boolean algebras.*

PROOF. Abbreviate Var $\boldsymbol{P}$ as the category $\mathfrak{V}$, and write the carrier P as $\{0, 1, \ldots, n-1\}$. Clearly, $\boldsymbol{P}$ has a lattice term-reduct, say a chain on P; hence it has BFC. In order to establish a categorical equivalence with Boolean algebras, we find in $\boldsymbol{P}$ an invertible idempotent term δ_1 relativizing $\{0, 1, \ldots, n-1\}$ to $\{0, 1\}$ (see Definition III.2.27). To that end define some characteristic functions, one for each i in P:

$$\delta_i(a) = \begin{cases} 1 & \text{if } a = i, \\ 0 & \text{otherwise;} \end{cases} \tag{3.2}$$

and an n-ary function t by

$$t(0, \ldots, 0, 1, 0, \ldots, 0) = i \quad \text{where 1 is at the } i\text{th argument.}$$

By primality these are all term-operations of $\boldsymbol{P}$, and hence

$$t\big(\delta_1(\delta_0(x)), \delta_1(\delta_1(x)), \ldots, \delta_1(\delta_{n-1}(x))\big) \approx x$$

is an identity of $\boldsymbol{P}$, and consequently also of $\mathfrak{V}$. Now the relativization $\delta_1(\boldsymbol{P})$ of $\boldsymbol{P}$ is term-equivalent to $\boldsymbol{B}_2$, the two-element Boolean algebra (see Definition II.1.8). So also is the varietal relativization $\delta_1(\mathfrak{V})$ term-equivalent to $\mathfrak{BooleAlg}$. Thus, Theorem III.2.28 gives the categorical equivalence, $\mathfrak{V} \simeq \mathfrak{BooleAlg}$. □

3.23. COROLLARY. *Any primal variety is dually equivalent to the category of Boolean spaces.*

4. Their Categories

Graduating from adjoint situations to categorical equivalences, while leaving complexes behind, we move on to categories of factorial braces and algebras with BFC. (As always we assume a fixed type of algebras.) The first will be the equivalence of $\mathfrak{FactorBraceRed}$, the category of all reduced factorial braces, with $\mathfrak{SheafBooleRed}$, the category of all reduced sheaves over Boolean spaces. The second equivalence will be between $\mathfrak{AlgBFC}$, the category of all algebras with Boolean factor congruences; and $\mathfrak{SheafBooleRedFt}$, the category of all reduced and factor-transparent sheaves over Boolean spaces. These last two categories are full subcategories of the previous two. These results were abstracted in [**Knoe92c**].

A simple example shows the need for reduced sheaves if we are to achieve categorical equivalence. Let the type be empty, that is, we work with sets. Consider the not reduced sheaf, $\mathcal{A} = \{a, b\} \uplus \{c\}$, over the base space $\{x, y\}$, where all the spaces are discrete. Then $\Gamma(\mathcal{A})$ has two global sections, that is, this algebra is a two-element set, with the two obvious factor congruences. This is a reduced factorial brace, but its sheaf $\Phi\big(\Gamma(\mathcal{A})\big)$ now has only two elements over a base space of one element, and so it is reduced. Thus we need to start with a reduced sheaf in order to return to it.

4.1. DEFINITION. The category $\mathfrak{FactorBraceRed}$ is the full subcategory of the category $\mathfrak{BooleBraceRed}$ restricted to factorial braces and their morphisms.

4.2. THEOREM. *The adjunction*

$$\langle \eta, \varepsilon \rangle \colon \Phi \dashv \Gamma \colon \langle \mathfrak{SheafBooleRed}, \mathfrak{BooleBraceRed} \rangle$$

of Theorem V.4.17, with $\mathfrak{BooleBraceRed}$ *restricted to* $\mathfrak{Factor}$–$\mathfrak{BraceRed}$, *is a categorical equivalence:* $\mathfrak{FactorBraceRed} \simeq \mathfrak{Sheaf}$–$\mathfrak{BooleRed}$.

PROOF. Use Theorem 1.8, which shows that η is a natural isomorphism, that is, $\eta(\langle \boldsymbol{A}, \boldsymbol{B} \rangle)$ is an isomorphism whenever $\langle \boldsymbol{A}, \boldsymbol{B} \rangle$ is a factorial brace. □

As these two categories are equivalent, so their skeletons are isomorphic. We now sum up Comer's results [**Comer71**] in categorical terms.

4.3. DEFINITION. The category $\mathfrak{AlgBFC}$ is the full subcategory of $\mathfrak{FactorBraceRed}$ with objects all factorial braces $\langle \boldsymbol{A}, \mathbf{Con}' \boldsymbol{A} \rangle$ where $\boldsymbol{A}$ has BFC.

Note well that a morphism in $\mathfrak{AlgBFC}$ will not be a single homomorphism of algebras, but rather a morphism of factorial braces, that is a pair $\langle \varphi, \upsilon \rangle$ of homomorphisms, $\varphi \colon \boldsymbol{A} \to \boldsymbol{A}^*$ and $\upsilon \colon \mathbf{Con}' \boldsymbol{A} \to \mathbf{Con}' \boldsymbol{A}^*$, such that $\varphi(\theta) \subseteq \upsilon(\theta)$ for all θ in $\mathbf{Con}' \boldsymbol{A}$.

4.4. DEFINITION. The category $\mathfrak{SheafBooleRedFt}$ is the full subcategory of $\mathfrak{SheafBooleRed}$ whose objects are factor-transparent.

Chap.	Algebras		Complexes		Sheaves
IV			$\mathfrak{Complex}$	$\dashv$	$\mathfrak{Sheaf}$
V	$\mathfrak{BooleBraceRed}$	$\dashv$	$\mathfrak{CompBooleRed}$	$\dashv$	$\mathfrak{SheafBooleRed}$
VI	$\mathfrak{FactorBraceRed}$		$\simeq$		$\mathfrak{SheafBooleRed}$
VI	$\mathfrak{AlgBFC}$		$\simeq$		$\mathfrak{SheafBooleRedFt}$
VII	$\mathfrak{UnitHalfShell}$		$\simeq$		$\mathfrak{SheafUnitHalfShell}$
VII	$\mathfrak{UnitShell}$		$\simeq$		$\mathfrak{SheafUnitShell}$
VIII	$\mathfrak{BaerStoneShell}$		$\simeq$		$\mathfrak{SheafShellIntegral}$
VIII	$\mathfrak{BaerRing}$		$\simeq$		$\mathfrak{SheafRingIntegral}$
Sect. III.4	$\mathfrak{BooleAlg}$		$\simeq$		$\mathfrak{BooleSpace}$

TABLE 1. Categories of Algebras Represented by Categories of Sheaves

Our categorical language allows us to put the notions of reduced and factor-transparent in perspective. It should be clear when everything is reduced and over Boolean spaces that Con$'\,\boldsymbol{\Gamma}(\boldsymbol{\mathcal{A}})$ always maps onto Clop $\boldsymbol{X}$, and its partial inverse, $U \mapsto \theta_U$, always maps injectively. These facts follow from the natural transformation ϵ being an isomorphism. The only thing standing in the way of full invertibility is the real possibility that not all factor congruences of $\boldsymbol{\Gamma}(\boldsymbol{\mathcal{A}})$ are utilized, as for example in the case of factorial braces. But, for algebras with BFC, the sheaves created from them are factor-transparent; so this possibility can not happen. Predecessors of the next theorem are [**Pier67**, theorem 6.6] and [**Comer72**, theorem 1.2].

4.5. THEOREM. *The categorical equivalence of Theorem 4.2, restricted to* $\mathfrak{AlgBFC}$ *and* $\mathfrak{SheafBooleRedFt}$*, is another categorical equivalence.*

PROOF. We have already proven in Theorem 3.15 that algebras with BFC go over to reduced and factor transparent sheaves and vice versa. Clearly this restriction is still a categorical equivalence. □

Table 1 summarizes most of the categories that have appeared so far, and some of those to come. Going down each column takes one from the most general to the most specific. Any category in a particular column is a full subcategory of all categories above it. In this chart, '$\dashv$' means that there is an adjunction between the two categories, and '$\simeq$' means a categorical equivalence. Categories of complexes are not listed after the first two lines since they are no longer needed; however such categories may easily be filled in by the reader. We will define the categories in the lower lines and explain their relationships in Sect. VII.6 and Sect. VIII.1. At the bottom are the classical Boolean algebras and Boolean spaces, which are not technically sheaves, but can be made into such if desired.

Categorical equivalence of varieties ought to imply equivalence of their categories of sheaves created from algebras in them. Some caution is needed since generally there may be many ways to create sheaves from one algebra $\boldsymbol{A}$. With BFC, however, there is the canonical way defined earlier: $\Phi(\boldsymbol{A}) = \boldsymbol{A} /\!\!/ \mathbf{Con}'\,\boldsymbol{A}$.

4.6. THEOREM. *If $\mathfrak{A}$ and $\mathfrak{B}$ are categorically equivalent varieties of algebras with* BFC, *then the categories of their sheaves $\boldsymbol{\mathcal{A}}$ and $\boldsymbol{\mathcal{B}}$ are also categorically equivalent.*

PROOF. In any sheaf coming from an algebra in $\mathfrak{A}$, replace its stalks, which must also be in $\mathfrak{A}$, by their corresponding algebras in $\mathfrak{B}$. Alternatively, for a functor $F: \mathfrak{A} \to \mathfrak{B}$, define $\overline{F}: \Phi(\mathfrak{A}) \to \Phi(\mathfrak{B})$ by catenating known functors:

$$\Phi\mathfrak{A} \xrightarrow{\Gamma} \mathfrak{A} \xrightarrow{F} \mathfrak{B} \xrightarrow{\Phi} \Phi\mathfrak{B}. \quad \square$$

Recall Theorem III.2.28, due to McKenzie [**McKe96**], that any categorical equivalence of varieties is a composition of three special equivalences: term-equivalence, matrix power and restriction by an idempotent and invertible term. Invoking this theorem leads immediately to three similar constructions for sheaves. For example, later in Sect. X.3 we will need matrix powers; the construction and proof is straightforward.

4.7. PROPOSITION. *For a sheaf $\boldsymbol{\mathcal{A}}$ and a positive integer k,*

$$\big(\Gamma(\boldsymbol{\mathcal{A}})\big)^{[k]} = \Gamma\big(\boldsymbol{\mathcal{A}}^{[k]}\big).$$

4.8. EXERCISE. Define the matrix power $\boldsymbol{\mathcal{A}}^{[k]}$ of a sheaf $\boldsymbol{\mathcal{A}}$ so that Proposition 4.7 is true. Hint: Leave the base space the same, and replace each stalk in $\boldsymbol{\mathcal{A}}$ by its matrix power. Less obvious is the topology for the new sheaf space.

VII
Shells

> In the first version of his *Treatise on Algebra* in 1830, Peacock defined symbolic algebra as 'the science which treats of the combinations of arbitrary signs and symbols by means of defined though arbitrary laws'. ... Peacock, however, did not avail himself of the 'arbitrary laws' of combination he advocated, ... In fact, he wrote in 1845 that 'I believe that no views of the nature of Symbolic Algebra can be correct or philosophical which made the selection of its rules of combination arbitrary and independent of arithmetic'. *Victor Katz* [1998, pp. 679–680] quoting *George Peacock*

In the spirit of Peacock's first assertion, we are now ready to specify some operations and identities in our algebras. But their choice is left mostly to the reader, contrary to Peacock's last assertion.

This chapter applies the previous results on representing algebras by sheaves to algebras called 'shells' that have one or two binary operations, among others, and two constants satisfying extremely weak identities – only those specifying how nullities and unities interact with the binary operations. It is a transition from the general theory, applicable to all algebras, to special classes of algebras in later chapters. The term 'shell' has two usages in this chapter: firstly, as a generic word in the title for a range of algebras with differing configurations of binary operations and constants; and secondly, as a specific term for the shells of Sect. 3 that have a particular fixed type. This chapter develops the tools that lead, in subsequent chapters,

A. Knoebel, *Sheaves of Algebras over Boolean Spaces*,
DOI 10.1007/978-0-8176-4642-4_VII,

to stronger hypotheses with stronger conclusions about representations by sheaves. The hypotheses imposed on algebras so far have been about their congruences, not always easily verifiable. But now the weak identities will typically be immediate to the applications.

Factor objects evolve throughout this chapter. Of the five kinds of internal factor objects, only congruences, bands and sesquimorphisms have so far been treated thoroughly. We gave independent definitions of these three, which led to Boolean algebras of them. Of necessity, the treatment of factor ideals was incomplete, and factor elements were only mentioned. But with the introduction of nullities and unities, we can now define factor ideals independently of the other factor objects; and, as the notion of shell is refined, factor elements will satisfy more and more identities, until they are characterized solely by identities.

The first section postulates at least one binary operation, which is treated like multiplication. With respect to it, 'sesqui-elements' are defined in analogy with factor sesquimorphisms, replacing them by left multiplication. But sesqui-elements do not capture all products. Their interest lies in their pointing the way to the existence of one-sided nullities and unities. These guarantee Boolean factor congruences, and hence canonical sheaf representations over Boolean spaces.

The next section makes these constants global in 'unitary half-shells'. Factor elements become sesqui-elements and they capture all products. Consequently, factor elements are in one-to-one correspondence with factor congruences and have a definition independent of the other factor objects. Each factor ideal is principal, generated by a factor element. When we relativize the operations of an algebra to a factor ideal, it becomes a homomorphic image of the algebra.

A 'unitary shell' in the third section has both a multiplication and an addition with a nullity and unity. Now all congruences, not just the factor congruences, are factored by products, and complementary factor elements are definable strictly with identities. Better yet, we can also define factor ideals independently: two such characterizations are given.

The fourth section rounds out this discussion by summarizing in a table the various operations that turn the five kinds of factor objects into isomorphic Boolean algebras. For the special classes of unitary rings and bounded lattices, we discover simpler alternative formulas for the Boolean operations of join, meet, and complement on their factor objects.

We prove in the next section that any algebra with Boolean factor congruences (from Chap. VI) is a subalgebra of a reduct of a unitary shell. 'Separable algebras' effect this proof. They generalize shells with more flexibility: the stalks of the sheaf constructed for this purpose may be unitary rings, bounded distributive lattices, or discriminator algebras.

The last section interprets categorically the results of the previous sections, as we did similarly for Chaps. IV–VI. This situation, at least formally, will be quite similar to that for rings, and, with all that has already been done, quite easy to achieve. We introduce the notion of 'conformal' homomorphism to restrict the homomorphisms to those that preserve factor elements. The theorem of this last section asserts that the category of shells with conformal homomorphisms is equivalent to the category of reduced sheaves over Boolean spaces.

Some of the theorems of this chapter were announced in [**Knoe72**], [**Knoe00**] and [**Knoe07b**].

1. Algebras with a Multiplication

Throughout this section we study algebras, $\boldsymbol{A} = \langle A; \times, \ldots, \omega, \ldots \rangle$, with one designated binary operation $\times$ and any number of other finitary operations ω. Examples are numerous: semilattices, lattices, semigroups and groups, possibly with operators, such as rings. Abbreviate the product: $ab = a \times b$.

In outline, we start with sesqui-elements, which are somewhat similar to factor elements but not equivalent to them initially. The first result of this section observes that each pair of complementary sesqui-elements decomposes the algebra as a product; but a counterexample shows that not every product arises in this way. A partial converse finds constants that point the way to sesquishells. The principal result states that sesquishells, defined by these constants, have BFC.

1.1. Definition. In an algebra, $\boldsymbol{A} = \langle A; \times, \ldots, \omega, \ldots \rangle$, a pair of elements e and e' are **complementary sesqui-elements** if:

(i) $e(ea) = ea$ and $e'(e'a) = e'a \qquad (a \in A)$;
(ii) $e(e'a) = e'(ea) \qquad (a \in A)$;
(iii) for each operation ω, including $\times$,
  (1) $e\omega(ea_1, ea_2, \ldots) = e\omega(a_1, a_2, \ldots) \qquad (a_1, a_2, \ldots \in A)$,
  (2) $e'\omega(e'a_1, e'a_2, \ldots) = e'\omega(a_1, a_2, \ldots) \qquad (a_1, a_2, \ldots \in A)$;
(iv) if $ea = eb$ and $e'b = e'b$ then $a = b \qquad (a, b \in A)$;
(v) for all a, b in A there exists an x in A such that $ea = ex$ and $e'b = e'x$.

A **sesqui-element** is such an element of $\boldsymbol{A}$ having a complement.

These clauses come straight from Definition II.2.10 of complementary sesquimorphisms. That is, if multiplication by a sesqui-element e is viewed as a function, $\mu_e(a) = ea$, then μ_e is a sesquimorphism as in Definition II.1.5.

Note that clause (iii) contains singular distributive laws needed to induce congruences. Clauses (iv) and (v) insure that each pair of complementary factor congruences creates a direct product. This last will be accomplished later by identities alone using the addition in shells. Notice also that (iv) insures uniqueness of the solution specified in (v).

Complementary sesqui-elements induce complementary factor congruences as the next proposition demonstrates. To formulate it we need two definitions: principal algebras and induced congruences. We do not try to define what would be meant by an ideal in this general setting, but only what is a principal ideal generated by a sesqui-element, which agrees with the complemented factor ideals of Sect. II.2,

1.2. DEFINITION. Let e be an element in an algebra, $\boldsymbol{A} = \langle A; \times, \ldots, \omega, \ldots \rangle$, with a designated multiplication $\times$. The **principal algebra** $e\boldsymbol{A}$ generated by e has the carrier,

$$eA = \{e \times a \mid a \in A\},$$

and relativized operations: $\omega^e = e\omega$. By this we mean that

$$\omega^e(a_1, a_2, \ldots) = e \times \omega(a_1, a_2, \ldots) \qquad (a_1, a_2, \ldots \in eA).$$

For example, in a bounded semilattice $\langle A; \times, 0, 1 \rangle$, we find that $e\boldsymbol{A}$ is a new bounded semilattice,

$$e\boldsymbol{A} = \langle eA; \times, 0, e \rangle$$

with a new unity e. A similar, but more involved example comes from a Boolean algebra $\langle A; +, \times, ', 0, 1 \rangle$ where all elements are sesqui-elements. The principal algebra becomes a new Boolean algebra,

$$e\boldsymbol{A} = \langle eA; +, \times, '^e, 0, e \rangle,$$

where $a'^e = ea'$ for a in eA.

Note for any sesqui-element e that in its principal algebra

$$ea = a \qquad (a \in eA).$$

Define its **induced congruence** θ_e by

$$a \; \theta_e \; b \text{ if } ea = eb \qquad (a, b \in A).$$

The next theorem correlates principal algebras with factor ideals. An origin is needed in part (d).

1.3. THEOREM. *Let $\boldsymbol{A}$ be an algebra $\langle A; \times, \ldots, \omega, \ldots \rangle$ with a designated binary operation $\times$. If e and e' are complementary sesqui-elements, then*

(a) *eA and $e'A$ are complementary factor ideals;*

(b) *the induced relations θ_e and $\theta_{e'}$ are a pair of complementary factor congruences;*

(c) *$\boldsymbol{A}/\theta_e \overset{\psi}{\cong} e\boldsymbol{A}$ and $\boldsymbol{A}/\theta_{e'} \overset{\psi'}{\cong} e'\boldsymbol{A}$, where $\psi(\frac{a}{\theta_e}) = ea$ and $\psi'(\frac{a}{\theta_{e'}}) = e'a$;*

(d) *$o/\theta_e = e'A$ and $o/\theta_{e'} = eA$, where $o = e(e'e)$;*

(e) *$\boldsymbol{A} \overset{\varphi}{\cong} e\boldsymbol{A} \times e'\boldsymbol{A}$, where $\varphi(a) = \langle ea, e'a \rangle$.*

PROOF. As actions, left multiplications by e and e' are sesquimorphisms, as noted earlier. So use Theorems II.2.12 and II.2.19. □

Part (d) of this theorem leads us to call eA the principal ideal generated by e. Not all products are captured in part (e), since an algebra with four elements and a single binary operation that is constant has many products, none of which can be created with sesqui-elements. Also, this theorem can not go full circle so that the resulting factor congruences always recreate the original sesqui-elements, since different pairs of sesqui-elements may yield the same pair of congruences. One reason for this we have already seen in Sect. II.1: different sesquimorphisms may yield the same congruences. But there is a more subtle reason, as seen next in a counterexample.

1.4. EXAMPLE. Here is a product decomposition of an algebra $\boldsymbol{A}$ with a multiplication in which the sesqui-elements may be chosen in many ways. Start with the algebra $\langle B; \times \rangle$ with the carrier, $B = \{o, t, 0, 1\}$, and one binary operation $\times$ given by the multiplication table:

$\times$	o	t	0	1
o	0	0	0	0
t	o	t	0	1
0	0	0	0	0
1	o	t	0	1

In effect, $\boldsymbol{B}$ has two one-sided unities. This would not be possible in an algebra with two-sided unities since they would absorb each other. The desired algebra $\boldsymbol{A}$ will be $\boldsymbol{B}^2$. Define some elements in $\boldsymbol{A}$:

$$e = \langle t, o \rangle, \qquad e' = \langle o, t \rangle,$$
$$f = \langle 1, 0 \rangle, \qquad f' = \langle 0, 1 \rangle.$$

Left to the reader is the verification that e and e' are complementary sesqui-elements, and also for the other pair f and f'. But their induced factor congruences are the same, and thus so are the corresponding factor bands. Clearly, the actions of e an f are the same, and likewise for e' and f'. It follows that $eA = fA$ and $e'A = f'A$. Thus, all the factor objects agree except the sesqui-elements. Incidentally, $ee' \neq e'e$.

1.5. EXERCISE. There is more subtlety in this example. It would seem that the origin of $\boldsymbol{A}$ for the complementary sesqui-elements e and e' should be the value $\langle 0, 0 \rangle$ of the terms in part (ii) of Definition 1.1. But the origin given by the proof of Proposition 1.7 below is $\langle o, o \rangle$. Explain this discrepancy.

1.6. EXERCISE. Theorem 1.3 has a partial converse. Consider two elements e and e' of an algebra, $\boldsymbol{A} = \langle A; \times, \ldots, \omega, \ldots \rangle$, that satisfy clauses (i) to (iii) in Definition 1.1. Prove that, if

$$\boldsymbol{A} \overset{\varphi}{\cong} e\boldsymbol{A} \times e'\boldsymbol{A} \quad \text{where} \quad \varphi(a) = \langle ea, e'a \rangle \quad (a \in A),$$

then e and e' are complementary sesqui-elements of $\boldsymbol{A}$.

The next proposition sets the stage for the algebras of the next section.

1.7. PROPOSITION. *There is a pair of complementary sesqui-elements e and e' in an algebra, $\mathbf{A} = \langle A; \times \ldots, \omega, \ldots \rangle$ if, and only if, there are elements o and t in A such that, for all a and b in A,*

$$o \times a = o \times b,$$
$$t \times a = a.$$

PROOF. $\Rightarrow$. As noted at the beginning of this section, there are corresponding sesquimorphisms given by $\mu(a) = ea$ and $\mu'(a) = e'a$. Let β be their corresponding factor band, as characterized in Theorem II.2.12a. Define $o = \beta(e', e)$ and $t = \beta(e, e')$.

Just as for sesquimorphisms in Proposition II.2.11, $e(e'(a)) = e(e'(b))$ for any a and b whatsoever. To establish the first equation, use β commuting with $\times$:

$$\begin{aligned} e(oa) &= \beta(e,e)\big(\beta(e',e)\beta(ea,e'a)\big) \\ &= \beta\big(e(e'(ea)), e(e(e'a))\big) \\ &= \beta\big(e(e'(eb)), e(e(e'b))\big) \\ &\;\;\vdots \\ &= e(ob). \end{aligned}$$

Similarly, $e'(oa) = e'(ob)$. Therefore, $oa = ob$, by (iv) of Definition 1.1.

For the second equation, note that

$$\begin{aligned} e(ta) &= \beta(e,e)\big(\beta(e,e')\beta(a,a)\big) \\ &= \beta\big(e(ea), e(e'a)\big) \\ &= \beta\big(e(ea), e'(ea)\big) \\ &= ea. \end{aligned}$$

and similarly $e'(ta) = e'(a)$. Therefore, $ta = a$.

$\Leftarrow$. Define the sesqui-elements as o and t. □

A corollary relates sesqui-elements to the factor elements of Sect. II.2; recall that the latter come from a factor band β and two designated elements, an origin o and a terminus t: thus $e = \beta(t, o)$ and $e' = \beta(o, t)$.

1.8. COROLLARY. *For any pair e and e' of complementary sesqui-elements of an algebra $\mathbf{A} = \langle A; \times \ldots, \omega, \ldots \rangle$, there are elements o and t of A such that e and e' are a pair of complementary factor elements of the doubly pointed algebra $\langle \mathbf{A}, o, t \rangle$.*

PROOF. Start with the first paragraph of the previous proof. □

Note that these elements o and t may well depend on the chosen sesqui-elements, as in Example 1.4. The sesquishells and unital half-shells, which are coming up, will each have a global origin and terminus. As sesquishells are a slight generalization of unital half-shells, the casual reader may cut directly to the next section.

1.9. DEFINITION. A **sesquishell**, $\boldsymbol{A} = \langle A; \times, 0, \ldots, \omega, \ldots\rangle$, is an algebra with a binary operation $\times$, a constant 0, and perhaps other operations ω, such that 0 is a **weak nullity**:

$$(0\mathsf{n}\times) \qquad 0 \times a = 0 \times b \qquad (a, b \in A).$$

It is said to be **unital** if there is another constant 1 such that

$$(1\mathsf{U}\times) \qquad 1 \times a = a \qquad (a \in A).$$

The term 'sesquishell' is ad hoc and will be used rarely after this section. We introduce it here to record the strongest conclusion from the weakest premise. Its prefix 'sesqui' comes from Proposition 1.7.

Many examples abound. However, there appears to be no naturally occurring binary operation that satisfies $(0\mathsf{n}\times)$ but not also the equation defining a half-shell:

$$(0\mathsf{N}\times) \qquad 0 \times a = 0 \qquad (a \in A).$$

(Half-shells are discussed in the next section.) For example, let $\boldsymbol{A}$ be a sesquishell, satisfying $(0\mathsf{n}\times)$, and set $z = 0 \times 0$. If $\times$ is associative, then z is a nullity of $\times$, that is, $(0\mathsf{N}\times)$ holds for z replacing 0. Alternatively, if $\times$ is commutative and $\boldsymbol{A}$ is unital $(1\mathsf{U}\times)$, then $(0\mathsf{n}\times)$ implies $(0\mathsf{N}\times)$. More comprehensively, a sesquishell is a half-shell iff $0 \times 0 = 0$.

Ross Willard[1] pointed out that such an algebra, satisfying only $(0\mathsf{n}\times)$ and $(1\mathsf{U}\times)$, must have Boolean factor congruences (BFC), as will be shown shortly. This has strong implications for the structure of an algebra, as seen in the previous chapter. This is remarkable, as all that is required among its operations is a binary operation $\times$ that has in its multiplication table one row that is a constant z and another that is the identity function, as shown in Fig. 1.

×	0	a	b	...	1
0	z	z	z	...	z
a					
b		any	thing		
⋮					
1	0	a	b	...	1

FIGURE 1. Multiplication in a sesquishell

[1]In a letter sent to the author in 1994, using a criterion in [**Will90**].

1.10. Theorem. *A unital sesquishell has Boolean factor congruences. Therefore, it is isomorphic to the unital sesquishell of all global sections of a reduced and factor-transparent sheaf over a Boolean space.*

Proof. In the next section, the proof of Theorem 2.2 and its supporting lemmas are so written as to hold more generally for unital sesquishells. □

2. Half-shells

Unital half-shells are defined, whose axioms slightly strengthen sesquishells. The factor elements now have a definition independent of the other factor objects. In fact, the Boolean algebra of factor elements in a unital half-shell is anti-isomorphic to the Boolean algebra of factor congruences. As unital half-shells have BFC, so they have a canonical sheaf representation.

In a unital half-shell, each factorization comes from a pair of complementary sesqui-elements, but the sesqui-elements are not unique. In more detail, sesqui-elements led to factor congruences in Sect. 1. In this section, factor congruences lead to sesqui-elements. But these processes may not go full circle, that is, these new sesqui-elements may not be the original ones. This will be rectified when the nullities and unities become two-sided later in this section.

2.1. Definition. A **half-shell**, $\boldsymbol{A} = \langle A; \times, 0, \ldots, \omega, \ldots \rangle$, is an algebra with a binary operation $\times$, a constant 0, and perhaps other operations ω such that

(0N×) $$0 \times a = 0 \qquad (a \in A).$$

The constant in (0N×) is called a **nullity** with respect to $\times$. If there is another constant 1 such that

(1U×) $$1 \times a = a \qquad (a \in A),$$

then $\boldsymbol{A}$ is said to be **unital**. The constant in (1U×) is called a **unity** with respect to $\times$.[2] More often than not, multiplication is abbreviated by juxtaposition: $a \times b = ab$.

Examples of unital half-shells are unital semigroups with nullity, and hence also unital rings and bounded lattices with operators. Other examples are primal algebras, relational algebras, and some classes of near-rings and semi-rings, as well as many other examples to be given later.

Unital half-shells have Boolean factor congruences. Before proving this, we review the concept of factor element and notions related to it. In these

[2] Elsewhere unities are called 'unit elements'.

half-shells, $\langle A; \times, 0, 1 \ldots, \omega, \ldots\rangle$, one normally assumes that complementary factor elements e and e' are to be figured with respect to the half-shell's nullity and unity, as the origin and terminus. Building on Sect. II.2, we thus have that

$$e = \beta(1,0) \quad \text{and} \quad e' = \beta(0,1),$$

where β is a factor band of the half-shell. Equivalently, for the complementary factor congruences θ and θ' corresponding to β, e and e' are the unique solutions to

$$0\,\theta'\,e\,\theta\,1 \quad \text{and} \quad 0\,\theta\,e'\,\theta'\,1. \tag{2.1}$$

(To prove equivalence, use Theorem II.2.5(b)(2).) Similarly, one usually assumes for sesquimorphisms μ and μ' of any half-shell that they fix the nullity: $\mu(0) = 0$ and $\mu'(0) = 0$. For any element e of $\boldsymbol{A}$, there is the induced relation θ_e on A given by:

$$a\,\theta_e\,b \text{ iff } ea = eb. \tag{2.2}$$

Under the right circumstances, θ_e will be a congruence of $\boldsymbol{A}$.

2.2. THEOREM. *A unital half-shell has factorable factor-congruences, that is, it has Boolean factor congruences.*

Before proving this theorem, we need two lemmas.

2.3. LEMMA. *Let θ and θ' be complementary factor congruences in a unital half-shell $\langle A; \times, 0, 1, \ldots, \omega, \ldots\rangle$, and let e and e' be the corresponding complementary factor elements given by (2.1). Then,*

(a) $\theta = \theta_e$ *and* $(\theta_e)' = \theta_{e'}$;
(b) $a\,\theta\,b$ *iff* $ea = eb$ $\quad (a, b \in A)$.

PROOF. (a) $\theta_e \subseteq \theta$. If $a\,\theta_e\,b$, then $ea = eb$. Since $e\,\theta\,1$, it follows that $a = 1a\,\theta\,ea = eb\,\theta\,1b = b$, and thus $a\,\theta\,b$.

$\theta \subseteq \theta_e$. If $a\,\theta\,b$, then $ea\,\theta\,eb$. Since $0\,\theta'\,e$, it follows that $ea\,\theta'\,0a = 0b\,\theta'\,eb$. Therefore, $ea\,(\theta \cap \theta')\,eb$; and thus $ea = eb$. Hence $a\,\theta_e\,b$.

(b) Use (a) and (2.2). □

2.4. LEMMA. *In a unital half-shell $\boldsymbol{A}$ with complementary factor congruences θ and θ' where $0\,\theta'\,e\,\theta\,1$ for some element e of A,*

$$\text{if } ca = cb, \text{ then } c(ea) = c(eb) \qquad (a, b, c \in A).$$

PROOF. This uses the same techniques as the previous lemma.

$$c(ea)\,\theta\,c(1a) = ca = cb = c(1b)\,\theta\,c(eb), \text{ and}$$
$$c(ea)\,\theta'\,c(0a) = c(0b)\,\theta'\,c(eb).$$

Since $\theta \cap \theta' = 0$, $c(ea) = c(eb)$. □

PROOF OF THEOREM 2.2. We verify Definition VI.2.11 of factorable factor-congruences:

$$(\eta \vee \theta) \cap (\eta \vee \theta') = \eta$$

for all factor congruences η, θ and θ' of $\boldsymbol{A}$ where θ and θ' are complementary.

$\subseteq$. Assume that

$$a\ (\eta \vee \theta)\ b \ \text{ and } \ a\ (\eta \vee \theta')\ b.$$

The first join means that there are $x_1, x_2, \ldots, x_n$ in A such that

$$a\ \eta\ x_1\ \theta\ x_2\ \eta\ \ldots\ \theta\ x_{n-1}\ \eta\ x_n\ \theta\ b. \tag{2.3}$$

Now there are e and d such that $0\ \theta'\ e\ \theta\ 1$ and $0\ \eta'\ d\ \eta\ 1$ for some complement η' of η. By Lemma 2.3, the alternating chain (2.3) translates into the chain

$$\mathrm{d}a = \mathrm{d}x_1,\ ex_1 = ex_2,\ \ldots,\ \mathrm{d}x_{n-1} = \mathrm{d}x_n,\ ex_n = eb.$$

Certainly $e(\mathrm{d}a) = e(\mathrm{d}x_1)$. And by Lemma 2.4, $e(\mathrm{d}x_1) = e(\mathrm{d}x_2)$, etc. Hence $e(\mathrm{d}a) = e(\mathrm{d}b)$, and thus $\mathrm{d}a\ \theta_e\ \mathrm{d}b$. Since $\theta_e = \theta$ by Lemma 2.3, we have that $\mathrm{d}a\ \theta\ \mathrm{d}b$. Similarly, from $a\ (\eta \vee \theta')\ b$, it follows that $\mathrm{d}a\ \theta'\ \mathrm{d}b$. Therefore, $\mathrm{d}a = \mathrm{d}b$ since $\theta \cap \theta' = 0$. So we may conclude what was desired: $a\ \eta\ b$.

$\supseteq$. Always true. □

That a unital half-shell has BFC is used implicitly from now on.

2.5. THEOREM. *Each unital half-shell* $\boldsymbol{A}$ *is represented by the reduced and factor-transparent sheaf,* $\mathcal{A} = \boldsymbol{A} /\!/ (\mathbf{Con}'\,\boldsymbol{A})$*, over the Boolean space,* $\boldsymbol{X} = \mathbf{Spec\,Con}'\,\boldsymbol{A}$*:*

$$\boldsymbol{A} \cong \boldsymbol{\Gamma}\big(\boldsymbol{A} /\!/ (\mathbf{Con}'\,\boldsymbol{A})\big).$$

PROOF. Use Theorems 2.2 and VI.3.15. □

Recall that this $\mathcal{A}$ is called the 'canonical sheaf' whenever $\mathbf{Con}'\,\boldsymbol{A}$ is a Boolean lattice. Many applications of this theorem will come in Chap. VIII.

2.6. EXERCISE. Consider a product $\prod_{i \in I} \boldsymbol{A}_i$ of unital half-shells $\boldsymbol{A}_i$ where I is infinite. An obvious sheaf representation is to take I as the base space X with the discrete topology; to the sheaf space $\biguplus_{i \in I} \boldsymbol{A}_i$ assign also the discrete topology. Notice that $\boldsymbol{X}$ is not compact. But by Theorem 2.2 this product has BFC; therefore, it ought to have a sheaf representation over a Boolean space. Explain this enigma. Hint: What are the prime ideals in an infinite power of a two-element Boolean algebra?

2.7. EXERCISE. Theorem 2.2 is important enough to prove it in three more ways, from the straightforward to the original, more involved method.

(a) Show that any unital half-shell has factorable factor-congruences by assuming that it has some factorization $\boldsymbol{A} \times \boldsymbol{A}'$. Consider another factorization of this product given by a factor congruence η, which is equal, by Theorem 2.11 below, to $\theta_{\langle e,e' \rangle}$ for some factor element $\langle e, e' \rangle$ in $\boldsymbol{A} \times \boldsymbol{A}'$. Then work with components of this product.

(b) Show that an algebra having factorable factor-congruences is equivalent to this condition of Bigelow and Burris [**BigBu90**] on any factorization of it, say $\boldsymbol{A}_1 \times \boldsymbol{A}_2$:

$$\text{if } \theta \in \mathrm{Con}'(\boldsymbol{A}_1 \times \boldsymbol{A}_2), \text{ then } \pi_1(\theta) \times 0_{\mathbf{Con}\,\boldsymbol{A}_2} \subseteq \theta,$$

where $\pi_1(\theta) = \{\langle \pi_1(a_1), \pi_1(a_2)\rangle | \langle a_1, a_2\rangle \in \theta\}$. Then use this to show that any unital half-shell has BFC.

(c) Fashion a proof based on Willard's in [**Will90**].

2.8. EXAMPLE. Although unital half-shells have BFC, that is, they have factorable factor-congruences, they need not have factorable congruences. To see this, let $\boldsymbol{A}$ be the bounded semilattice $\langle A; \wedge, 0, 1\rangle$:

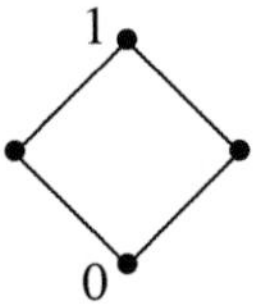

Let θ and θ' be the complementary factor congruences decomposing $\boldsymbol{A}$ as the product of two-element semilattices. Further, let η be the congruence gathering the bottom three elements together; clearly η is not a factor congruence. Now $(\eta \vee \theta) = 1 = (\eta \vee \theta')$. So $(\eta \vee \theta) \cap (\eta \vee \theta') = 1 \neq \eta$. Therefore, η is not congruence-factorable, and this counterexample is established. But Theorem 3.2 of the next section reveals that this is no longer possible in unital shells, which also have an addition.

Theorem VI.3.15 tells us that the canonical sheaf created from an algebra $\boldsymbol{A}$ with BFC is reduced. But in the case of unital half-shells the reduction is more severe: each stalk has at least two elements, as shown next. This was in the original definition of 'reduced'. Recall that the induced relation $\theta(a, b)$ is the smallest congruence relating a and b.

2.9. PROPOSITION. *Let $\boldsymbol{A}$ be a unital half-shell.*

(a) *Then, $\theta(0, 1) = 1_{\mathbf{Con}\,\boldsymbol{A}}$.*

(b) *Consequently, the supremum of any proper ideal of congruences of $\boldsymbol{A}$ is a proper congruence of $\boldsymbol{A}$.*

(c) *As a result, no stalk is trivial in the canonical sheaf representation of Theorem 2.5.*

PROOF. (a). Let a and b be any elements of A. Then,

$$a = 1a \;\theta(0,1)\; 0a = 0 = 0b \;\theta(0,1)\; 1b = b.$$

(b). Suppose that Θ is a proper ideal of $\mathbf{Con}'\,\boldsymbol{A}$. By way of contradiction suppose that $\bigvee\Theta = 1_{\mathbf{Con}\,\boldsymbol{A}}$. Then $0 \bigvee\Theta\, 1$. Since $\bigvee\Theta = \bigcup\Theta$, by Proposition V.1.5, there is a θ in Θ such that $0\,\theta\,1$. So $\theta = 1_{\mathbf{Con}\,\boldsymbol{A}}$ by (a). Hence, it is in Θ, which implies that Θ is improper.

(c) This follows immediately from (b). □

Here is a useful proposition.

2.10. PROPOSITION. *For e and e' complementary factor elements in a unital half-shell $\boldsymbol{A}$ and for η a factor congruence of $\boldsymbol{A}$, the following hold.*

(a) $e\,\eta\,1$ *iff* $\theta_e \subseteq \eta$ *iff* $0\,\eta\,e'$.

(b) $\theta_e = \theta(e,1)$.

(c) $\theta_e = \theta(0,e')$.

PROOF. As (b) and (c) are immediate consequences of (a), we need only prove the latter. As $e\,\theta_e\,1$ and $0\,\theta_e\,e'$, we need only prove two implications within (a).

$e\,\eta\,1 \Rightarrow \theta_e \subseteq \eta$. Suppose that $e\,\eta\,1$ and $a\,\theta_e\,b$; then $ea = eb$. Therefore,

$$a = 1a\,\eta\,ea = eb\,\eta\,1b = b,$$

and hence $a\,\eta\,b$.

$0\,\eta\,e' \Rightarrow \theta_e \subseteq \eta$. Suppose that $0\,\eta\,e'$ and $a\,\theta_e\,b$; then $ea = eb$. Hence,

$$e'a\,\eta\,0a = 0 = 0b\,\eta\,e'b.$$

Thus, $e'a\,\eta\,e'b$. Since $1\,\theta_{e'}\,e'$,

$$a = 1a\,\theta_{e'}\,e'a\,\eta\,e'b\,\theta_{e'}\,1b = b,$$

and hence $a\,(\eta \vee \theta_{e'})\,b$. By Theorem 2.2, $\boldsymbol{A}$ has BFC, and so by Theorem VI.3.9, η is factorable: $\eta = (\eta \vee \theta_{e'}) \cap (\eta \vee \theta_e)$. Therefore, $a\,\eta\,b$. □

Factor elements fit into products of unital half-shells as central idempotents do in unital rings.

2.11. THEOREM. *Let θ and θ' be complementary factor congruences in a unital half-shell $\langle A; \times, 0, 1 \ldots, \omega, \ldots\rangle$, and let β, μ, μ', e and e' be the corresponding factor band, complementary factor sesquimorphisms, and complementary factor elements, that is, $e = \beta 10$ and $e' = \beta 01$. Then,*

(a) *e and e' are complementary sesqui-elements;*

(b) *eA and $e'A$ are complementary factor ideals;*

(c) $\boldsymbol{A}/\theta \overset{\psi}{\cong} e\boldsymbol{A}$ *and* $\boldsymbol{A}/\theta' \overset{\psi'}{\cong} e'\boldsymbol{A}$, *where* $\psi(a/\theta) = ea$ *and* $\psi'(a/\theta') = e'a$;

(d) $0/\theta = e'\boldsymbol{A}$ *and* $0/\theta' = e\boldsymbol{A}$;

(e) $\boldsymbol{A} \overset{\varphi}{\cong} e\boldsymbol{A} \times e'\boldsymbol{A}$, *where* $\varphi(a) = \langle ea, e'a\rangle$;

(f) *moreover, these factor elements e and e' multiply as the sesquimorphisms μ and μ' act:*

$$ea = \mu(a) = \beta(a,e) \quad \textit{and} \quad e'a = \mu'(a) = \beta(a,e') \qquad (a \in A).$$

PROOF. Recall from Lemma 2.3 that $\theta = \theta_e$ and $(\theta_e)' = \theta_{e'}$.

(a) We need to verify axioms (i)–(v) of Definition 1.1 for sesqui-elements. We can demonstrate (i)–(iii) by showing that both sides are related by θ as well as θ'. By virtue of Theorem 2.18 of Chap. VI this amounts to verifying these equations when the sesqui-elements are just 0 or 1.

To verify axiom (iv) assume that $ea = eb$ and $e'a = e'b$. These amount to saying that $a\,\theta\,b$ and $a\,\theta'\,b$. Since $\theta \cap \theta' = 0$, it follows that $a = b$.

For any a and b in A, since $\theta \circ \theta' = 1$, there is an x such that $a\ \theta\ x\ \theta'\ b$. Therefore, $ea = ex$ and $e'b = e'x$. This proves (v).

Parts (b)–(e) of this theorem follow from (a) and Theorem 1.3.

(f) From the defining properties of factor bands and Theorem II.2.12,

$$\mu(a) = \beta(a,0) = \beta(1a,0a) = \beta(1,0)\,\beta(a,a) = ea, \quad \text{and}$$
$$\mu(a) = \beta(a,0) = \beta(a,\beta(1,0)) = \beta(a,e). \qquad \square$$

There is a subtlety in this theorem: it is possible for different sesquielements to have the same action. However, the next corollary extends the one-to-one correspondences, begun in Theorem II.2.12 for bands, congruences, and sesquimorphisms, to the remaining two factor objects, elements, and ideals.

2.12. COROLLARY. *Let* $\boldsymbol{A}$ *be a unital half-shell.*

(a) *Each factor element in* $\boldsymbol{A}$ *comes from a unique congruence, and consequently from a unique factor band.*
(b) *Likewise, each factor ideal in* $\boldsymbol{A}$ *uniquely determines its corresponding factor congruence.*
(c) *Thus, factor elements and ideals have unique complements.*

PROOF. In view of Sect. II.2, it suffices to work with the sesquimorphisms of a half-shell $\boldsymbol{A}$. Let $\langle \mu, \mu' \rangle$ and $\langle \nu, \nu' \rangle$ be pairs of complementary sesquimorphisms with respective pairs of factor elements, $\langle e, e' \rangle$ and $\langle f, f' \rangle$, such that $e = f$. Then by the previous theorem, for all a in A,

$$\mu(a) = ea = fa = \nu(a).$$

Since complements are unique in the Boolean algebra $\mathbf{Sesq}'\,\boldsymbol{A}$, also $\mu' = \nu'$.

Uniqueness of factor ideals follows from Proposition VI.2.8 $\square$

In unital half-shells, factor elements satisfy more identities than in sesquishells. Even more factor identities will appear as the concept of a shell is further refined. Already idempotence appears, but not centrality, $ea = ae$, which depends on the constants 0 and 1 being two-sided, as in ring theory.

2.13. PROPOSITION. *In a unital half-shell* $\boldsymbol{A}$, *any elements* a, b *and factor element* e *satisfy*

$$ee = e,$$
$$e(ab) = (ea)b = (ea)(eb),$$
$$e0 = 0,$$
$$e1 = e.$$

Complementary factor elements e *and* e' *satisfy*

$$ee' = 0 = e'e.$$

Two factor elements e *and* f *satisfy*

$$ef = fe.$$

PROOF. Use Theorem VI.2.18, which tests the validity of identities by substituting 0s and 1s for the factor elements. □

Of course, the factor elements also satisfy the clauses of Definition 1.1 since they are sesqui-elements by Theorem 2.11.

2.14. EXERCISE. Show that in a sesquishell the factor identity $e(ab) = (ea)b$ may fail for a sesqui-element e. Hint: create a sesquishell having a regression of zeros, $0a = z_1$ and $z_ia = z_{i+1}$. Show also that the factor identity $ea = ae$ may fail, even in a unital half-shell.

Recall from Sect. VI.2 that, in an algebra $\boldsymbol{A}$ with Boolean factor congruences, the factor bands and the factor sesquimorphisms form Boolean algebras, isomorphic or anti-isomorphic to one another: $\mathbf{Con}'\,\boldsymbol{A} \cong \mathbf{Band}'\,\boldsymbol{A} \overset{\text{anti}}{\cong} \mathbf{Sesq}'\,\boldsymbol{A}$. In unital half-shells we have in the next theorem also the Boolean algebra of factor elements, $\mathbf{Elem}'\,\boldsymbol{A}$, which is anti-isomorphic to $\mathbf{Con}'\,\boldsymbol{A}$.

We are tempted to call $\mathrm{Elem}'\,\boldsymbol{A}$ the 'center', following Birkhoff [**Birk67**, p. 67], who uses it to pick out those elements identifying a factorization of a bounded partial order. But this word is often used in a different sense to mean the set of elements commuting with all others, as by Kurosh [**Kuro63**, p. 105]. For this reason we avoid it.

2.15. THEOREM. *Let $\boldsymbol{A}$ be a unital half-shell $\langle A; \times, 0, 1, \ldots, \omega, \ldots\rangle$.*

(a) *The set* $\mathrm{Elem}'\,\boldsymbol{A}$ *of factor elements forms a Boolean algebra,*

$$\mathbf{Elem}'\,\boldsymbol{A} = \langle \mathrm{Elem}'\,\boldsymbol{A}; \vee, \times, ', 0, 1\rangle,$$

in which $e \vee f = (e' \times f')'$.

(b) $\mathbf{Elem}'\,\boldsymbol{A} \overset{\text{anti}}{\cong} \mathbf{Con}'\,\boldsymbol{A}$, *with the maps given in each direction by:*

$$e \mapsto \theta_e \quad \text{where } a\,\theta_e b \quad \text{iff} \quad ea = eb;$$
$$\theta \mapsto e_\theta \quad \text{where } 0\,\theta'\,e_\theta\,\theta\,1.$$

PROOF. (a) In view of the factor identities above and Proposition V.1.2, we need only prove that $\mathrm{Elem}'\,\boldsymbol{A}$ is a semilattice and it is a Boolean subsemilattice of itself, that is,

$$ef' = 0 \quad \text{iff} \quad ef = e \qquad (e, f \in \mathrm{Elem}'\,\boldsymbol{A}).$$

We do this with the help of axiom (iv) in Definition 1.1.

(b) As a one-to-one correspondence has been established in Corollary 2.12, it suffices to show that $\theta_e \circ \theta_f = \theta_{ef}$, or equivalently:

$$0\,(\theta_e \circ \theta_f)'\,ef\,(\theta_e \circ \theta_f)\,1.$$

(Remember that factor congruences commute since $\boldsymbol{A}$ has BFC.) To establish the first relationship, realize by DeMorgan's laws that

$$(\theta_e \circ \theta_f)' = \theta_e' \cap \theta_f' = \theta_{e'} \cap \theta_{f'}.$$

But $0\,\theta_{e'}\,ef$ since $e'0 = 0 = e'ef$ by the identities just proven. And similarly $0\,\theta_{f'}\,ef$. The second relationship is established by taking f to be the intermediary in the composition. □

Factor elements can now be defined independently of the other factor objects.

2.16. PROPOSITION. *In a unital half-shell* $\boldsymbol{A}$, *the elements* e *and* e' *are a pair of complementary factor elements if, and only if,*

(a) $ee = e$ *and* $e'e' = e'$ $\quad (a \in A)$;
(b) $e1 = e$ *and* $e'1 = e'$;
(c) $ee' = 0$ *and* $e'e = 0$ $\quad (a \in A)$;
(d) $e(ab) = (ea)b$ *and* $e'(ab) = (e'a)b$ $\quad (a, b \in A)$;
(e) *for each operation* ω, *including* $\times$,
  (1) $e\omega(ea_1, ea_2, \ldots) = e\omega(a_1, a_2, \ldots)$ $\quad (a_1, a_2, \ldots \in A)$,
  (2) $e'\omega(e'a_1, e'a_2, \ldots) = e'\omega(a_1, a_2, \ldots)$ $\quad (a_1, a_2, \ldots \in A)$;
(f) *if* $ea = eb$ *and* $e'b = e'b$ *then* $a = b$ $\quad (a, b \in A)$;
(g) *for all* a, b *in* A *there is an* x *in* A *such that* $ea = ex$ *and* $e'b = e'x$.

PROOF. $\Rightarrow$. By Definition 1.1, Theorem 2.11 and Proposition 2.13.
$\Leftarrow$. Let e and e' be elements satisfying (a)–(g). These clauses imply that e and e' are complementary sesqui-elements. By Theorem 1.3, the induced relations θ_e and $\theta_{e'}$ of (2.2) are complementary factor congruences. It suffices to establish (2.1) that $0\,\theta_{e'}\,e\,\theta_e\,1$ and $0\,\theta_e\,e'\,\theta_{e'}\,1$. Now

$$e'0 = e'(ee') = (e'e)e' = 0e' = 0 = e'e \quad \text{and} \quad ee = e = e1.$$

Hence, by (2.2), $0\,\theta_{e'}\,e\,\theta_e\,1$. Thus, e is a factor element, and likewise e'. □

The factor identity, e0 = 0, is not needed in this characterization, although it holds in unital half-shells and is derived toward the end of the proof.

2.17. COROLLARY. *Let* e *and* e' *be a pair of complementary factor elements in a unital half-shell* $\boldsymbol{A}$. *Then*

$$a \in eA \quad \textit{iff} \quad ea = a \quad \textit{iff} \quad e'a = 0 \qquad (a \in A).$$

2.18. EXERCISE. Show that in a unital half-shell $\boldsymbol{A}$ the Boolean operations on factor elements in Elem$'\,\boldsymbol{A}$ may all be defined independently:

(1) $e \wedge f = e \times f$;
(2) e' is the unique f such that for all g, $eg = 0$ iff $fg = g$;
(3) $e \vee f = (e' \wedge f')'$.

2.19. EXERCISE. Pierce [**Pier67**, p. 15] defines a *reduced* sheaf of unital rings as a sheaf $\langle \boldsymbol{\mathcal{R}}, \pi, \boldsymbol{X} \rangle$ such that

(i) $\boldsymbol{X}$ is a Boolean space, and
(ii) if $\varepsilon \in \text{Elem}'\,\Gamma(\boldsymbol{\mathcal{R}})$, then for all x in X, either $\varepsilon(x) = 0$ or $\varepsilon(x) = 1$.

Generalize this to unital half-shells as a characterization of our restricted 'reduced' and 'factor-transparent' of Definition VI.3.14. Compare this with proposition 2.2 of [**Comer71**].

Specifically, for a global section σ of the canonical sheaf $\langle \boldsymbol{\mathcal{A}}, \pi, \boldsymbol{X} \rangle$ of a unital half-shell $\boldsymbol{A}$, prove that σ is a factor element of $\Gamma(\boldsymbol{\mathcal{A}})$ if, and only if,

$$\text{(2.4)} \qquad \text{either } \sigma(x) = 0 \text{ or } \sigma(x) = 1 \qquad (x \in X).$$

Hint: use two facts for a factor element σ, that $0\,\theta'_\sigma\,\sigma\,\theta_\sigma\,1$, and either $\theta_\sigma \subseteq \bigvee P$ or $\theta'_\sigma \subseteq \bigvee P$ for a prime ideal P. Vaggione solved this exercise when it was originally a problem [**Vagg10**].

Beware still that not all sesqui-elements are factor elements (see Example 1.4). As seen before, different pairs of complementary sesqui-elements might lead to the same factor congruences. In the extreme, there might be vagrant pairs of complementary sesqui-elements that are not directly obtainable from any complementary factor congruences.

This will be resolved when the nullity and unity become two-sided. Factor elements and sesqui-elements will be one and the same when the corresponding sesquimorphisms preserve 0; and the sesqui-elements will be uniquely determined by the factorizations they create. And these elements will satisfy more identities.

2.20. DEFINITION. A **two-sided half-shell** $\langle A; \times, 0 \ldots, \omega, \ldots \rangle$ has a designated binary operation and constant such that

$$\text{(0N×N0)} \qquad 0 \times a = 0 = a \times 0 \qquad (a \in A).$$

It is called **unital** if there is another constant 1 such that

$$\text{(1U×U1)} \qquad 1 \times a = a = a \times 1 \qquad (a \in A).$$

2.21. THEOREM. *Let* $\boldsymbol{A}$ *be a two-sided unital half-shell in which complementary sesqui-elements are anchored at the nullity*

$$ee' = 0 = e'e.$$

Then any sesqui-element of $\boldsymbol{A}$ *is a factor element, and vice versa.*

PROOF. Let e and e' be such a pair of complementary sesqui-elements in a unital two-sided half-shell $\langle A; \times, 0, 1 \ldots, \omega, \ldots \rangle$, and let θ_e and $\theta_{e'}$ be their factor congruences:

$$\text{(2.5)} \quad a\,\theta_e\,b \text{ iff } ea = eb \quad \text{and} \quad a\,\theta_{e'}\,b \text{ iff } e'a = e'b \qquad (a, b \in A).$$

Let f and f' be the factor elements derived from them:

$$\text{(2.6)} \qquad 0\,\theta_{e'}\,f\,\theta_e\,1 \text{ and } 0\,\theta_e\,f'\,\theta_{e'}\,1.$$

By Theorem 1.3, θ_e and $\theta_{e'}$ are complementary factor congruences. Hence, the solutions f and f' to (2.6) are unique. If e and e' were also to solve (2.6), then the old and new factor elements would be equal. Now

$$e'0 = 0 = e'e \text{ and } ee = e = e1.$$

Hence, by (2.5), $0\,\theta_{e'}\,e\,\theta_e\,1$. Therefore, $e = f$ and likewise $e' = f'$.
Theorem 2.11a takes care of the vice versa. □

Note how essential, in this proof and the succeeding ones, is the two-sidedness of the constants. Next, even more factor identities hold for two-sided half-shells than for one-sided ones. Finally appearing, remarkably rather late, is the commutative factor identity, $ea = ae$, which is central to the definition of factor elements in noncommutative unital rings, but which could not have appeared earlier since it does not hold generally in one-sided half-shells.

2.22. PROPOSITION. *In a two-sided unital half-shell* $\mathbf{A}$ *any pair of complementary factor elements* e *and* e' *satisfy the identities and formulas of Definition 1.1 and Proposition 2.13, and in addition now these:*

(a) $ea = ae$ *and* $e'a = ae'$ $\quad (a \in A)$;
(b) $a(eb) = (ae)b$ *and* $a(e'b) = (ae')b$ $\quad (a, b \in A)$;
(c) $a(be) = (ab)e$ *and* $a(be') = (ab)e'$ $\quad (a, b \in A)$;
(d) *for each operation* ω *of* $\mathbf{A}$, *including* $\times$,

$$\omega(a_1e, a_2e, \ldots)e = \omega(a_1, a_2, \ldots)e \qquad (a_1, a_2, \cdots \in A),$$
$$\omega(a_1e', a_2e', \ldots)e' = \omega(a_1, a_2, \ldots)e' \qquad (a_1, a_2, \cdots \in A);$$

(e) $ea = 0$ *if, and only if,* $e'a = a$ $\quad (a \in A)$.

PROOF. Use Theorem VI.2.18. □

These identities and the previous ones are not independent. For example, the middle associative law (b) follows from (a), (c) and the left-sided associative law given in Proposition 2.13. A characterization of factor elements with only factor identities will have to wait till addition is appended in the next section. Look ahead to Theorem 3.4 and Exercise 3.12.

With two-sidedness comes a duality of factor identities. Each factor identity previously established in one-sided unital half-shells has a mirror image now holding in two-sided unital half-shells. By a mirror image is meant the reversal of all products in a factor identity. Thus, the mirror image of $e(xy) \approx (ex)y$ in Proposition 2.16d is $(yx)e \approx y(xe)$.

3. Shells

Another binary operation supplements the multiplication and constants of the previous section. It is written additively, with 0 acting as its unity. This is enough to ensure that products factor all congruences in unital shells, not merely factor congruences, as in the previous section. Factor identities alone can now characterize factor elements; this gives a common extension of their classical definitions in rings and lattices.

We describe factor ideals in two new ways, independent of the other factor objects. These simplify when the unital shell is two-sided and distributive. This fills out the roster of independent definitions for all five factor objects. Along the way we look at factor ideals as principal ideals generated by factor elements.

3.1. DEFINITION. A **shell**, $\boldsymbol{A} = \langle A; +, \times, 0, \ldots, \omega, \ldots\rangle$, is an algebra with two binary operations, $+$ and $\times$, a constant 0, and perhaps other operations ω such that for all a in A,

(0N×) $$0 \times a = 0,$$

(0U+U0) $$0 + a = a = a + 0.$$

If there is another constant 1 such that

(1U×) $$1 \times a = a,$$

then $\boldsymbol{A}$ is said to be **unital**.

Again ω is a generic operation beyond those special operations already given; there may be many of these additional operations, or none at all. The constant 0 is called a **nullity** in axiom 0N×, and 1 is a **unity** in axiom 1U×. These terms reflect the function of the constant rather than what symbol represents it. The constant in axiom 0U+U0 is still called a 'unity' with respect to + even though it is designated 0.

However, it is useful to have the convention for algebras with two designated binary operations and two constants, such as rings $\langle R; +, \times, 0, 1\rangle$ or bounded lattices $\langle L; \vee, \wedge, 0, 1\rangle$, that the second operation, $\times$ or $\wedge$, determines the name: 0 for the 'nullity' and 1 for the 'unity'.

One may interpret these shell operations in many different ways. For example, 1 and 0 may play their usual roles as in number theory, or they may represent truth and falsity in logic, or they may even be maximal and minimal elements in a lattice. The binary operations similarly have widely ranging interpretations: the first operation + may be the summing of magnitudes, the disjunction (OR'ing) of logical quantities, the joining of lattice elements, or the union of sets, according to need; and the other operation × can subsume the multiplication of numbers, the conjunction of truth values (AND'ing), the meet of elements in lattices, and the intersection of sets, as desired. Thus, shells encompass both unital rings and bounded lattices as well as expansions and generalizations of them, and yet, as will be seen, there is a significant common theory.[3]

3.2. THEOREM. *A unital shell has factorable congruences; consequently, it has Boolean factor congruences.*

---

[3]Vaggione generalizes shells by replacing the nullity and unity by sequences of unary operations [**Vagg96**]. For the passage between his shells and those of the author, see [**Knoe00**].

PROOF. By Definition VI.2.11 it suffices to prove that

$$\eta = (\eta \vee \theta) \cap (\eta \vee \theta')$$

for all η in Con $\boldsymbol{A}$ and complementary θ and θ' in Con$'$ $\boldsymbol{A}$. One direction of inclusion, $\subseteq$, is always true. To prove the other, assume that a and b are related by the right side. Then $a\ (\eta \vee \theta)\ b$ and $a\ (\eta \vee \theta')\ b$. So there must exist sequences of elements $x_1, x_2, \ldots, x_m$ and $y_1, y_2, \ldots, y_n$ in $\boldsymbol{A}$ related alternately by these relations:

$$a\, \eta\, x_1\, \theta\, x_2\, \eta\, x_3 \cdots x_{m-1}\, \eta\, x_m\, \theta\, b \quad \text{and} \quad a\, \eta\, y_1\, \theta'\, y_2\, \eta\, y_3 \cdots y_{n-1}\, \eta\, y_n\, \theta'\, b.$$

By repeating the last elements and congruences, these sequences may be made the same length, $m = n$. Let e and e' be the complementary factor elements defined by the factor band β determined by θ and θ':

$$e = \beta(1,0) \quad \text{and} \quad e' = \beta(0,1).$$

Then the relations above become, by Lemma 2.3:

$$x_1\, \theta\, x_2 \ \text{ iff } \ ex_1 = ex_2; \quad y_1\, \theta'\, y_2 \ \text{ iff } \ e'y_1 = e'y_2; \ \text{ etc.}$$

Now $e'y_1\, \theta\, e'y_2$ since $e(e'y_1) = (ee')y_1 = 0y_1 = 0y_2 = (ee')y_2 = e(e'y_2)$, by Proposition VI.2.13. Thus $ex_1{+}e'y_1\ \theta\ ex_2{+}e'y_2$, and so forth. Therefore,

$$a = ea{+}e'a\ \ \eta\ \ ex_1{+}e'y_1\ \ (\theta \cap \theta')\ \ ex_2{+}e'y_2\ \ \eta\ \ ex_3{+}e'y_3 \cdots$$
$$\cdots ex_{n-1}{+}e'y_{n-1}\ \ \eta\ \ ex_n{+}e'y_n\ \ (\theta \cap \theta')\ \ eb{+}e'b = b.$$

Since $\theta \cap \theta'$ is equality, it follows that $a\, \eta\, b$.

It has BFC by Theorem VI.3.9. □

One may well wonder what fortunate consequences might follow from a shell having factorable congruences. Of course, it follows that each unital shell $\boldsymbol{A}$ is represented by a reduced and factor-transparent sheaf over a Boolean space, as in Theorem VI.3.15: $\boldsymbol{A} \cong \boldsymbol{\Gamma}\big(A /\!\!/ (\mathbf{Con}'\, A)\big)$. But more might be true: for example, see Problem VI.3.13. Also, how general might the Boolean algebra of factor congruences be in a shell, as suggested next.

3.3. PROBLEM. Show that for any algebra $\boldsymbol{A}$ with BFC there is a unital shell $\boldsymbol{B}$ such that $\mathbf{Con}'\, \boldsymbol{A} \cong \mathbf{Con}'\, \boldsymbol{B}$. Hint: consider Boolean algebras. More concretely, could this still be true if one insists that $A = B$ and $\mathbf{Con}'\, \boldsymbol{A} = \mathbf{Con}'\, \boldsymbol{B}$.

The concept of unital shells is sufficiently strong to support a characterization of factor elements solely in terms of factor identities.

3.4. THEOREM. *Two elements e and e' of a unital shell $\boldsymbol{A}$ are complementary factor elements if, and only if, for all a, a_i and b in A:*

$$ee = e, \qquad e'e' = e';$$
$$ee' = 0, \qquad e'e = 0;$$
$$e0 = 0, \qquad e'0 = 0;$$
$$e1 = e, \qquad e'1 = e';$$
$$e(ab) = (ea)b, \qquad e'(ab) = (e'a)b;$$
$$e(ab) = (ea)(eb), \qquad e'(ab) = (e'a)(e'b);$$
$$a = ea + e'a, \qquad a = e'a + ea;$$
$$e(a+b) = ea + eb, \qquad e'(a+b) = e'a + e'b;$$
$$e\omega(ea_1, ea_2, \ldots) = e\omega(a_1, a_2, \ldots), \quad e'\omega(e'a_1, e'a_2, \ldots) = e'\omega(a_1, a_2, \ldots)$$

(for all operations ω besides $+, \times, 0$ and 1).

The import of this last clause is that the last two factor identities should hold for all additional operations ω, but their truth for $+, \times, 0$ and 1 follows readily from the previous clauses.

PROOF. $\Rightarrow$. Use Theorem VI.2.18 to verify these factor identities.
$\Leftarrow$. The clauses of Proposition 2.16 follow readily except possibly its (g); to prove it, set $x = ea + e'b$. □

What a lot of factor identities! Most are trivially satisfied in traditional algebras. We now realize why those classical identities of centrality come out the way they do. Examples illuminate. Because rings are defined by many identities, it is necessary to require only a few additional ones in order to single out those elements that factor the ring. In a commutative ring with unity, factor elements are just the idempotents. In going from commutative ring theory to noncommutative, factor elements become central idempotents – they must be idempotents commuting with all elements of the ring. Additionally, if the ring is nonassociative, the factor elements will not only be central idempotents, as already expected, but they must associate with all other elements. If further, the ring were not to be even distributive, then the factor elements would also have to distribute over all the other elements of the ring.

As one might guess, if a bounded lattice is not distributive, then factor elements must distribute over everything else, and also the factor identity, $a = ea + e'a$, must hold. Why one distributive law, $e(a+b) = ea + eb$, is necessary and not others, such as $a(e+b) = ae + ab$, is now understood: the former is satisfied when we substitute 0 or 1 for e but not in the latter (see Theorem VI.2.18). More factor identities are given in Proposition 2.22.

What we lack in general, and perhaps appear to need in the theory of factor elements in shells, is subtraction. Surprisingly, this is not a deficiency. All that we need is a 'complement' f of e satisfying $e \times f = 0$ and $e + f = 1$.

Such conditions do not need to be explicitly stated for unital rings since the potential complement f is indeed a subtraction, $f = 1 - e$. As a consequence, the expression of the Boolean operations on factor elements in terms of the shell operations varies according to the kind of algebra; for example, Boolean join is the usual join in lattices, but in rings it is $e \vee f = (e'f')' = e + f - ef$. Section 4 has more such expressions.

3.5. PROBLEM. For what classes of algebras does the set of all factor identities characterize the factor elements? When is there a basis of factor identities from which the remaining may be derived by a 'factor logic' analogous to equational logic? In these questions, the class may be that of one of the sections of this chapter, or a common class such as rings, groups or lattices. Ask similar questions about factor formulas (see Problem VI.2.19).

Factor ideals may now be defined independently of the other factor objects. Remember from Sect. II.2 that a pair of complementary factor ideals are given in a general algebra $\boldsymbol{A}$ with an origin o by images of a factor band: $\beta(A, o)$ and $\beta(o, A)$, or equivalently as images of the corresponding factor sesquimorphisms. Recall that complementary factor sesquimorphisms always capture the notion of inner direct product, whereas in general there may be ambiguity with complementary factor ideals (see the discussion before Definition II.2.22). But, as will be proven, factor ideals in unital shells have their own unique unities; thus there is no ambiguity in the product they specify within the shell. Inner direct products can also be identified in unital shells with complementary factor elements, $e = \beta(1, 0)$ and $e' = \beta(0, 1)$, that generate the complementary factor ideals eA and $e'A$ with unities e and e'. See Theorem 2.11.

Here are several characterizations of pairs of complementary factor ideals as inner products in unital shells, sometimes two-sided, the last one requiring distributivity of $\times$ over $+$. These are reminiscent of older definitions of inner products in terms of ideals in classical algebras. We assume in a shell that all ideals have 0 as their origin. For readability we write a potential complement of the factor ideal I as J rather than I'.

Another but temporary notion of ideal is phrased in the traditional language of ring theory. It is given the French spelling 'idéal' to distinguish it from the kernels $0/\theta$ of congruences θ. Its definition will vary over several sections according to the needs of a particular section. For now, we define it as follows.

3.6. DEFINITION. In a shell $\boldsymbol{A}$ a subset I of A is an **idéal** if

(i) $0 \in I$.
(ii) If $i_1, i_2 \in I$, then $i_1 + i_2 \in I$.
(iii) If $i \in I$ and $a \in A$, then $ia \in I$.

3.7. THEOREM. *In a unital shell,* $\mathbf{A} = \langle A; +, \times, 0, 1 \ldots, \omega, \ldots \rangle$, *two subsets* I *and* J *are complementary factor ideals if, and only if,*

(a) I *and* J *are idéals;*
(b) *each* a *in* A *is a unique sum of some* a_I *in* I *and some* a_J *in* J:

$$a = a_I + a_J;$$

(c) *for each* a *in* A,

$$a_I = 1_I a \quad \text{and} \quad a_J = 1_J a;$$

where 1_I *and* 1_J *come from (b).*
(d) *whenever* $i_1,\ i_2 \in I$ *and* $j_1,\ j_2 \in J$, *then*

$$(i_1 + j_1) + (i_2 + j_2) = (i_1 + i_2) + (j_1 + j_2);$$

(e) *whenever* $i_1,\ i_2 \in I$ *and* $j_1,\ j_2 \in J$, *then*

$$(i_1 + j_1)(i_2 + j_2) = i_1 i_2 + j_1 j_2;$$

(f) *whenever* $i_1, i_2, \ldots \in I$ *and* $j_1, j_2, \ldots \in J$, *then for the operations* ω,

$$\omega(i_1 + j_1, i_2 + j_2, \ldots) = 1_I \omega(i_1, i_2, \ldots) + 1_J \omega(j_1, j_2, \ldots).$$

The operations ω of condition (f) of this theorem need not cover $+$, $\times$, 0 and 1; but if they do, then (d) and (e) may be omitted. Note that (c) and (f) have a touch of factor elements in 1_I and 1_J. By left multiplication, these become factor sesquimorphisms, thereby filling out the inner direct product as given in Definition II.2.22. Before proving this theorem, we establish a preliminary result.

3.8. LEMMA. *Assume that* I *and* J *are subsets of a unital shell* $\mathbf{A}$ *that satisfy conditions (a)–(e) of Theorem 3.7. Decompose* 1 *uniquely as* $e = 1_I$ *and* $e' = 1_J$. *We conclude the following for all* a *and* b *in* A.

(1) $a = ea + e'a$.
(2) $e(a + b) = ea + eb$.
(3) $a \in I$ *iff* $ea = a$ *iff* $e'a = 0$.
(4) $e(ea) = ea$ *and* $e'(e'a) = e'a$.
(5) $I = eA$ *and* $J = e'A$.

PROOF. (1) By unique decomposition, (b) and (c) of Theorem 3.7.
(2) There are the unique decompositions:

$$a + b = (a + b)_I + (a + b)_J, \quad \text{and}$$
$$a + b = (a_I + a_J) + (b_I + b_J) = (a_I + b_I) + (a_J + b_J) \quad \text{by (d)}.$$

Finish by (c).
(3) By the repeated use of unique decompositions.
(4) Use (3).
(5) Use (3) and (4). □

With regard to conclusion (5) of this lemma, we say that e **generates** I and e' generates J.

Proof of Theorem 3.7. $\Leftarrow$. Assume that I and J are subsets that satisfy (a)–(f). Define $e = 1_I$, $e' = 1_J$, and a binary operation β on A by $\beta(a,b) = ea + e'b$. We verify the three axioms (II.2.1)–(II.2.3) that define a factor band β, freely using the preceding lemma. For the first,

$$\beta(a,a) = ea + e'a = a.$$

For the second,

$$\begin{aligned}
\beta(\beta(a,b),\beta(c,d)) &= e(ea + e'b) + e'(ec + e'd)\\
&= (e+0)(ea + e'b) + (0 + e')(ec + e'd)\\
&= \big(e(ea) + 0(e'b)\big) + \big(0(ec) + e'(e'd)\big)\\
&= (ea + 0) + (0 + e'd)\\
&= ea + e'd\\
&= \beta(a,d).
\end{aligned}$$

For the third, which is similar to the second but more involved,

$$\begin{aligned}
&\beta\big(\omega(a_1,a_2,\ldots),\omega(b_1,b_2,\ldots)\big)\\
&\quad= e\omega(ea_1 + e'a_1, ea_2 + e'a_2,\ldots)\\
&\qquad + e'\omega(eb_1 + e'b_1, eb_2 + e'b_2,\ldots)\\
&\quad= e\big(e\omega(ea_1,ea_2,\ldots) + e'\omega(e'a_1,e'a_2,\ldots)\big)\\
&\qquad + e'\big(e\omega(eb_1,eb_2,\ldots) + e'\omega(e'b_1,e'b_2,\ldots)\big)\\
&\quad= e\omega(ea_1,ea_2,\ldots) + e'\omega(e'b_1,e'b_2,\ldots)\\
&\quad= \omega(ea_1 + e'b_1, ea_2 + e'b_2,\ldots)\\
&\quad= \omega\big(\beta(a_1,b_1),\beta(a_2,b_2),\ldots\big).
\end{aligned}$$

Now I is the ideal determined by β since $I = eA = eA + e'0 = \beta(A,0)$, the last by Theorem 3.4.

$\Rightarrow$. For any pair of complementary factor ideals I and $J = I'$ in $\mathbf{A}$, we establish clauses (i)–(iii) in the definition of idéal and conclusions (b)–(f). Associated with I and J are complementary factor congruences θ and θ', and complementary factor elements e and e', as developed in Sect. II.2, Theorem 2.11, and the preceding sections.

(i) This is true since 0 is the origin.

(ii) Since $i_1, i_2 \in I$, then $ei_1 = i_1$ and $ei_2 = i_2$. Hence, $i_1 + i_2 = ei_1 + ei_2 = e(i_1 + i_2)$. That is, $i_1 + i_2 \in I$.

(iii) $ia = (ei)a = e(ia)$; therefore, $ia \in I$.

(b) It is already known from Theorem 3.4 that $a = ea + e'a$. Uniqueness follows since, if also $a = i + j$ for i in I and j in J, then $ea = e(i+j) = ei + ej = ei = i$, because $ej = 0$, and by Corollary 2.17.

(c) From the proof for (b).

(d) Premultiply each side by e and then e', and distribute over the sums, losing terms.

(e) and (f) are similar to (d). □

3.9. COROLLARY. *Let $\boldsymbol{A}$ be a unital shell with complementary factor ideals, I and J. The elements, $e \in I$ and $e' \in J$, given as the unique solution to $e + e' = 1$, are the complementary factor elements that generate I and J. Consequently, they satisfy all the factor identities of Theorem 3.4.*

PROOF. Two representations of 0 have components in I and J:

$$0 = 0 + 0 \text{ and } 0 = e0 + e'0.$$

By uniqueness, $e0 = 0$ and $e'0 = 0$. Similarly, $e1 = e$ and $e'1 = e'$. With the factor band β defined in the proof of the preceding theorem, one concludes that $\beta(1,0) = e1 + e'0 = e$ and likewise, $\beta(0,1) = e'$. Thus, e and e' satisfy the definition of factor elements coming from a factor band, as reviewed at the beginning of Sect. 2. □

We give several simple consequences of Theorem 3.7. For some of those we need the merger of 'two-sided', 'unital' and 'shell' of Definitions 2.20 and 3.1.

3.10. DEFINITION. An algebra $\langle A; +, \times, 0, 1, \ldots, \omega, \ldots\rangle$ is a **two-sided unital shell** if it satisfies, for all a in A,

(0U+U0) $$0 + a = a = a + 0;$$

(0N×N0) $$0 \times a = 0 = a \times 0;$$

(1U×U1) $$1 \times a = a = a \times 1.$$

3.11. PROPOSITION. *Let I and J be a pair of complementary factor ideals of a unital shell $\boldsymbol{A}$. Assume that $a \in A$, $i \in I$, and $j \in J$. Conclude that*

(a) $$i + j = j + i, \quad \textit{and}$$

(b) $$(i + j)a = ia + ja.$$

If $\boldsymbol{A}$ is two-sided, then also

(c) $$ij = 0, \quad \textit{and}$$

(d) $$a(i + j) = ai + aj.$$

PROOF. (a) Substitute $i_1 = 0$ and $j_2 = 0$ into (d) of Theorem 3.7.
(b) Use that theorem again:

$$\begin{aligned}(i + j)a &= (i + j)(a_I + a_J)\\ &= ia_I + ja_J\\ &= (ia_I + 0a_J) + (0a_I + ja_J)\\ &= (i + 0)(a_I + a_J) + (0 + j)(a_I + a_J)\\ &= ia + ja.\end{aligned}$$

(c) Substitute $i_2 = 0$ and $j_1 = 0$ into (e) of Theorem 3.7.
(d) Redo (b) on the other side. □

3.12. EXERCISE. (a) For a two-sided unital shell $\boldsymbol{A}$ (Definition 3.10), prove directly from Theorem 3.4 and independently of Theorem VI.2.18 that

$$ea = ae \qquad (e \in \mathrm{Elem}'\boldsymbol{A}, a \in A).$$

(b) Find a one-sided unital shell in which the factor identity of (a) fails.

The next theorem will introduce complex addition and multiplication into the characterization of inner products:

$$I + J = \{i + j \mid i \in I \text{ and } j \in J\}, \text{ and}$$
$$I \times J = \{i \times j \mid i \in I \text{ and } j \in J\}.$$

Using the previous theorem and proposition, we easily prove next that multiplication and intersection of factor ideals are synonymous.

3.13. PROPOSITION. *Let $\boldsymbol{A}$ be a unital shell.*

(a) *Whenever e_1 and e_2 are factor elements of $\boldsymbol{A}$, then*

$$(e_1 e_2)A = e_1 A \cap e_2 A.$$

(b) *Whenever I and J are factor ideals of a two-sided unital shell, then*

$$I \times J = I \cap J.$$

∞∞∞∞

In analogy with factor congruences, we would like to identify factor ideals with those ideals I for which there is another ideal J such that $I + J = A$ and $I \cap J = \{0\}$. We say that an element 1_I is a **unity** of a subset I of a shell $\boldsymbol{A}$ if, whenever $i \in I$, then $1_I i = i$.

3.14. THEOREM. *Two subsets, I and J, of a two-sided unital shell $\boldsymbol{A}$ are a pair of complementary factor ideals if, and only if,*

(a) *I and J are idéals;*
(b) *$I + J = A$;*
(c) *$I \cap J = \{0\}$;*
(d) *I and J have unities 1_I and 1_J such that, for all a and b in A,*

$$1_I(a + b) = 1_I a + 1_I b, \text{ and}$$
$$1_J(a + b) = 1_J a + 1_J b;$$

(e) *For all i in I, j in J and a in A,*

$$(i + j)a = ia + ja, \text{ and}$$
$$a(i + j) = ai + aj.$$

(f) *for all the operations ω (including $+$ and $\times$) with the unities 1_I and 1_J of part (d), whenever $i_1, i_2, \ldots \in I$ and $j_1, j_2, \ldots \in J$, then*

$$\omega(i_1 + j_1, i_2 + j_2, \ldots) = 1_I \omega(i_1, i_2, \ldots) + 1_J \omega(j_1, j_2, \ldots).$$

PROOF. We show the equivalence of these conditions with those of Theorem 3.7. To make this clear, prefix the conditions of Theorem 3.7 with 'A' and those of the present theorem with 'B'. We need prove only the equivalence of (Ab)–(Ae) with (Bb)–(Be) in the presence of the other conditions, which are the same.

$\Rightarrow$. Assume (Aa)–(Af).

(Bb). This is clear from (Ab).

(Bc). Clearly $0 \in I \cap J$. Now suppose $a \in I \cap J$. Then the equations,

$$a = a + 0 = 0 + a,$$

give two representations of a. Hence, $a = 0$ since such representations are unique.

(Bd) It has already been established that factor ideals have unities that are factor elements (Theorem 2.11 and Corollary 2.17).

(Be). This is Proposition 3.11b,d.

$\Leftarrow$. Assume (Ba)–(Bf).

(Ab). Using (Bb), define e in I and f in J by $e + f = 1$. Then by (Be), for any a in A,

$$a = 1a = (e + f)a = ea + fa.$$

This representation is unique, for if $a = i + j$ with $i \in I$ and $j \in J$, and the conditions of the proposition are freely applied, then

$$ea = e(i + j) = ei + ej = ei + 0 = ei + fi = (e + f)i = 1i = i.$$

Likewise, $fa = j$.

(Ac) Since $a = 1a = (1_I + 1_J)a = 1_I a + 1_J a$, then $1_I a = a_I$ from (Ab).

(Ad) and (Ae). Use (Bf). □

When the distributive law is assumed, conditions (d) and (e) of this theorem vanish and (f) may be simplified, as in the next corollary. When $\boldsymbol{A}$ is a unital ring, distributivity is automatic, and all three are redundant; idéals are now the same as the traditional ideals of ring theory.

3.15. COROLLARY. *Suppose $\boldsymbol{A}$ is a two-sided unital shell in which $\times$ distributes over $+$:*

$$a(b + c) = ab + ac, \quad \text{and} \quad (a + b)c = ac + bc \qquad (a, b, c \in A).$$

Two subsets I and J of A form an inner direct product of factor ideals if, and only if,

(a) *I and J are idéals;*

(b) *$I + J = A$;*

(c) *$I \cap J = \{0\}$; and*

(d) *I has a unity 1_I and J has a unity 1_J such that $1_I + 1_J = 1$, and for all operations ω,*

$$1_I\omega(1_I a_1, 1_I a_2, \ldots) = 1_I\omega(a_1, a_2, \ldots) \qquad (a_1, a_2, \ldots \in A), \quad \text{and}$$
$$1_J\omega(1_J a_1, 1_J a_2, \ldots) = 1_J\omega(a_1, a_2, \ldots) \qquad (a_1, a_2, \ldots \in A).$$

PROOF. $\Rightarrow$. Use Theorem 3.14.

$\Leftarrow$. Assume(a)–(d). These imply the clauses of Theorem 3.14, of which only (f) requires much proof. For it, with distributivity, (c) and (d),

$$\begin{aligned} 1_I\omega(i_1+j_1, i_2+j_2, \ldots) &= 1_I\omega(1_I(i_1+j_1), 1_I(i_2+j_2), \ldots) \\ &= 1_I\omega(1_Ii_1+1_Ij_1, 1_Ii_2+1_Ij_2, \ldots) \\ &= 1_I\omega(1_Ii_1, 1_Ii_2, \ldots) \\ &= 1_I\omega(i_1, i_2, \ldots). \end{aligned}$$

A similar derivation holds for premultiplication by 1_J. By distributivity,

$$\begin{aligned} \omega(i_1+j_1, i_2+j_2, \ldots) &= 1_I\omega(i_1+j_1, i_2+j_2, \ldots) \\ &\quad + 1_J\omega(i_1+j_1, i_2+j_2, \ldots) \\ &= 1_I\omega(i_1, i_2, \ldots) + 1_J\omega(j_1, j_2, \ldots). \end{aligned}$$ □

3.16. COUNTEREXAMPLE. In this corollary distributivity is essential. To see otherwise, let $\boldsymbol{A}$ be the nondistributive lattice $\boldsymbol{N}_5$.

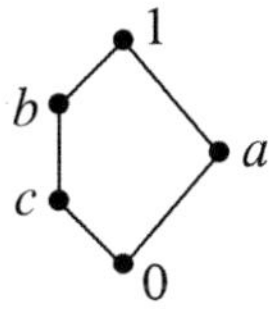

Look at the ideals, $I = \{0, a\}$ and $J = \{0, b, c\}$. They satisfy (a), (b), and (c); but they do not form an inner direct product.

Although not a shell, a multiplicative counterexample is $\boldsymbol{S}_3$, the permutation group on the three letters α, β, and γ. Its subgroups, $I = \{(), (\alpha\beta)\}$ and $J = \{(), (\alpha\beta\gamma), (\alpha\gamma\beta)\}$, satisfy (b) and (c) of Corollary 3.15, but they are not an inner product.

This section closes with several exercises about factor objects in two other fragments of shells. It is instructive to fit these into our framework. We assume in these exercises that the sesquimorphisms preserve 0.

The algebras of Jónsson and Tarski [**JónTa47**] are close to shells, but sufficiently different not to have a canonical sheaf representation, and therefore they will play a minor role in this book.

3.17. DEFINITION. A **J-T algebra** is an algebra

$$\langle A; +, 0, \ldots, \omega, \ldots \rangle \text{ of type } \langle 2, 0, \ldots, n, \ldots \rangle$$

such that, for all a in A,

(0U+U0) $$0 + a = a = a + 0,$$

and for all operations ω of $\boldsymbol{A}$,

$$\omega(0, 0, \ldots, 0) = 0.$$

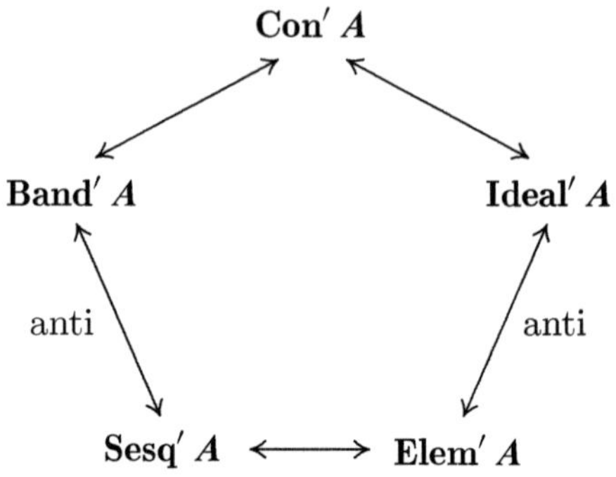

FIGURE 2. Isomorphisms between Boolean Algebras of Factor Objects for a Two-sided Unital Shell $\boldsymbol{A}$.

4.2. PROPOSITION. *If I and J are factor ideals in a two-sided unital shell $\boldsymbol{A}$, then*

(a) $$I \vee J = I + J;$$

(b) $$I \wedge J = I \cap J; \textit{ and}$$

(c) $$I' = \{a \in A \mid ai = 0 \textit{ for all } i \textit{ in } I\}.$$

PROOF. (a) Clearly, by the bijection between ideals and congruences already established in Proposition VI.2.8, it suffices to prove that

$$\frac{0}{\theta} + \frac{0}{\eta} = \frac{0}{\theta \vee \eta},$$

for any factor congruences θ and η.

$\subseteq$. If $a \in 0/\theta$ and $b \in 0/\eta$, then $(a + b)\,\theta\,(0 + b)\,\eta\,(0 + 0) = 0$, and $a + b \in 0/(\theta \vee \eta)$.

$\supseteq$. Suppose that $a \in 0/(\theta \vee \eta)$. Since $\theta \vee \eta = \theta \circ \eta$ by commutativity, there is an x such that $0\,\theta\,x$ and $x\,\eta\,a$. Therefore,

$$0 = e_\theta 0 = e_\theta x \quad \text{and} \quad e_\eta x = e_\eta a.$$

So, by appropriate multiplications, we get $0 = e_\eta e_\theta a$ by Proposition 2.13. Hence,

$$a = e'_\theta a + (e'_\eta e_\theta a + e_\eta e_\theta a) = e'_\theta a + e'_\eta e_\theta a.$$

Since $e'_\theta A = 0/\theta$, etc., by Theorem 2.11, then $e'_\theta a \in 0/\theta$ and $e'_\eta e_\theta a \in 0/\eta$. Therefore,

$$a = e'_\theta a + e'_\eta e_\theta a \in \frac{0}{\theta} + \frac{0}{\eta}.$$

(b) It is trivial to verify that

$$\frac{0}{\theta} \cap \frac{0}{\eta} = \frac{0}{\theta \cap \eta}.$$

$I_\theta = 0/\theta$ $\theta \mapsto I_\theta$ $\mathbf{Con}'\,\boldsymbol{A} \longleftrightarrow \mathbf{Ideal}'\,\boldsymbol{A}$ $\theta_I \leftarrow\!\shortmid I$ $\theta_I =$ the unique θ in $\mathrm{Con}'\,\boldsymbol{A}$ with $I = 0/\theta$
e_I is the unity of I' $I \mapsto e_I$ $\mathbf{Ideal}'\,\boldsymbol{A} \overset{\text{anti}}{\longleftrightarrow} \mathbf{Elem}'\,\boldsymbol{A}$ $I_e \leftarrow\!\shortmid e$ $I_e = e'A$
$\mu_e(a) = ea$ $e \mapsto \mu_e$ $\mathbf{Elem}'\,\boldsymbol{A} \longleftrightarrow \mathbf{Sesq}'\,\boldsymbol{A}$ $e_\mu \leftarrow\!\shortmid \mu$ $e_\mu = \mu(1)$
$\beta_\mu(a,b) = c$ iff $\mu(c) = \mu(a)$ and $\mu'(c) = \mu'(b)$ $\mu \mapsto \beta_\mu$ $\mathbf{Sesq}'\,\boldsymbol{A} \overset{\text{anti}}{\longleftrightarrow} \mathbf{Band}'\,\boldsymbol{A}$ $\mu_\beta \leftarrow\!\shortmid \beta$ $\mu_\beta(a) = \beta(a,0)$
$a\,\theta_\beta\,b$ iff $\beta(a,b) = b$ $\beta \mapsto \theta_\beta$ $\mathbf{Band}'\,\boldsymbol{A} \longleftrightarrow \mathbf{Con}'\,\boldsymbol{A}$ $\beta_\theta \leftarrow\!\shortmid \theta$ $\beta_\theta(a,b) = c$ iff $a\,\theta\,c$ and $c\,\theta'\,b$

TABLE 1. Formulas for the Isomorphisms.

	Join	Meet	Complement
Con′ A	$\theta \vee \eta = \theta \circ \eta$	$\theta \wedge \eta = \theta \cap \eta$	$\theta' =$ unique η with $\theta \cap \eta = 0$ & $\theta \circ \eta = 1$
Ideal′ A	$I \vee J = I + J$	$I \wedge J = I \cap J$	$I' =$ annihilator of I
Elem′ A	$e \vee f = (e'f')'$	$e \wedge f = ef$	$e' =$ unique f with $ef = 0$ & $e + f = 1$
Sesq′ A	$\mu \vee \nu = (\mu'\nu')'$	$\mu \wedge \nu = \mu \circ \nu$	$\mu' =$ unique ν with $\mu\nu = 0$ & $\mu + \nu = 1$
Band′ A	$(\alpha \vee \beta)(a,b)$ $= \alpha(\beta(a,b),b)$	$(\alpha \wedge \beta)(a,b)$ $= \alpha(a,\beta(a,b))$	$\alpha'(a,b) = \alpha(b,a)$

TABLE 2. Boolean Operations on Factor Objects in Two-sided Unital Shells.

(c) $\subseteq$. Recall from Proposition 3.13 and Theorem 3.14 that $I'I = \{0\}$. $\supseteq$. If a belongs to the right side, then $ae_I = 0$. Hence $a = ae_I' = ae_{I'}$. Consequently $a \in I'$. □

In the next proposition, by $\mu\nu = 0$ is meant composition $(\mu \circ \nu)(a) = 0$, and by $\mu + \nu = 1$, addition computed pointwise: $(\mu + \nu)(a) = \mu(a) + \nu(a) = a$.

4.3. PROPOSITION. *Let $\boldsymbol{A}$ be a two-sided unital shell.*

(a) *For a factor element e of $\boldsymbol{A}$, its complement e' is the unique f such that*
$$ef = 0, \text{ and } e + f = 1.$$
(b) *For a factor sesquimorphism μ of $\boldsymbol{A}$, its complement μ' is the unique ν such that*
$$\mu\nu = 0, \text{ and } \mu + \nu = 1.$$

PROOF. (a) The only thing not proven already in previous sections is uniqueness. So suppose that f satisfies the two equations. By Theorem 3.4,
$$f = ef + e'f = 0 + e'f = e'e + e'f = e'(e + f) = e'1 = e'.$$
(b) This follows from (a) by the correspondence, $\mu(a) = ea$, in Theorem 2.11(f). □

In certain common algebras in Table 3, the Boolean operations on factor elements are expressed in terms of the operations of the algebras. Propositions 4.4 and 4.5 verify these formulas.

$e \vee f$	$= e + f - ef$	(unital rings)
	$= e + f$	(bounded distributive lattices)
e'	$= 1 - e$	(unital rings)

TABLE 3. Formulas for the Boolean operations on factor elements in special shells.

4.4. PROPOSITION. *In a two-sided unital shell $\boldsymbol{A}$ in which addition is idempotent ($a + a = a$),*
$$e \vee f = e + f \quad (e, f \in \text{Elem}'\ A).$$

PROOF. It suffices to show that the right side, $e + f$, together with the complement of the left side, $e'f'$, are complementary factor elements. This means verifying the identities of Theorem 4.3(a):
$$(e + f)e'f' = 0, \text{ and } (e + f) + e'f' = 1.$$
This is done most efficiently by testing them with 0 and 1 replacing the factor elements, as in Theorem VI.2.18. □

4.5. PROPOSITION. *In a unital ring that is not necessarily associative,*

(a) $$e \vee f = e + f - ef \qquad (e, f \in \text{Con}'\ \boldsymbol{A});$$
(b) $$e' = 1 - e \qquad (e \in \text{Con}'\ \boldsymbol{A}).$$

PROOF. We prove (b) first, since (a) depends on (b). Part (b) is true since $e' + e = 1$ in rings.

For (a) let L represent the left side and R the right. We demonstrate that the complement of L is the complement of R by showing that $L' + R = 1$ and $L'R = 0$, as in Theorem 4.3. First,

$$\begin{aligned} L' + R &= (e \vee f)' + (e + f - ef) \\ &= e'f' + (e + f - ef) \\ &= (1 - e)(1 - f) + (e + f - ef) \\ &= 1 - f - e + ef + e + f - ef \\ &= 1. \end{aligned}$$

Secondly, in a similar fashion, with the distributive law, we see that

$$\begin{aligned} L'R &= (e \vee f)'(e + f - ef) \\ &= 0. \end{aligned}$$

Thus $e \vee f = e + f - ef$. □

We summarize these special results by saying that the meet operation for factor objects in shells is the obvious operation already available at any of the five levels: product of elements, intersection of ideals, composition of sesquimorphisms, intersection of congruences, and composition of bands. But we cannot always describe the other two operations so easily, and in general they float free of obvious formulas.

Why were the choices made that led to the unappealing appearance of anti-isomorphisms in Fig. 2: **Band′ *A*** $\overset{\text{anti}}{\longleftrightarrow}$ **Sesq′ *A*** and **Ideal′ *A*** $\overset{\text{anti}}{\longleftrightarrow}$ **Elem′ *A***? These could have been avoided by interchanging joins and meets so that all links in this figure become isomorphisms; after all, any Boolean algebra is anti-isomorphic to itself. For example, '$\mu \wedge \nu$' might be notated '$\mu \vee \nu$' for sesquimorphisms, but this would run counter to the meet of factor elements in ring theory being the multiplication of central idempotents. Or mutatis mutandis, the Boolean operations on factor bands could be redefined, but this would run afoul of the operations on congruences. And these would be artificial. Strong convention dictates how these Boolean operations are named.

At one end of the spectrum of factor objects, the congruences of an algebra have a natural partial ordering of inclusion, and it is conventional to define the join as the least upper bound, which becomes composition for commuting factor congruences. A factor sesquimorphism μ corresponding to a factor congruence θ ought to be one of those attached to the congruence as developed in Sect. II.1:

$$a \; \theta_\mu \; b \quad \text{iff} \quad \mu(a) = \mu(b).$$

Quite naturally one expects compositions of congruences to correspond to compositions of sesquimorphisms, consequently the join of sequimorphisms would seem to be composition. But beware!

At the other end of the spectrum, factor elements comprise all the elements in Boolean rings, where conventionally multiplication is meet in their term-equivalents as Boolean algebras. As premultiplication by factor elements is the same as the action of the corresponding factor sequimorphisms:

$$(ef)a = e(fa) = \mu_e(\mu_f(a)) = (\mu_e \circ \mu_f)(a),$$

the meet of factor elements should correspond to the meet of factor sequimorphisms. But look how compositions clash: the join of factor congruences has become the meet of factor elements,

$$\theta_e \vee \theta_f = \theta_e \circ \theta_f = \theta_{ef} = \theta_{e \wedge f}.$$

This settles it for factor congruences, ideals, sesquimorphisms and elements, leaving only factor bands to consider, where the choice was a toss-up; it was set to keep company with congruences, which unfortunately runs afoul of the composition of sesquimorphisms:

$$(\mu \wedge \nu)(a) = \beta_\mu(\beta_\nu(a,0),0) = (\beta_\mu \vee \beta_\nu)(a,0).$$

Their isomorphisms and anti-isomorphisms are set down in Theorems II.2.5, II.2.12, II.2.19 and VII.2.11.

4.6. EXERCISE. One way to reverse the anti-isomorphisms just discussed is to reverse the factor elements coming from factor congruences. Redefine them by the relationships, $0 \; \theta \; e \; \theta' \; 1$ and $0 \; \theta' \; e' \; \theta \; 1$. Trace this new definition through the other factor objects, making appropriate adjustments in other definitions. What are its advantages and disadvantages?

5. Separator Algebras

While unital shells have Boolean factor congruences and also have factor elements that determine all the other factor objects, BFC alone is not sufficient to support factor elements. This short section presents 'separator algebras', which generalize shells, but nevertheless have many of their factorization properties. The main result of this section embeds any algebra with BFC into a separator algebra so that within the latter there are factor elements realizing the factorizations of the former [**Knoe02**]. In other words, we adjoin imaginary factor elements to an algebra with BFC.

5.1. DEFINITION. A **separator** algebra is an algebra $\boldsymbol{A}$ having a quaternary term-operation q and term-constants 0 and 1 satisfying, for all a and b in A,

$$q(1,0,a,b) = a, \quad \text{and}$$
$$q(0,1,a,b) = b.$$

A **separator** variety is an equational class in which there are terms $q, 0, 1$ satisfying these identities.

Examples are unital rings and bounded lattices: $q(a,b,c,d) = ac + bd$, and more generally unital shells. In a sense, so are discriminator algebras. A **discriminator** algebra is an algebra $\boldsymbol{A}$ with a quaternary term-operation t satisfying, for all a, b, c and d in A,

$$t(a,b,c,d) = c \quad \text{if} \quad a = b, \quad \text{and}$$
$$t(a,b,c,d) = d \quad \text{if} \quad a \neq b.$$

A discriminator algebra is also called **quasi-primal**

5.2. PROPOSITION. *Any discriminator algebra $\boldsymbol{A}$ is polynomially equivalent to a separator algebra, but not conversely.*

PROOF. When $\boldsymbol{A}$ is nontrivial pick two distinct elements 0, 1 of A, and define a polynomial $q(w,x,y,z) = t(0,x,y,z)$. Then q separates.

For a counterexample, let the separator $q(w,x,y,z)$ be the positive term $wy \vee xz$ of the two-element lattice on $\{0,1\}$, and consider the separator algebra $\langle \{0,1\}; q, 0, 1\rangle$. The operation q is 'positive' in the sense that q is monotonic in each argument, and so is each term composed from it; but this cannot be true of the discriminator term. □

'Discriminator' is an old notion; the newer term 'separator' was coined for this book in analogy with it. Chajda [**Cha94**] defined the concept of **T(0,1)-algebras**, which are equivalent to separator algebras. This means that the tolerance[4] generated by two constants 0 and 1 is the universal relation. For background on quasiprimal algebras and discriminator varieties, consult [**Pixl96**, sect. 2] and [**BurSa81**, chap. IV, sects. 9, 10].

As Keimel and Werner [**KeiWe74**] have shown, any variety generated by a finite discriminator algebra $\boldsymbol{A}$ is dually equivalent to the category of Hausdorff sheaves over Boolean spaces with their simple stalks being the subalgebras of $\boldsymbol{A}$ (see also Bulman-Fleming and Heinrich Werner [**BulWe77**]). We seek something similar for separator algebras; but the stalks remain unknown in general, as already seen for unital shells, a subclass of them.

5.3. PROPOSITION. *Any separator algebra has Boolean factor congruences, and hence has a canonical sheaf representation.*

PROOF. By Theorem VI.3.2, it suffices to prove that any two factor bands β and γ of a separator algebra $\boldsymbol{A}$ commute: $\beta(\gamma ab, \gamma cd) = \gamma(\beta ac, \beta bd)$. To this end, define the corresponding complementary factor elements: $e = \beta 10$ and $e' = \beta 01$. For all a and b in A, because factor bands commute with term operations,

$$\begin{aligned}\beta ab &= \beta\big(q(1,0,a,b), q(0,1,a,b)\big)\\ &= q(\beta 10, \beta 01, \beta aa, \beta bb)\\ &= q(e, e', a, b).\end{aligned}$$

[4] A tolerance is a reflexive, symmetric relation that preserves the operations.

Using this, we prove that

$$\begin{aligned}\beta(\gamma ab, \gamma cd) &= q(e, e', \gamma ab, \gamma cd) \\ &= q(\gamma ee, \gamma e'e', \gamma ab, \gamma cd) \\ &= \gamma\big(q(e, e', a, c), q(e, e', b, d)\big) \\ &= \gamma(\beta ac, \beta bd).\end{aligned}$$

□

5.4. EXERCISE. (a) Show that the converse of this proposition is false, that is, find an algebra with BFC that does not have a separator term.
(b) Show that separator algebras are more general than unitary shells, that is, find a separator algebra that is not term-equivalent to a unital shell.

Diego Vaggione [**Vagg96a**] both generalizes Proposition 5.3 and sharpens it. His 'Pierce variety' generalizes the notion of separator variety, and in his theorem 5, he proves that any Pierce variety has factorable congruences. In his theorem 8, varieties with factorable congruences in which the stalks of the canonical sheaf are directly indecomposable are characterized as certain kinds of Pierce varieties.

Now we represent any algebra with BFC within a separator algebra, thereby creating factor elements for the original algebra.

5.5. THEOREM. *For any algebra,* $\boldsymbol{A} = \langle A; \ldots, \omega, \ldots \rangle$, *with Boolean factor congruences, there is a separator algebra,* $\boldsymbol{Q} = \langle Q; q, 0, 1, \ldots, \omega, \ldots \rangle$, *whose type expands that of* $\boldsymbol{A}$ *by a quaternary operation* q *and constants 0 and 1, such that*

(a) $\boldsymbol{A}$ *is a subalgebra of the reduct* $\langle Q; \ldots, \omega, \ldots \rangle$, *and*
(b) *for any factor band* β *of* $\boldsymbol{A}$ *there are complementary factor elements* e *and* e' *of* $\boldsymbol{Q}$ *such that*

$$\beta(a, b) = q(e, e', a, b) \quad (a, b \in A). \tag{5.1}$$

PROOF. (a) Recall Theorem VI.3.15 that an algebra $\boldsymbol{A}$ with BFC has a reduced and factor-transparent sheaf $\langle \mathcal{A}, \pi, \boldsymbol{X} \rangle$ over the Boolean space $\boldsymbol{X}$ with stalks $\boldsymbol{A}_x$. The algebra $\boldsymbol{\Gamma}(\mathcal{A})$ of global sections is isomorphic to $\boldsymbol{A}$, and will be used in place of it. Let $\boldsymbol{P}$ be the product of the stalks: $\prod_{x \in X} \boldsymbol{A}_x$. Clearly $\boldsymbol{\Gamma}(\mathcal{A})$ is a subalgebra of $\boldsymbol{P}$. On each stalk impose a separator q with constants 0 and 1. This can be done in many ways; for example, by adding lattice operations with bounds 0 and 1. Extend q by co-ordinates to all of P, and designate $\boldsymbol{Q}$ as the separator algebra $\langle P; q, 0, 1, \ldots, \omega, \ldots \rangle$ expanding $\boldsymbol{P}$.

(b) Consider a factor band β of $\boldsymbol{\Gamma}(\mathcal{A})$ with associated complementary factor congruences θ and θ'. By factor-transparency there are clopen subsets, $U = U_\theta$ and $U' = U_{\theta'}$, of $\boldsymbol{X}$ such that for all global sections σ and τ in $\Gamma(\mathcal{A})$

$$\sigma\,\theta\,\tau \quad \text{iff} \quad U \subseteq \sigma{:}\tau, \quad \text{and} \quad \sigma\,\theta'\,\tau \quad \text{iff} \quad U' \subseteq \sigma{:}\tau.$$

By Corollary VI.3.16,

$$\beta(\sigma,\tau) = \sigma|U \cup \tau|U' \qquad (\sigma,\tau \in \Gamma(\mathcal{A})).$$

Use this formula to extend β from $\Gamma(\mathcal{A})$ to all of the product P. By manipulating patching, one shows that β is a factor band of $\boldsymbol{Q}$. (The band β commutes with q since q is defined pointwise in $\boldsymbol{Q}$ just as the original operations ω are.)

Now define

$$e = 1|U \cup 0|U' \quad \text{and} \quad e' = 0|U \cup 1|U'.$$

Since $e = \beta 10$ and $e' = \beta 01$, it follows that e and e' are complementary factor elements of $\boldsymbol{Q}$. Equation (5.1) is verified:

$$\begin{aligned} q(e,e',\sigma,\tau) &= q(\beta 10, \beta 01, \beta\sigma\sigma, \beta\tau\tau) \\ &= \beta\big(q(1,0,\sigma,\tau), q(0,1,\sigma,\tau)\big) \\ &= \beta\sigma\tau. \end{aligned}$$

□

5.6. Corollary. *In the previous theorem, the stalks may be chosen to have a reduct that is a unital ring or a bounded distributive lattice.*

Proof. In the previous proof, the stalks are easily chosen to be whatever kind of separator algebras one wants, including bounded chains. □

We close this section with some exercises and problems.

5.7. Exercise. Show that the definition of separator variety may be simplified to the following when the variety has commuting congruences. There is a ternary term t and constants 0 and 1 satisfying these identities:

$$\begin{aligned} &t(0,x,y) \approx x, \quad \text{and} \\ &t(1,x,y) \approx y. \end{aligned}$$

5.8. Exercise. Show that taking a subalgebra is essential to the truth of Theorem 5.5. That is, find an algebra with BFC that is not a reduct of a separator algebra with the same carrier and factors.

5.9. Exercise. Let $\langle A; q, 0, 1, \ldots, \omega, \ldots\rangle$ be a separator algebra. Prove that two elements e and f of A are complementary factor elements if, and only if, they satisfy these axioms when $a,b,c,\ldots \in A$:

(i) $q(e,f,a,a) = a$,
(ii) $q\big(e,f,a,q(e,f,b,c)\big) = q(e,f,a,c) = q\big(e,f,q(e,f,a,b),c\big)$,
(iii) $q(e,f,1,0) = e$ and $q(e,f,0,1) = f$,
(iv) $q\big(e,f,\omega(a_1,a_2,\ldots),\omega(b_1,b_2,\ldots)\big) = \omega\big(q(e,f,a_1,b_1),q(e,f,a_2,b_2),\ldots\big)$.

Hint. For the 'only if' direction, first prove that $q(e,f,a,b) = \beta(a,b)$ for the factor band associated with e and f.

5.10. PROBLEM. As a counterpart to the previous exercise, find an axiomatic definition for factor ideals in separator algebras.

5.11. PROBLEM. Redo Sect. 4 for separator algebras.

6. Categories of Shells

We specialize further the categories of algebras and their sheaves defined in the last section of the previous chapter. When the algebras with Boolean factor congruences are restricted to be unital half-shells, their morphisms in the category $\mathfrak{UnitHalfShell}$ have an especially appealing form. This is the first order of investigation. The other is its mate, the category $\mathfrak{SheafUnitHalfShell}$ of reduced and factor-transparent sheaves of unital half-shells. These two categories are equivalent, and the equivalence resembles that for rings, as outlined by Pierce [**Pier67**]. As usual, we assume a fixed type of unital half-shells $\langle A; \times, 0, \ldots, \omega, \ldots \rangle$, with perhaps also $+$ being one of the operations.

6.1. DEFINITION. A homomorphism $\varphi : \boldsymbol{A} \to \boldsymbol{B}$ between unital half-shells is **conformal** if $\varphi(e)$ is a factor element whenever e is a factor element. The category $\mathfrak{UnitHalfShell}$ consists of unital half-shells as objects and conformal homomorphisms as morphisms.

That factor elements do not automatically go over to factor elements under arbitrary homomorphisms is seen in this example of bounded lattices.

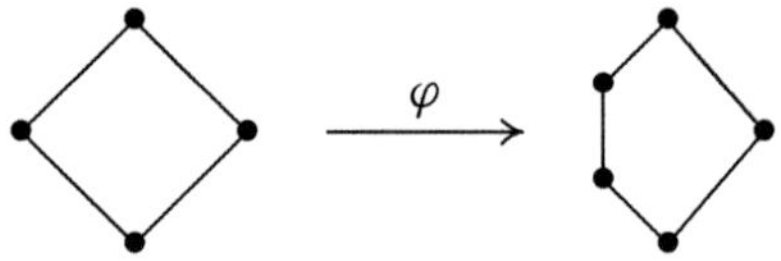

$\boldsymbol{M}_2$ $\boldsymbol{N}_5$

This injection φ takes elements of the first lattice to their obvious counterparts in the second, missing one element in the right lattice. They are all factor elements in the first, but only the top and bottom are now factor elements in the second. Strict inclusion of the factor elements of one shell into another is possible, as seen in the left inverse $\widetilde{\varphi}$ of the φ above, where the two left middle elements coalesce.

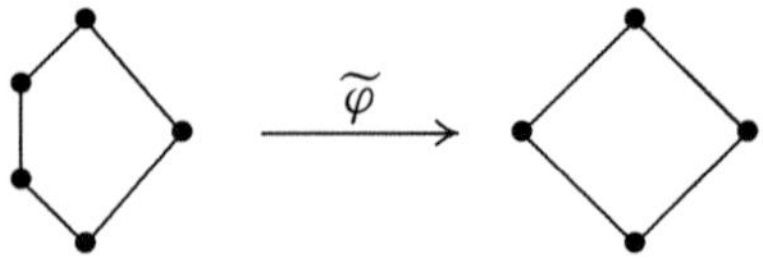

$\boldsymbol{N}_5$ $\boldsymbol{M}_2$

As the objects of the categories $\mathfrak{UnitHalfShell}$ and $\mathfrak{AlgBFC}$ are presented differently, the first as algebras and the second as special Boolean braces, it is necessary to reconcile these in the next proposition.

6.2. PROPOSITION. *The category* $\mathfrak{UnitHalfShell}$ *is isomorphic to a full subcategory of* $\mathfrak{AlgBFC}$. *The isomorphism takes a unital half-shell to itself, and a conformal homomorphism* $\varphi: \boldsymbol{A} \to \boldsymbol{B}$ *goes to the pair* $\langle \varphi, \upsilon \rangle$ *where the homomorphism of Boolean algebras,* $\upsilon: \mathbf{Con}' \boldsymbol{A} \to \mathbf{Con}' \boldsymbol{B}$, *is induced by* φ:

$$\upsilon(\theta_e) \equiv \theta_{\varphi(e)} \qquad (e \in \mathrm{Elem}' \ A).$$

PROOF. Right away, one can see that $\mathfrak{UnitHalfShell}$ is a category since identity maps are conformal and conformality of homomorphisms is preserved by composition. Next, one must show that $\langle \varphi, \upsilon \rangle$ is a morphism in $\mathfrak{AlgBFC}$ whenever $\varphi: A \to B$ is a conformal homomorphism of these shells. It is easy to check that $\varphi(\theta_e) \subseteq \upsilon(\theta_e)$ in Definition V.4.1. Note that φ preserves the complementation $'$ of factor elements since the equations defining complementation of factor elements are preserved by φ. Thus φ preserves the Boolean operations of $\mathbf{Elem}' \boldsymbol{A}$. From Theorem 2.15b we know that $\mathbf{Elem}' \ A$ is anti-isomorphic to $\mathbf{Con}' \boldsymbol{A}$; in turn υ preserves the Boolean operations in $\mathbf{Con}' \boldsymbol{A}$ and is a Boolean homomorphism.

To establish fullness of the embedding, we must show that, whenever $\boldsymbol{A}$ and $\boldsymbol{B}$ are such shells and $\langle \varphi, \upsilon \rangle$ is a morphism in $\mathfrak{AlgBFC}$, then $\varphi: A \to B$ is a conformal homomorphism in $\mathfrak{UnitHalfShell}$. It suffices to prove that $\varphi(e) \in \mathrm{Elem}' \boldsymbol{B}$ when $e \in \mathrm{Elem}' \boldsymbol{A}$. Consider any factor congruence θ_e determined by the factor element e. As a factor element, e satisfies $0 \, \theta_{e'} \, e \, \theta_e \, 1$. Since υ takes factor congruences into factor congruences, there must be a factor element d in B such that $\upsilon(\theta_e) = \theta_d$.

We need to establish that $d = \varphi(e)$. Because υ is a homomorphism, it preserves complementation: $\upsilon(\theta_{e'}) = \theta_{d'}$. Recall from Definition V.4.1 that $\varphi(\theta_e) \subseteq \upsilon(\theta_e)$. Using these facts, together with how φ acts on congruences, we have these relationships between elements and congruences.

$$\begin{array}{ccccc} 0 & \theta_{d'} & d & \theta_d & 1 \\ & \cup\!| & & \cup\!| & \\ 0 & \varphi(\theta_{e'}) & \varphi(e) & \varphi(\theta_e) & 1 \end{array}$$

Thus, $0 \, \theta_{d'} \, \varphi(e) \theta_d 1$. By uniqueness of the middle components, $\varphi(e) = d$. Therefore, $\varphi(e) \in \mathrm{Elem}' \boldsymbol{B}$. □

6.3. DEFINITION. The category $\mathfrak{SheafUnitHalfShell}$ is the full subcategory of $\mathfrak{SheafBooleRedFt}$ consisting of sheaves whose stalks are unital half-shells.

Since unital half-shells have BFC, it follows for every unital half-shell that its canonical sheaf, which is reduced and factor-transparent, is in $\mathfrak{SheafBooleRedFt}$. So we may proceed to restrict the categorical equivalences of previous chapters. The next theorem needs an isomorphism for the reason given before Proposition 6.2.

6.4. Theorem. *The categories of* $\mathfrak{UnitHalfShell}$ *and* $\mathfrak{SheafUnitHalfShell}$ *are equivalent; more precisely, when restricted to isomorphic copies of these categories, the adjunction of Theorem VI.4.2,*

$$\langle \eta, \varepsilon \rangle : \Phi \dashv \Gamma : \langle \mathfrak{SheafBooleRedFt}, \mathfrak{AlgBFC} \rangle,$$

is still a categorical equivalence.

Proof. Note that unital half-shells are defined equationally. Hence the stalks of the sheaf representation of a unital half-shell are again unital half-shells, since they are homomorphic images. Thus, when $\boldsymbol{A}$ is in $\mathfrak{UnitHalfShell}$, so $\Phi(A)$ is in $\mathfrak{SheafUnitHalfShell}$. Conversely, it is clear that a sheaf $\boldsymbol{\mathcal{A}}$ of unital half-shells goes back to a unital half-shell $\Gamma(\boldsymbol{\mathcal{A}})$. Now use Theorem VI.4.5. □

While this theorem extends Stone's representation theorem for Boolean algebras, we note two things. First, in the category of unital half-shells, not all homomorphisms are allowed, only the conformal ones. Second, the category of reduced and factor-transparent sheaves of unital half-shells over Boolean spaces is certainly more involved than just the category of Boolean spaces themselves. One reason for this is that different sheaves may be the same topologically, both in their base and sheaf spaces, and yet be different algebraically. For example, compare the three sheaves coming from the bounded lattice, $\boldsymbol{M}_2 = \boldsymbol{C}_2{}^2$, from the unital ring $\mathbb{Z}_2{}^2$ and from the mixed shell $\boldsymbol{C}_2 \times \mathbb{Z}_2$. More generally, compare the squares of any two non-isomorphic product-indecomposable algebras of the same cardinality. It is also conceivable that there are sheaves of shells over the same base space but with different sheaf spaces.

We will apply Theorem 6.4 in Chap. VIII.

Adding a binary operation $+$ as one of the generic operations ω with 0 as its unity creates two new categories, $\mathfrak{UnitShell}$ and $\mathfrak{SheafUnitShell}$ as full subcategories of $\mathfrak{UnitHalfShell}$ and $\mathfrak{SheafUnitHalfShell}$, respectively. The next corollary is a generalization of theorem 6.6 of [**Pier67**].

6.5. Corollary. *The categories* $\mathfrak{UnitShell}$ *and* $\mathfrak{SheafUnitShell}$ *are equivalent.*

We noted earlier that a homomorphism does not always take factor elements to factor elements. However, when a homomorphism of unital shells is surjective, then it is conformal, as shown next.

6.6. Proposition. *If a homomorphism,* $\varphi \colon \boldsymbol{A} \to \boldsymbol{B}$, *of unital shells is surjective, then*

$$\varphi(\mathrm{Elem}'\,\boldsymbol{A}) \subseteq \mathrm{Elem}'\,\boldsymbol{B}.$$

PROOF. This inclusion is true since all the equations defining factor elements in Theorem 3.4 easily go over from $\boldsymbol{A}$ to $\boldsymbol{B}$. □

A consequence of this observation is that, in a subdirect representation, the canonical homomorphisms from a unital shell onto its quotients are conformal.

VIII
Baer–Stone Shells

> Von Neumann's major contribution to lattice theory centered around his concepts of a 'continuous geometry' and of a 'regular ring'.
>
> *Garrett Birkhoff* [**Birk58**, p. 50]

Out of the concept of von Neumann regular ring comes the more general notion of Baer ring. Out of the concept of Boolean algebra comes the more general notion of Stone lattice. A further common generalization of these two notions we label a 'Baer–Stone shell'.

This chapter presents several applications of the theory developed earlier. For every Baer–Stone half-shell $\boldsymbol{A}$ (that is two-sided and unital) there is a reduced and factor-transparent sheaf $\boldsymbol{\mathcal{A}}$ over a Boolean space with the stalks having no divisors of zero such that $\boldsymbol{A} \cong \boldsymbol{\Gamma}(\boldsymbol{\mathcal{A}})$. With all that has been done in previous chapters, the proof is relatively short. The bulk of the work in the first section lies in showing that every regular congruence is integral. Just as the results of the previous chapters may be cast into categories, so we restate this as the equivalence of two categories.

In the second section, there are two more applications. Each von Neumann regular, commutative and unital ring is isomorphic to the ring of all global sections of a sheaf of fields over a Boolean space. And every biregular ring is so represented by a sheaf with simple stalks over a Boolean space. These results also go over to half-shells when their terms are refined appropriately.

A. Knoebel, *Sheaves of Algebras over Boolean Spaces*,
DOI 10.1007/978-0-8176-4642-4_VIII,

To avoid continual repetition, some of the adjectives modifying shell, like 'two-sided', 'unital' and others, will be omitted occasionally in the middle of proofs. Throughout the chapter, Sect. VII.2 is employed extensively.

1. Integrality

Annihilators and divisors of zero are the focus of this section. The adjective 'Baer–Stone' means that each annihilator of a half-shell is a factor ideal. After formulating several lemmas and propositions, we prove Theorem 1.13 that two-sided unital Baer–Stone half-shells $\boldsymbol{A}$ are representable by sheaves with stalks having no divisors of zero. By adding a loop to the operations, the sheaf becomes a Boolean product.

1.1. Definition. In a half-shell, $\boldsymbol{A} = \langle A; \times, 0, \ldots, \omega, \ldots \rangle$, we define the **right annihilator** of an element a of A by

$$a^{\perp} = \{b \in A \mid ab = 0\};$$

and the **left annihilator** by

$$^{\perp}a = \{b \in A \mid ba = 0\}.$$

So to speak, $^{\perp}a$ gives all one-sided quotients $0/a$, and $a^{\perp}$ gives all quotients on the other side, $a\backslash 0$.

Here is a notation for the **principal ideal** generated by an arbitrary element a in a half-shell $\boldsymbol{A}$:

$$[a] \text{ is the smallest ideal } 0/\theta \quad \text{for which} \quad a\,\theta\,0 \qquad (\theta \in \operatorname{Con}\boldsymbol{A}).$$

This is equivalent to $[a]$ being the intersection of all ideals containing a:

$$[a] = \bigcap \left\{ \frac{0}{\theta} \;\middle|\; a\,\theta\,0 \text{ and } \theta \in \operatorname{Con}\boldsymbol{A} \right\}.$$

Do not confuse this notation with $[a]_x$ in Sect. IV.2 for the component of a in the xth stalk of a sheaf. We state a crucial property of $[a]$.

1.2. Proposition. *If e is a factor element in a unital half-shell $\boldsymbol{A}$, then $[e] = eA$.*

Proof. $[e] \subseteq eA$. Suppose $a \in \bigcap\{0/\theta \mid \theta \in \operatorname{Con}\boldsymbol{A}$ and $e\,\theta\,0\}$. Since $e\,\theta_{e'}\,0$, it follows that $a \in 0/\theta_{e'}$. Therefore, $a \in eA$, by Theorem VII.2.11d.

$eA \subseteq [e]$. If $a \in eA$, then $ea = a$. For any θ in $\operatorname{Con}\boldsymbol{A}$ for which $e\,\theta\,0$, we have that $a = ea\,\theta\,0a = 0$. Hence, a belongs to the intersection of all such $0/\theta$. □

1.3. Definition. A half-shell $\boldsymbol{A}$ has the **Baer–Stone** property if each right annihilator of a single element is a factor ideal and likewise for any left annihilator.

Since factor ideals are generated by unique factor elements in a unital half-shell, being Baer–Stone is equivalent to these two conditions together:

$$\forall a \in A\ \exists a^* \in \mathrm{Elem}'\,\boldsymbol{A}\ (a^\perp = [a^*]),$$
$$\forall a \in A\ \exists {}^*a \in \mathrm{Elem}'\,\boldsymbol{A}\ ({}^\perp a = [{}^*a]).$$

Examples of Baer–Stone shells are commutative Baer rings[1] [**Kist63**, pp. 40, 45] and Stone lattices[2] [**Birk67**, p. 130]. In fact, a bounded distributive lattice is a Stone lattice iff it is a Baer–Stone shell in our sense [**GrätSc57**, theorem 3]. Specific examples of these are Boolean lattices and the ring of all n-by-n matrices over a field. This concept was also studied in semigroups [**Keim71a**]. The five-element modular lattice M_3 is an example of a unital half-shell that is not Baer–Stone.

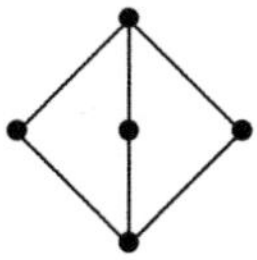

$\boldsymbol{M}_3$

The Venn diagram of Fig. 1 illustrates the relationships between the various notions. The intersection contains only one-element algebras.

We show next for certain Baer–Stone half-shells that $a^* = {}^*a$; this yields our previously defined complements of factor elements.

1.4. PROPOSITION. *Let* $\boldsymbol{A}$ *be a two-sided unital half-shell that is Baer–Stone, and let* a *be an element determining factor elements* a^* *and* ${}^*a$*, as they are given after Definition 1.3. Then,*

(a) $$a^* = {}^*a \qquad (a \in A),$$

(b) $$e^* = e' = {}^*e \quad (e \in \mathrm{Elem}'\,\boldsymbol{A}).$$

PROOF. (a) First $0 = ({}^*a)a = a({}^*a)$ since ${}^*a \in \mathrm{Elem}'\,\boldsymbol{A}$. Hence, ${}^*a \in a^\perp$, which is a^*A. Since a^*, as a factor element, is the unity of the factor ideal it generates, we conclude that ${}^*a = ({}^*a)(a^*)$. Likewise, $a^* = (a^*)({}^*a) = ({}^*a)(a^*) = {}^*a$, by commutativity of factor elements.

(b) To show first that $e^\perp = [e']$, we reason this way for any b in A:

$$b \in e^\perp \text{ iff } eb = 0 \text{ iff } e'b = b \text{ iff } b \in [e'],$$

---

[1]Kaplansky's [**Kapl68**, p. 3] original definition of a Baer ring differs from that of Kist's and Hofmann's [**Hofm72**, pp. 327–8], and is both more and less than that of being Baer–Stone in our sense. Kaplansky requires that the annihilator of each subset of $\boldsymbol{R}$ be a principal ideal generated on one side by an idempotent, which nevertheless need not be central. In a short personal history near the beginning of his book on rings of operators, Kaplansky [**Kapl68**] writes that this is in honor of Reinhold Baer's [**Baer52**] early use of the condition.

[2]This term arose when Grätzer and Schmidt [**GrätSc57**] answered a question of Stone.

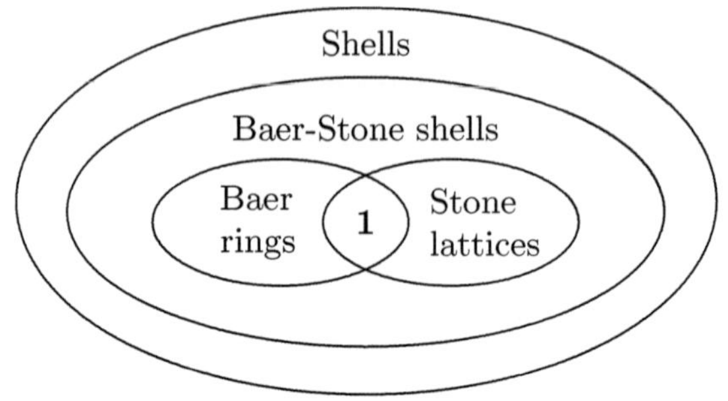

FIGURE 1. Special Shells.

courtesy of Proposition VII.2.22. Then the generators of the factor ideals, $e^{\perp}$ and $[e']$, must be equal: $e^* = e'$. □

So let us agree, under these circumstances, to replace the stars for left and right complements by the prime, thereby extending the notation′ to all elements. Thus, if a is in A, then a' is the unique factor element such that

$$a^{\perp} = [a'] = {}^{\perp}a.$$

It follows that for any factor element e in a Baer–Stone, two-sided, unital half-shell,

$$[e] = (e')^{\perp}.$$

This discussion shows that annihilators are synonymous with factor ideals.

1.5. PROPOSITION. *In a two-sided unital half-shell that is Baer–Stone,*

$$\{a^{\perp} \mid a \in A\} = \text{Ideal}' \boldsymbol{A}.$$

Being Baer–Stone is equivalent to a superficially stronger condition. First, extend the annihilator from single elements to arbitrary subsets B of A:

$$B^{\perp} = \{a \in A \mid ba = 0 \text{ for all } b \text{ in } B\} = \bigcap_{b \in B} b^{\perp}.$$

1.6. PROPOSITION. *A two-sided unital half-shell* $\boldsymbol{A}$ *is Baer–Stone if, and only if, each annihilator of any finite subset* F *of* A *is a factor ideal.*

PROOF. We know by the preceding proposition that $b^{\perp} \in \text{Ideal}' \boldsymbol{A}$. But the latter is a Boolean algebra and hence closed to finite intersections. □

Here is a proposition giving six equivalent formulations of annihilation.

1.7. PROPOSITION. *In a two-sided unital half-shell that is Baer–Stone, these statements are equivalent for all pairs of elements* a *and* b *in* A*:*

(a) $ab = 0$,
(b) $ba = 0$,
(c) $a'b = b$,
(d) $a''b = 0$,

(e) $a \in [b']$,
(f) $b \in [a']$.

PROOF. We apply Corollary VII.2.17, which states that for complementary factor element e and e' of $\boldsymbol{A}$,

$$b \in [e] \quad \text{iff} \quad eb = b \quad \text{iff} \quad e'b = 0 \qquad (b \in A).$$

By the definition of Baer–Stone,

$$ab = 0 \quad \text{iff} \quad b \in [a'] \quad \text{iff} \quad a'b = b \quad \text{iff} \quad a''b = 0.$$

since $a' \in \operatorname{Elem}' \boldsymbol{A}$. Thus equations (a), (f), (c), and (d) are equivalent. So are the remaining dual equations since $a^{\perp} = {}^{\perp}a$. □

In this next corollary, the third part has a tricky proof, reminiscent of bouncing back and forth in a Galois connection, up to three times. It is crucial to proving that every regular congruence of a shell is integral.

1.8. COROLLARY. *In a Baer–Stone, two-sided and unital half-shell $\boldsymbol{A}$, these three statements always hold for all a, b in A and e in* $\operatorname{Elem}' \boldsymbol{A}$*:*

(a) $a \in [e]$ *if, and only if,* $a'' \in [e]$*;*
(b) $a \in [a'']$*;*
(c) $a''b'' = (ab)''$.

PROOF. We use Proposition 1.7 and its proof repeatedly.
(a) $a \in eA$ iff $ae' = 0$ iff $a''e' = 0$ iff $a'' \in eA$.
(b) Substitute a'' for e in part (a).
(c) We show separately that each side is equal to $a''b''(ab)''$. First, from the proposition and the fact that $ab(ab)' = 0$, it follows that $a''b(ab)' = 0$, and hence $a''b''(ab)' = 0$, since Boolean elements commute and associate by Proposition VII.2.22. Therefore, $a''b''(ab)'' = a''b''$.
Second, the equalities,

$$a'(ab)=(a'a)b=0b=0$$

imply that $a'(ab)''=0$. Hence $a''(ab)''=(ab)''$. Similarly, $b''(ab)''=(ab)''$. Therefore,

$$a''b''(ab)'' = a''(ab)'' = (ab)''.$$

Putting everything together, we find that $a''b'' = a''b''(ab)'' = (ab)''$. □

1.9. DEFINITION. A nontrivial half-shell $\boldsymbol{A}$ is said to have **no divisors of zero** if, for all a and b in A,

$$\text{whenever } ab = 0, \text{ then } a = 0 \text{ or } b = 0.$$

Synonymously, $\boldsymbol{A}$ is an **integral** half-shell.

This definition should remind one of integral domains. In the case of noncommutative rings, some say instead that $\boldsymbol{A}$ is 'dense', 'entire', or 'completely prime'. Any chain with a nullity, thought of as a lattice, has no divisors of zero. In fact, a lattice, with a least element 0, has no divisors of zero if, and only if, 0 is meet-irreducible, or equivalently, if, and only if, the lattice is weakly irreducible. (See [**Birk44**] for definitions.)

We extend the notion of integrality to ideals and congruences. Recall that an ideal I of a half-shell $\boldsymbol{A}$ is the 0-coset of some congruence θ of $\boldsymbol{A}$, that is, $I = 0/\theta$.

1.10. Definition. A proper ideal I of a half-shell $\boldsymbol{A}$ is **integral** if,

$$\text{whenever } ab \in I, \text{ then } a \in I \text{ or } b \in I \qquad (a, b \in A).$$

A congruence θ of a half-shell $\boldsymbol{A}$ is **integral** if its ideal $0/\theta$ is integral.

To capture divisors of zero before a quotient is taken motivates this definition. We had hoped to avoid introducing a new term by a straightforward adoption of the classical notion of prime ideal in ring theory.[3] Unfortunately, only in commutative ring theory do these two notions coincide. In noncommutative ring theory some prime ideals are not integral. For an example, see [**Hung74**, p. 133, ex. 9b].

This terminology parallels commutative ring theory in that, in an integral domain, the smallest ideal $\{0\}$ is integral by our definition. More generally, we observe that a half-shell is **integral** iff the zero ideal $\{0\}$ is; that is, the half-shell has no divisors of zero. In another parallel, just as a ring is semisimple if it is a subdirect product of simple rings, we say that a half-shell is **semi-integral** if it is a subdirect product of integral half-shells.

1.11. Proposition. *In any half-shell A, a congruence θ of A is integral if, and only if, A/θ is integral.*

Proof. This is a straightforward application of the definitions. □

Recall from Definition VI.3.3 that a congruence θ of a unital half-shell $\boldsymbol{A}$ is 'regular' if there is an ideal N of $\mathbf{Con}'\,\boldsymbol{A}$ such that $\theta = \bigvee N$. Its ideal $0/\theta$ is also called 'regular'. Remember too that a 'maximal regular congruence' is maximal only with regard to all regular congruences. Maximal regular congruences are those coming from prime ideals of factor congruences, as demonstrated in Proposition VI.3.4.

1.12. Proposition. *In a Baer–Stone, two-sided and unital half-shell, each maximal regular congruence is integral, and so is each maximal regular ideal.*

Proof. Let θ be a maximal regular congruence, with $\boldsymbol{P}$ being the prime ideal of $\mathbf{Con}'\,\boldsymbol{A}$ such that $\theta = \bigvee \boldsymbol{P}$. Suppose $ab \in 0/\theta$. Since

[3]An ideal $\boldsymbol{P}$ of a ring is *prime* if, whenever $\boldsymbol{JK} \subset \boldsymbol{P}$ for some ideals $\boldsymbol{J}$ and $\boldsymbol{K}$, then $\boldsymbol{J} \subset \boldsymbol{P}$ or $\boldsymbol{K} \subset \boldsymbol{P}$. Here $\boldsymbol{JK}$ is complex multiplication.

$\bigvee P = \bigcup P$ by Theorem V.1.5, we then have that $ab\,\eta\,0$ for some η in P. By Proposition VII.2.11d, $ab \in [e_\eta{}']$. Therefore, by Corollary 1.8a,c,

$$a''b'' = (ab)'' \in [e_\eta{}'].$$

By the anti-isomorphism of **Elem**$'$ $\boldsymbol{A}$ and **Ideal**$'$ $\boldsymbol{A}$ in Sect. VII.4,

$$[a''] \cap [b''] = [a''b''] \subseteq [e_\eta{}'].$$

Since $e_\eta{}'A \in 0/P$ where $0/P = \{0/\theta \mid \theta \in P\}$, and $0/P$ is an ideal of ideals,

$$[a''] \cap [b''] \in \frac{0}{P}.$$

By the primeness of this ideal,

$$[a''] \in \frac{0}{P} \text{ or } [b''] \in \frac{0}{P}.$$

Then by Corollary 1.8 again and Proposition VI.3.4a,

$$a \in [a''] \subseteq \bigvee \frac{0}{P} = \frac{0}{\bigvee P} = \frac{0}{\theta},$$

or similarly $b \in 0/\theta$. □

The next theorem applies the sheaf representation of unital half-shells to yield stalks with no divisors of zero. Earlier versions are found in [**Knoe72**, **Knoe00**]. Swamy [**Swam80**] proved it for semigroups.

1.13. THEOREM. *Every two-sided unital half-shell that is Baer–Stone is isomorphic to the unital half-shell of all global sections of a reduced and factor-transparent sheaf over a Boolean space of half-shells without divisors of zero.*

PROOF. Recall Theorem VII.2.5: each unital half-shell is isomorphic to the algebra $\boldsymbol{\Gamma}(\boldsymbol{\mathcal{A}})$ of all global sections of a reduced and factor-transparent sheaf $\boldsymbol{\mathcal{A}}$ with stalks over a Boolean space, $\boldsymbol{X} =$ **Spec Con**$'$ $\boldsymbol{A}$. In this sheaf each stalk is a quotient $\boldsymbol{A}/\theta$ by a maximal regular congruence θ, which is the supremum of a prime ideal of **Con**$'$ $\boldsymbol{A}$. Since we have just proven in Proposition 1.12 that maximal regular congruences in $\boldsymbol{A}$ are integral, the quotients have no divisors of zero. □

When Theorem 1.13 is applied to Stone lattices, we may compare it to their representation by subdirect products. Balbes and Horn [**BalHo70**] prove the theorem in this setting. A Stone lattice, by traditional definition, is any pseudo-complemented distributive lattice $\boldsymbol{A}$ in which

$$a' \vee a'' = 1 \qquad (a \in A).$$

In their theorem 2, Grätzer and Schmidt [**GrätSc57**] tell us that any bounded distributive lattice is a Stone lattice in this sense if, and only if, it is Baer–Stone in our sense. Grätzer [**Grät69**] has shown that, up to isomorphism, there are only two subdirectly irreducible Stone lattices: the two- and three-element chains, $\boldsymbol{C}_2$ and $\boldsymbol{C}_3$. In our representation the stalks may have more than three elements; nevertheless, the stalks will have no divisors of zero See also [**Grät98**, pp. 112–] and [**Prie74**].

We can relate direct indecomposability to integrality. This last theorem shows that a Baer–Stone half-shell that is unital and two-sided is directly indecomposable if, and only if, it is integral.

More generally, we can only prove for an arbitrary unital half-shell that, if it is integral, then it is directly indecomposable. For suppose, on the contrary, that it is decomposable. Then there are non-zero factor elements e and e' for which $ee' = 0$. Hence, the half-shell has divisors of zero.

The bounded lattice $\boldsymbol{M}_5$ is a counterexample to the converse. See [**BalHo70**] for more results of this kind about Stone lattices.

We now entertain the question about when the Baer–Stone property leads to a Boolean product. It appears that we must insist on having a group so that we can translate sections about. Actually, the full strength of a group is not needed; a loop will do.

Recall from [**Bruck58**] that a **loop** is a half-shell $\langle A; +, 0\rangle$ satisfying these axioms:

(i) $\forall a, b \in A\ \exists! x \in A\ (x + a = b)$,
(ii) $\forall a, b \in A\ \exists! y \in A\ (a + y = b)$,
(iii) $\forall a \in A\ (a + 0 = a = 0 + a)$.

Although written additively, the operation $+$ need not be commutative. As in group theory, the x and y in the first two axioms may be different for the same a and b. These axioms about unique existence define new binary operations:

$$b / a = x \quad \text{and} \quad a \backslash b = y.$$

We would like to use the sign for difference '$-$' to go along with '$+$', but two kinds of subtraction are needed, so we adopt the two slants that are almost horizontal. In group theory, the inverse operations would be written multiplicatively as ba^{-1} and $a^{-1}b$, respectively; but we must be more discriminating about inverses in loop theory, as the next example shows.

A non-commutative loop on the set $\{0, 1, 2, 3, 4\}$ that is not a group is given by its Cayley table in Fig. 2. Unlike groups, loops may have inverses that are not two-sided. For example, $0/3 = 1 \neq 4 = 3\backslash 0$.

$+$	0	1	2	3	4
0	0	1	2	3	4
1	1	2	4	0	3
2	2	3	0	4	1
3	3	4	1	2	0
4	4	0	3	1	2

FIGURE 2. Cayley Table of a loop with unequal inverses.

The following laws do hold in loops:

$$(b/a) + a = b,$$
$$a + (a\backslash b) = b,$$
$$a/a = 0 = a\backslash a,$$
$$a/0 = a = 0\backslash a.$$

The theory of loops can be given a strictly equational definition; this tells us that subloops and homomorphic images are again loops.

With the concept of a loop in hand, we now turn to strengthening Theorem 1.13; we state it as two logical equivalences to make clearer where clopen sets play their part. Recall that a Boolean product is a sheaf over a Boolean space in which each equalizer is not only open but also closed; equivalently by Corollary V.3.1, a Boolean product is a Hausdorff sheaf over a Boolean space.

1.14. PROPOSITION. *Consider a half-shell,* $\boldsymbol{A} = \langle A; \times, 0, 1, \ldots, \omega, \ldots\rangle$, *that is unital and two-sided. Let* $\mathcal{A}$ *be the resulting sheaf* $\boldsymbol{A} /\!\!/ \mathbf{Con}' \boldsymbol{A}$ *over the Boolean space,* $\boldsymbol{X} = \mathbf{Spec}\,\mathbf{Con}'\,\boldsymbol{A}$, *constructed as in Sect. V.2.*

(a) $\boldsymbol{A}$ *is Baer–Stone if, and only if, each stalk is integral and the equalizer* $\sigma{:}0$ *is clopen for any global section* σ *of* $\boldsymbol{\Gamma}(\boldsymbol{A})$.
(b) *Assume there is a term* $+$ *of* $\boldsymbol{A}$ *such that the algebra* $\langle A; +, 0\rangle$ *is a loop. Then* $\boldsymbol{A}$ *is Baer–Stone if, and only if,* $\boldsymbol{A}$ *is a Boolean product of two-sided unital half-shells that are integral.*

PROOF. (a) $\Rightarrow$. Integrality has already been proven in Theorem 1.13. In order to demonstrate clopeness, let σ be any global section of $\boldsymbol{\Gamma}(\mathcal{A})$. Since $\boldsymbol{A}$ is Baer–Stone and hence isomorphic to $\boldsymbol{\Gamma}(\mathcal{A})$, there is a global section ε that is a factor element of $\boldsymbol{\Gamma}(\mathcal{A})$ and such that $\sigma^{\perp} = [\varepsilon]$. Because any stalk is integral, it is indecomposable as a product; hence, for any x in the base space, $\varepsilon(x)$ is 0 or 1, and so $\varepsilon{:}0 = \varepsilon'{:}1$. Therefore, both $\varepsilon{:}0$ and $\varepsilon{;}0$ are clopen. Since each stalk of $\mathbf{A}$ has no divisors of zero, it follows that

$$\sigma{;}0 \cap \varepsilon{;}0 = \emptyset.$$

To establish that these sets are complementary, it suffices to show also that

$$\sigma{;}0 \cup \varepsilon{;}0 = X.$$

To that end suppose, by way of contradiction, that $x_0 \in X$ but that $\sigma(x_0) = 0 = \varepsilon(x_0)$. Since $\sigma{:}0$ and $\varepsilon{:}0$ are open, there is a basic clopen subset U of $\sigma{:}0 \cap \varepsilon{:}0$ containing x_0. Let ρ be a global section such that $\rho(x_0) \neq 0$; this exists by Proposition VII.2.9(c) and by the interpolation property stated in Proposition V.2.4(b). Define a new global section τ by patching:

$$\tau(x) = \begin{cases} \rho(x) & \text{if } x \in U, \\ 0 & \text{if } x \in X \sim U. \end{cases}$$

Then $\sigma(x)\tau(x) = 0$ for all x in X, and so $\sigma\tau = 0$. Hence $\tau \in \sigma^{\perp}$. But $\tau(x_0) = \rho(x_0) \neq 0 = \varepsilon(x_0)$. Therefore, $\tau \neq \varepsilon\tau$, meaning that $\tau \notin [\varepsilon]$, thereby violating **A** being Baer-Stone.

Thus, $\sigma;0$ and $\varepsilon;0$ are complementary open subsets of X. Since $\sigma;0$ is always closed, $\sigma:0$ is clopen.

$\Leftarrow$. Let σ be any global section in $\boldsymbol{\Gamma}(\mathcal{A})$. We define a new global section $\tau_\circ\colon X \to \mathcal{A}$ by patching over clopen subsets of $\boldsymbol{X}$: $\tau_\circ = 0|(\sigma;0) \cup 1|(\sigma:0)$. Otherwise stated,

$$\tau_\circ(x) = \begin{cases} 0 & \text{if } \sigma(x) \neq 0, \\ 1 & \text{if } \sigma(x) = 0. \end{cases}$$

It is straightforward to check that

$$\sigma^{\perp} = \{\tau \in \boldsymbol{\Gamma}(\mathcal{A}) \mid \tau_\circ{:}0 \subseteq \tau{:}0\},$$

since the stalks have no zero divisors. Clearly, $\tau_\circ \in \mathrm{Elem}'\,\boldsymbol{\Gamma}(\mathcal{A})$ since $\tau_\circ$ and $\tau_\circ{}'$ satisfy all the clauses of Proposition VII.2.16 defining a factor element. Most importantly, $\sigma^{\perp} = [\tau_\circ]$, confirming Baer–Stone:

$$\tau \in \sigma^{\perp} \text{ iff } \tau_\circ{:}0 \subseteq \tau{:}0 \text{ iff } \tau_\circ\tau = \tau \text{ iff } \tau \in [\tau_\circ].$$

(b) $\Rightarrow$. Since $\boldsymbol{A} \cong \boldsymbol{\Gamma}(\mathcal{A})$ for a Baer-Stone $\boldsymbol{A}$, it suffices to work with global sections in $\boldsymbol{\Gamma}(\mathcal{A})$ to prove that $\mathcal{A}$ is Hausdorff. We need only show that each equalizer is closed. For any two global sections, σ and τ, it is easy to see from the loop structure that $\sigma:\tau = (\sigma/\tau):0$. Hence, we have reduced the problem to showing that $\rho:0$ is closed for any global section ρ. But this is (a).

$\Leftarrow$. See (a) again. □

This result is given for unital rings by Peercy [**Peer70**, theorem 2.2.7] and Hofmann [**Hofm72**, pp. 327–328]. See also Krauss and Clark [**KraCl79**, pp. 93–96], for a different approach to some of the results of this section when the Baer–Stone shells are rings.

These special shells fit into the hierarchy of categories mentioned at the end of Sect. VI. To save space, we omit the adjectives 'two-sided' and 'unital' from their names. Let **BaerStoneShell** be the category of all Baer-Stone shells that are two-sided and unital, with conformal homomorphisms as the morphisms; and let **SheafShellIntegral** be the category of sheaves over Boolean spaces of two-sided unital shells that are integral. Then these are full subcategories of **UnitHalfShell** and **SheafUnitHalfShell**, respectively; and, therefore, by the theorems of this section and Sect. VII.6, they are categorically equivalent: **BaerStoneShell** $\simeq$ **SheafShell–Integral**.

This immediately specializes to the classical equivalences of **BaerRing** and **SheafRingIntegral**, the categories of Baer rings and sheaves of integral rings over Boolean spaces, respectively, where the rings are assumed to have two-sided unities. In symbols, **BaerRing** $\simeq$ **SheafRingIntegral**.

These further specialize to the classical duality between Boolean algebras and Boolean spaces. Return to the end of Sect. VI.4 for a chart of all the categories to see how these new ones fit in with the old ones.

1.15. PROBLEM. In Baer–Stone unital half-shells, the existence of a unique factor element e generating the annihilator of an element a leads to a unary operation: $a \mapsto a^*$, where $a^* = e$ when $a^\perp = [e]$. Introduce this new operation $'$ into the type of Baer–Stone half-shells. Then conformal homomorphisms are the same as homomorphisms. Study the equational class generated by these new algebras; it would include Baer–Stone lattices and rings. What is a simple set of defining identities? Are there half-shells satisfying these identities that are not Baer–Stone? To get started, look up 'Baer*-rings'.

2. Regularity

This section addresses von Neumann regularity, strong regularity and biregularity. For a ring $\boldsymbol{R}$ to be **regular**[4] in the sense of von Neumann means that

$$\forall a \in R\, \exists r \in R\ (ara = a).$$

We start with a modern version with sheaves of a classical result.

2.1. THEOREM. *Each von Neumann regular, commutative and unital ring $\boldsymbol{R}$ is Baer–Stone. Therefore, $\boldsymbol{R}$ is isomorphic to the ring of all global sections of a reduced and factor-transparent sheaf of fields over a Boolean space.*

PROOF. Assume that a ring $\boldsymbol{R}$ satisfies the hypotheses. Let a be an element of R with a pseudoinverse r, that is, $ara = a$. One can verify that $1 - ar$ is an idempotent, and thus it is a factor element in a commutative ring.

To show that R is Baer–Stone, it suffices to prove that $a^\perp = [1 - ar]$. In one direction of inclusion, if $b \in a^\perp$, then $ab = 0$, and hence

$$b(1 - ar) = b - bar = b - 0r = b;$$

thus, $b \in [1-ar]$. In the other direction, if $b \in [1-ar]$, that is, $b(1-ar) = b$, then $b - arb = b$, and so $arb = 0$; hence, $ab = arab = arba = 0$.

That the stalks are fields follows readily from Theorem 1.13, which proves that the stalks are both integral and regular. The solution in a stalk to $ax = 1$ when $a \neq 0$ is found in the equation $ara = a$, since then $a(ra-1) = ara - a = 0$ and hence $ra = 1$. □

[4]This concept differs from the regular ideals and congruences studied previously.

Without sheaves, this theorem is due to Köthe [**Köthe30**] and [**McCoy38**].

With the adjective 'commutative' removed from its statement, it becomes false. Put another way, in noncommutative ring theory, 'regular' and 'Baer–Stone' are incomparable notions. On the one hand, we know that the ring $\boldsymbol{M}_2(\mathbb{Q})$ of 2 by 2 matrices with rational entries is regular and simple, so the right annihilator of, say, $\left[\begin{smallmatrix}0&1\\0&0\end{smallmatrix}\right]$ cannot be a two-sided ideal; thus $\boldsymbol{M}_2(\mathbb{Q})$ is not a Baer–Stone ring. In short, since $\left[\begin{smallmatrix}0&1\\0&0\end{smallmatrix}\right]\left[\begin{smallmatrix}0&1\\0&0\end{smallmatrix}\right] = 0$, clearly $\boldsymbol{M}_2(\mathbb{Q})$ has divisors of zero. On the other hand, we know that the ring $\mathbb{Z}$ of integers is trivially a Baer–Stone ring since there are no divisors of zero; and yet the only elements with pseudo-inverses are 0 and 1, so $\mathbb{Z}$ is not regular. Therefore, any attempt to generalize Theorem 2.1 directly to noncommutative rings is doomed.[5]

But there are several ways to adapt Theorem 2.1 to some noncommutative rings; we mention the class of **strongly regular** rings. These are rings satisfying the property:

$$\forall a \in R\, \exists r \in R\ (aar = a),$$

which is equivalent to regularity when the ring is commutative. In general, any strongly regular, unital ring is both regular and biregular (Definition 2.2, but this is nontrivial to show.[6] The stalks in the sheaf representation of strongly regular, unital rings are division rings, in analogy with commutative regular rings where the stalks are fields.

Something similar in spirit is found in **near-rings**, $\langle A; +, \times, 0\rangle$, where $\langle A; +, 0\rangle$ is a group (not necessarily Abelian), $\times$ is a semigroup, and $\times$ distributes over $+$ on one side. These are useful in co-ordinatizing certain geometric planes and doubly transitive groups. In analogy with the set of all endomorphisms of an Abelian group forming a ring in a natural way, the set of all endomorphisms of a group, not necessarily Abelian, generates a near-ring. Stronger hypotheses allow one to prove more about the stalks: any von Neumann regular, unital near-ring without nilpotents is isomorphic to a subdirect product of near-fields. This is theorem 9.158 of [**Pilz83**, p. 346], which has references to the discoverers of this theorem.

We end this chapter with a representation theorem for some shells with a definition superficially similar to that of Baer–Stone shells. Our motivation for this is the paper of [**Szeto77**] on biregular near-rings, which in turn was motivated by the papers of Arens and Kaplansky [**AreKa48**] and Dauns and Hofmann [**DauHo66**], but whose sheaf is a Boolean product and different from ours.

---

[5] See the comments at the beginning of Sect. XII.5 for related notions and references.

[6] See [**AreKa48**] and [**Ledb77**]. See also [**ForMc46**] and [**Goode79**, p. 35].

2.2. Definition. A **biregular** half-shell $\boldsymbol{A}$ is a half-shell when every principal ideal is a factor ideal. If the half-shell has a unity, this amounts to:

$$\forall a \in A\ \exists e \in \mathrm{Elem}'\,\boldsymbol{A}\ ([a] = [e]).$$

(Remember that $[a]$ is the smallest ideal $0/\theta$ containing a.)

This is a direct generalization of what it means for a unitary ring to be biregular: each principal (two-sided) ideal is generated by a central idempotent.[7] The theorem coming up will apply to biregular shells in any class satisfying two rather broad conditions [**Knoe91a, Knoe00**]. These conditions are satisfied by rings, and more generally by shells in which the addition is a loop, which was defined in Sect. 1.

2.3. Theorem. *Consider a class $\mathfrak{C}$ of unital half-shells in which:*

(i) *Any homomorphic image of an algebra in $\mathfrak{C}$ is again in the class*

(ii) *Whenever θ is a congruence of an algebra in $\mathfrak{C}$ and $\theta \neq 0$, there is some non-zero element a such that $a\,\theta\,0$.*

Then any biregular shell $\boldsymbol{A}$ in $\mathfrak{C}$ is isomorphic to the unital half-shell of all global sections of a reduced and factor-transparent sheaf of simple unital half-shells over a Boolean space.

Proof. We use Theorem VII.2.5 to represent $\boldsymbol{A}$ by a reduced and factor-transparent sheaf. All that remains is to show that the stalks are simple.[8] A generic stalk is $\boldsymbol{A}/\zeta$ where ζ is the supremum of a prime ideal of factor congruences, that is, $\zeta = \bigvee P$ with $P \in \mathrm{Spec}\,\mathbf{Con}'\,\boldsymbol{A}$. By way of contradiction, assume that this stalk is not simple. With the isomorphism Theorem II.1.3 we know that $\mathbf{Con}(\boldsymbol{A}/\zeta)$ is isomorphic to the lattice of intermediate congruences: $\{\eta \in \mathrm{Con}\,\boldsymbol{A} \mid \eta \supseteq \zeta\}$. By nonsimplicity there exists an intermediate congruence η of $\boldsymbol{A}$ such that $\zeta \subset \eta \subset 1$. Condition (i) implies that $\boldsymbol{A}/\zeta$ satisfies (ii). And (ii) implies that there is an element a in A such that $a\,\eta\,0$ but *not* $a\,\zeta\,0$.

Define a new proper ideal of $\mathbf{Con}'\,\boldsymbol{A}$ as

$$N = \{\theta \in \mathrm{Con}'\,\boldsymbol{A} \mid \theta \subseteq \eta\}.$$

Corollary VII.2.10 characterizes members of this set by factor elements:

$$e\,\eta\,0 \ \text{ iff } \ \theta_{e'} \subseteq \eta \qquad (e \in \mathrm{Elem}'\,\boldsymbol{A}). \tag{2.1}$$

In particular, this logical equivalence also holds for ζ.

[7]However, when the shell is a lattice, this is nothing new. For, as Mai Gehrke pointed out, any biregular bounded lattice $\boldsymbol{A}$ is a Boolean lattice. This is true since $[a] = \{x | x \leq a\}$ in a lattice. Thus $[a] = [e]$ implies $a = e$.

[8]Consult the book [**Diers86**] for a strongly categorical approach to sheaves with simple stalks over Boolean spaces.

Now biregularity asserts the existence of a factor element e connected to the a found earlier: $[a] = [e]$. From Corollary VII.2.17 it follows that $a = ea$. Consequently,

$$a\,\theta\,0 \text{ iff } e\,\theta\,0 \qquad (\theta \in \operatorname{Con}\boldsymbol{A}),$$

because

$$a\,\theta\,0 \text{ iff } [a] \subseteq \frac{0}{\theta} \text{ iff } [e] \subseteq \frac{0}{\theta} \text{ iff } e\,\theta\,0.$$

So $e\,\eta\,0$ but not $e\,\zeta\,0$; and thus $\theta_{e'} \subseteq \eta$ but $\theta_{e'} \not\subseteq \zeta$, by (2.1). Hence, $\theta_{e'} \in N$ and $\theta_{e'} \notin P$. Thus $N \supsetneqq P$; so P is not maximal. Therefore, our assumption of nonsimplicity is wrong. The stalk $\boldsymbol{A}/\zeta$ must be simple. □

2.4. Problem. Can hypotheses (i) and (ii) be eliminated from the theorem just proven?

2.5. Problem. Try to convert the class of biregular unital half-shells to a variety by adding a unary operation, $a \mapsto a^*$, to the type, where a^* is the unique factor element e such that $[a] = [e]$. What is a simple set of defining identities? Are there half-shells satisfying these identities that are not biregular?

Another variant on regularity are the unital near-rings of [**Szeto77a**] in which each element is a nontrivial power of itself. Unitary near-rings are unital half-shells and hence Theorem VII.2.5 may be applied. He proved independently of this that any such near-ring is bijectively representable by sheaves of near-fields over Boolean spaces.

Still another variant are the biregular rings lacking a unity. They are represented in [**BurSa81**, p. 163] by sheaves not over Boolean spaces, but over locally compact, totally disconnected, Hausdorff spaces, sometimes called 'locally Boolean spaces'. The stalks are simple unital rings.

There are even more ways to represent biregular rings by sheaves; these are referenced in Sect. XII.5. Biregular semigroups were represented by Keimel [**Keim70**].

IX

Strict Shells

The ring $\mathbb{Z}$ of integers has many desirable properties, stimulating much research in more general systems. Among them are cancellation and no divisors of zero:

$$(0.1) \qquad md = nd \quad \text{implies} \quad m = n \quad \text{if} \quad d \neq 0;$$

$$(0.2) \qquad mn = 0 \quad \text{implies} \quad m = 0 \quad \text{or} \quad n = 0.$$

In a commutative ring these conditions are equivalent. In a half-shell, the first always implies the second, but the three element chain $\boldsymbol{C}_3$ that is a semilattice satisfies the second but not the first. So we concentrate on the more general notion (0.2) of having no divisors of zero, as we did in Chap. VIII.

There the hypothesis that shells be Baer–Stone is rather restrictive. Now we replace this by less limiting conditions involving implications between equations, which are usually easy to verify. For example, commutative rings without any nilpotent elements properly include Baer rings. Not surprisingly, to obtain a theorem comparable to Theorem VIII.1.13, we must weaken its conclusion: the algebra can now be represented only as a subalgebra of the algebra of all global sections of a sheaf over a Boolean space, rather than as an isomorphic image. Most importantly, the stalks will still have no divisors of zero. They are quotients by congruences induced by annihilators.

The first section introduces the basic notions and proves the theorem just summarized. This is done in the context of algebras that have only a multiplication and one constant, a nullity, which we call strict half-shells. When

A. Knoebel, *Sheaves of Algebras over Boolean Spaces*,
DOI 10.1007/978-0-8176-4642-4_IX,

there is neither another operation of addition nor a unity of multiplication, life is difficult but not impossible. We must assume the half-shells have no nilpotents and satisfy a condition called null-symmetry. These assumptions are implications between products that are zero.

The second section starts by exploring the consequence of having no divisors of zero when the sheaf is a Boolean product: the enveloping half-shell is then Baer–Stone. When the sheaf space is extremal disconnected, the enveloping half-shell is completely Baer–Stone. This section continues with an equivalent characterization of no divisors of zero. It concludes by showing that the assumed implications of Sect. 1 imply all other implications of null products.

The third section adds a unity to the half-shell; this makes for stronger and simpler conclusions. It then uses these results to prove a similar theorem when addition is included, that is, for shells themselves. However, for this to go over, it is necessary to make additional assumptions about the shells: addition is a loop operation and multiplication distributes over it. As a corollary we apply this theorem to clusters.

These last results show that the type of the algebra is important. We obviously need a multiplication and a nullity in order to define annihilators. But, since the congruences are induced by the annihilators, additional operations may foul these congruences and cause them to be too big. Therefore, more operations cannot be added without also adding more identities. We close this last section by comparing various notions of regularity in ring theory.

There is another way to look at this chapter and put it into context. We can make many different choices in identifying congruences that form a Boolean subsemilattice of **Con** $\boldsymbol{A}$. In Chaps. VII and VIII, it was factor congruences. In this chapter, it will be congruences induced by annihilators. In Sect. XI.2, it will be factor elements of **Con** $\boldsymbol{A}$, that is, congruences that are factor elements in the *lattice* of congruences.

1. Nilpotents and Null-symmetry

Null-symmetry is a new condition on products that are the nullity, which is needed to prove the results of this chapter, where the multiplication is not assumed to be commutative and associative. When these two laws are present, null-symmetry automatically follows and so its presence is masked in earlier presentations.[1] Roughly, null-symmetry means that if a product

[1]See, for example,the papers of Keimel [**Keim71a**], Kennison [**Kenn76**] and Kist [**Kist63, Kist69**]. See also those of Hofmann [**Hofm72**, p. 330] and Krauss and Clark [**KraCl79**, p. 97].

of elements of the half-shell is the nullity, than so is any rearrangement of the elements in the product.

We reintroduce annihilators and show that they form a Boolean subsemilattice of ideals. The minimal congruences obtained by relating all elements of each ideal provide a parallel Boolean subsemilattice of congruences. With this, a sheaf over a Boolean space is constructed, and under suitable hypotheses it is proven to have stalks with no zero divisors.

1.1. DEFINITION. A **strict half-shell** $\langle A, \times, 0\rangle$ with one binary operation $\times$ and a constant 0, with no other operations, that satisfies

$$(0\mathsf{N}\times) \qquad 0 \times a = 0 \qquad (a \in A).$$

It is a **strict unital half-shell** if it is an algebra $\langle A, \times, 0, 1\rangle$ that also satisfies

$$(1\mathsf{U}\times) \qquad 1 \times a = a \qquad (a \in A).$$

They are **two-sided** if $a \times 0 = 0$ for both, and $a \times 1 = a$ for the second. As before we abbreviate $\times$ by juxtaposition: $ab = a \times b$.

This chapter will invoke various combinations of these hypotheses, postulated to hold in a strict half-shell $\boldsymbol{A}$ for all a, b, c and d in A:

(I) if $a^n = 0$, then $a = 0 \quad (n \in \mathbb{Z}^+)$;

(II) if $ab = 0$, then $ba = 0$;

(IIIa) $a(bc) = 0$ if, and only if, $(ab)c = 0$;

(IIIb) if $a(bc) = 0$, then $a(cb) = 0$;

(IV) if $(ab)(cd) = 0$, then $(ac)(bd) = 0$.

In (I), we are assuming association to the left: $a^n = \big(((aa)a)\cdots a\big)a$.

The following are typical consequences of clauses (I)–(IV):

$$\begin{aligned}
&\text{if } ab = 0, \ \text{ then } \ b(ac) = 0;\\
&\text{if } a(ab) = 0, \ \text{ then } \ b(ba) = 0;\\
&\text{if } (ab)(cd) = 0, \ \text{ then } \ (ac)(db) = 0;\\
&\text{if } \big(a\big(a((be)c)\big)\big)d = 0, \ \text{ then } \ a\big(b(c(de))\big) = 0;\\
&\text{if } \big(a\big(a\big(a\cdots(aa)\ldots\big)\big)\big) = 0, \ \text{ then } \ a = 0.
\end{aligned}$$

Some of these may not be as easy to prove as they look; one has to exercise caution here. More complicated implications of null products will follow but only with some work. Even more unexpectedly, postulate IV does not appear to follow from the previous ones.

Higher-order implications and their relationships will be addressed in Sect. 2. The casual reader may assume all these hypotheses for the principal theorems of Sects. 1 and 3, and skip the latter part of Sect. 2 on the axiomatics of such conditions.

For the time being, in order to focus our discussion, we need to attach some names to various combinations of these clauses.

1.2. DEFINITION. When (I) is satisfied in a half-shell, we say that it has no nilpotents, a **nilpotent** being any non-zero element some power of which is zero. For the remaining postulates, we say that a half-shell is **null-symmetric** whenever (II), (IIIa), (IIIb) and (IV) all hold. By a **null product** we mean any product like these that is equal to zero.

Normally, nilpotence will appear in the presence of the other implications; therefore, even though we are not assuming the associative law, any association within a null power of an element typically implies that the element is the nullity.

Two examples of half-shells satisfying all these conditions are: semilattices with a minimum element, and nullital commutative semigroups without nilpotents.

A commutative ring satisfying (I) is called **semi-prime**. Classical ring theory characterizes a semi-prime ideal as an intersection of prime ideals [**McCoy64**, p. 69]. That an ideal is prime in the commutative case assures us that the quotient ring will have no zero divisors. In the noncommutative rings, the concepts of primeness and lack of zero divisors diverge. For example, the ring $\boldsymbol{M}_2(\mathbb{Q})$ of all 2-by-2 matrices over the rational numbers is simple, and hence prime. But if

$$A = \begin{pmatrix} 0 & 0 \\ 0 & 1 \end{pmatrix}, \text{ and } B = \begin{pmatrix} 0 & 1 \\ 0 & 0 \end{pmatrix},$$

then $AB = 0$. Thus the ring has zero divisors. Also note that $BA \neq 0$; so (II) is violated [**Hung74**, p. 133, ex. 9]; hence, $\boldsymbol{M}_2(\mathbb{Q})$ is not null-symmetric.

Incidentally, axiom II makes any half-shell $\boldsymbol{A}$ two-sided:

$$0a = 0 = a0 \qquad (a \in A).$$

Our working hypotheses of null-symmetry and no nilpotents, that is, clauses I to IV, are at the level of elements, and, with fewer operations and identities satisfied, they are of necessity more involved. This new class of algebras, not quite in keeping literally with noncommutative ring theory, but nevertheless in the spirit of commutative ring theory, is justified by the theorem to come.

In anticipation of breaking half-shells down by means of sheaf constructions, we make a few comments about the nature of these clauses. It would be convenient in quotient half-shells to be able to use the implications between equations that define null-symmetry. Unfortunately, implications are not necessarily preserved by homomorphism. In other words, universal sentences of this form define only a quasivariety, not a variety. Put another way, quasivarieties are closed to the construction of subdirect products, but a subdirect decomposition of an algebra in a quasivariety may produce factors outside the quasivariety.

We may now state precisely the main result of this section. For it, we need a name for sheaves whose base space is dually isomorphic to a complete Boolean algebra. Such spaces are called **extremally disconnected** in Definition III.4.3: the closure of any open set is again open.

1.3. THEOREM. *Any null-symmetric strict half-shell $\boldsymbol{A}$ without nilpotents is isomorphic to a subalgebra of the half-shell $\boldsymbol{\Gamma}(\mathcal{A})$ of all global sections of a sheaf $\mathcal{A}$ over an extremally disconnected Boolean space with stalks that have no divisors of zero.*

The proof is long and splits naturally into three parts, involving a variety of constructions, subordinate propositions and lemmas. First, we single out annihilators of subsets of the carrier and show that they form a complete Boolean semilattice of ideals. Second, from these annihilators we create congruences, which in turn will form a complete Boolean subsemilattice of $\mathbf{Con}\,\boldsymbol{A}$. By Theorem V.2.1, it will follow readily that $\boldsymbol{A}$ is represented by a subalgebra of a sheaf over a Boolean space. The third point to prove is that the stalks of this sheaf have no zero divisors. The whole proof is drawn together after Proposition 1.14.

First, we define idéals of strict half-shells, which will turn out to be kernel ideals of Sect. II.1, and which are slightly different from the idéals of Sect. VII.3 since now there is no addition. This new notion is a stepping stone to annihilators, and will disappear in a few pages.

1.4. DEFINITION. An **idéal** of a half-shell is a subset I of $\boldsymbol{A}$ such that:

(i) $0 \in I$, and

(ii) if $a \in A$ and $i \in I$, then $ai \in I$ and $ia \in I$.

Denote the set of all idéals of a half-shell $\boldsymbol{A}$ by Idéal A.

In a half-shell $\boldsymbol{A}$, when I and J are idéals of it, the intersection $I \cap J$ and union $I \cup J$ are also idéals. Thus, the set of idéals is a bounded lattice:

$$\textbf{Idéal}\,A = \langle \text{Idéal}\,\boldsymbol{A}; \cup, \cap, \{0\}, A\rangle.$$

1.5. DEFINITION. For any subset B of a half-shell $\boldsymbol{A}$, define the **right annihilator** of B to be

$$B^{\perp} = \{a \in A \mid \text{ if } b \in B \text{ then } ba = 0\}.$$

The **left annihilator** is naturally

$$^{\perp}B = \{a \in A \mid \text{ if } b \in B \text{ then } ab = 0\}.$$

1.6. PROPOSITION. *In a half-shell $\boldsymbol{A}$, the relation, $ab = 0$, is a polarity, and hence we obtain a Galois connection in which the operation of left annihilation is an anti-isomorphism of the complete lattices of all right and left annihilators, and right annihilation is its inverse. We can conclude immediately that whenever $B, C \subseteq A$ and $B_k \subseteq A$ $(k \in K)$:*

(a) $(^{\perp}(B^{\perp}))^{\perp} = B^{\perp}$;

(b) *if* $B \subseteq C$, *then* $B^{\perp} \supseteq C^{\perp}$;

(c) $\bigcap_{k\in K}(B_k^{\perp}) = \left(\bigcup_{k\in K} B_k\right)^{\perp}$;
(d) $\{B^{\perp} \mid B \subseteq A\} = \{B \subseteq A \mid ({}^{\perp}B)^{\perp} = B\}$.
Also true are the dual statements in which right and left annihilation are interchanged.

PROOF. Polarities and Galois connections are discussed in Birkhoff's book on lattices [**Birk67**]. □

With condition (II) in the definition of null-symmetry, we no longer need to distinguish on which side we are annihilating:

$$B^{\perp} = {}^{\perp}B \qquad (B \subseteq A).$$

From now on we need use only the right 'perp'. Hence the notation,

$$\mathrm{Ideal}^{\perp}\,\boldsymbol{A} = \{B^{\perp} \mid B \subseteq A\},$$

is justified for the set of annihilators in a null-symmetric half-shell $\boldsymbol{A}$. Thus, we find that $B^{\perp}$ is also an idéal:

$$\mathrm{Ideal}^{\perp}\,\boldsymbol{A} \subseteq \mathrm{Idéal}\boldsymbol{A}.$$

The following lemmas and propositions prove what we need to know about these lattices.

1.7. LEMMA. *For idéals I, J of a half-shell $\boldsymbol{A}$ without nilpotents,*

$$I \cap J = \{0\} \textit{ if, and only if, } IJ = \{0\},$$

where $IJ = \{ij \mid i \in I \text{ and } j \in J\}$.

PROOF. $\Rightarrow$. Assume $I \cap J = \{0\}$. Any element of IJ is of the form ij where $i \in I$ and $j \in J$. Since I and J are idéals, it follows that $ij \in I$ and $ij \in J$. Therefore, $ij \in I \cap J$. Hence $ij = 0$.

$\Leftarrow$. Assume that $IJ = \{0\}$ and that $a \in I \cap J$. Then $aa \in IJ$. Thus $aa = 0$. Since $\boldsymbol{A}$ has no nilpotents, $a = 0$. □

1.8. PROPOSITION. *Let $\boldsymbol{A}$ be a null-symmetric half-shell without nilpotents.*

(a) *The algebra,* $\mathbf{Ideal}^{\perp}\,\boldsymbol{A} = \langle \mathrm{Ideal}^{\perp}\,\boldsymbol{A};\, \cap, {}^{\perp}, 0\rangle$, *is a Boolean subsemilattice of the semilattice* **Idéal** $\boldsymbol{A}$.
(b) *The two semilattices,* $\mathbf{Ideal}^{\perp}\,\boldsymbol{A}$ *and* **Idéal** $\boldsymbol{A}$, *are both complete lattices, but their suprema may differ. For a family $\{B_k \mid k \in K\}$ of annihilators, we have that*

(1) $$\bigvee_{k\in K} B_k = \bigcup_{k\in K} B_k \qquad (\textstyle\bigvee \textit{ operates in } \mathbf{Idéal}\ \boldsymbol{A}),$$

(2) $$\bigsqcup_{k\in K} B_k = \Big(\bigcap_{k\in K}(B_k^{\perp})\Big)^{\perp} \qquad (\textstyle\bigsqcup \textit{ operates in } \mathbf{Ideal}^{\perp}\,\boldsymbol{A}),$$

(3) $$\bigvee_{k\in K} B_k \subseteq \bigsqcup_{k\in K} B_k.$$

PROOF. (a) We should first convince ourselves that $\text{Ideal}^{\perp} \boldsymbol{A}$ is closed to the operations, $\cap$ and $^{\perp}$. Closure to these comes from the nature of a Galois connection. The set $\{0\}$ is the annihilator of A for the following reason. If $b \in A^{\perp}$, then $ab = 0$ for all a in A; in particular, $bb = 0$ and so $b = 0$, by the lack of nilpotents. Thus $A^{\perp} = \{0\}$.

The remaining thing to prove is the Boolean part (Definition V.1.1):

$$I \cap J^{\perp} = \{0\} \text{ iff } I \subseteq J \qquad (I, J \in \text{Ideal}^{\perp} \boldsymbol{A}).$$

$\Rightarrow$. Suppose that $I \cap J^{\perp} = \{0\}$ and $i \in I$. Then $IJ^{\perp} = \{0\}$ by Lemma 1.7. Hence, $ik = 0$ for all k in $J^{\perp}$. Thus, $i \in J^{\perp\perp} = J$.

$\Leftarrow$. On the supposition that $I \subseteq J$ and $a \in I \cap J^{\perp}$, it follows that $a \in I \subseteq J$ and $a \in J^{\perp}$. Therefore, $aa \in JJ^{\perp}$. But $JJ^{\perp} = \{0\}$, and hence $a = 0$ since there are no nilpotents.

(b) Completeness of these semilattices follows immediately from the definitions of idéal and annihilator. Hence, they have joins.

(1) The join $\bigvee_{k \in K} B_k$ is defined to be the smallest idéal containing all the B_k, and the union $\bigcup_{k \in K} B_k$ is an idéal.
(2) The join $\bigsqcup_{k \in K} B_k$ is defined to be the smallest annihilator containing all the B_k, and in a Galois connection this is $\left(\bigcup_{k \in K} B_k\right)^{\perp\perp}$. An application of $^{\perp}$ to part (c) of Proposition 1.6 completes this argument.
(3) There are more idéals than annihilators. □

Trying to prove the next lemma motivated much of the preceding discussion. It is needed at a crucial step in order to prove that the stalks in the sheaf to be constructed will have no divisors of zero. Note the similarity to the earlier Lemma VIII.1.8c.

1.9. LEMMA. *Whenever a and b are elements in a null-symmetric half-shell $\boldsymbol{A}$ without nilpotents, then*

$$(ab)^{\perp\perp} = a^{\perp\perp} \cap b^{\perp\perp},$$

where we have simplified the notation: $a^{\perp} = \{a\}^{\perp}$.

PROOF. $\subseteq$. First, we show that $a^{\perp} \subseteq (ab)^{\perp}$. If $x \in a^{\perp}$, then $xa = 0$, and also $(xa)b = 0$. So by null-symmetry $x(ab) = 0$, and hence $x \in (ab)^{\perp}$. Thus $a^{\perp} \subseteq (ab)^{\perp}$, and likewise also $b^{\perp} \subseteq (ab)^{\perp}$. Therefore, with a reversal of inclusions, $(ab)^{\perp\perp} \subseteq a^{\perp\perp} \cap b^{\perp\perp}$,.

$\supseteq$. Since $\textbf{Ideal}^{\perp} \boldsymbol{A}$ is a Boolean subsemilattice, it suffices to prove:

$$a^{\perp\perp} \cap b^{\perp\perp} \cap (ab)^{\perp} = \{0\}.$$

Suppose that $x \in a^{\perp\perp} \cap b^{\perp\perp} \cap (ab)^{\perp}$. We will show that $x = 0$. Setting out what membership in each annihilator means, we write down that:

(i) $xy = 0 \qquad (y \in a^{\perp})$;
(ii) $xz = 0 \qquad (z \in b^{\perp})$;
(iii) $x(ab) = 0$.

The given logical implications between various null products will be used freely throughout. From (iii), by definition, $xa \in b^{\perp}$. Hence, from (ii) now, $x(xa) = 0$. Thus $xx \in a^{\perp}$. Next from (i), $x(xx) = 0$. Therefore, $(xx)(xx) = 0$, and so $xx = 0$, and finally $x = 0$, since there are no nilpotents. □

The second part of the proof of Theorem 1.3 involves converting these annihilator ideals into congruences so that the machinery of Chap. V may be started up. The obvious choice is the minimal congruence $\theta(I)$ that relates all elements of an idéal I. For half-shells this is the extremely simple and constructive Rees congruence [**Petr77**, p. 9] [**ChaEL03**, p. 187]. Strictness of the half-shells now becomes essential.

1.10. PROPOSITION. *Let I and J be idéals of a strict half-shell $\boldsymbol{A}$.*

(a) *The minimal congruence $\theta(I)$ is characterized by the condition:*

$$a \ \theta(I) \ b \text{ if, and only if, } a, b \in I \text{ or } a = b \qquad (a, b \in A).$$

(b) *We may recover any idéal from its corresponding congruence:*

$$I = \frac{0}{\theta(I)}.$$

(c) *It follows that $\theta(I) = \theta(J)$ if, and only if, $I = J$.*
(d) *If $I \subseteq J$, then $\theta(I) \subseteq \theta(J)$.*
(e) $\theta(I) \cap \theta(J) = \theta(I \cap J)$.

PROOF. (a) Check that the right side of the equivalence defines a congruence whenever I is an idéal. Clearly, it is minimal; so it is $\theta(I)$.
(b) This is clear.
(c) Use (b).
(d) This follows directly from the definition of $\theta(I)$.
(e) $\subseteq$. Split this inclusion into the cases defining $\theta(I)$.
$\supseteq$. Use part (d). □

1.11. DEFINITION. The set of all congruences associated with annihilators in a half-shell is designated:

$$\mathrm{Con}^{\perp} \boldsymbol{A} = \left\{ \theta(I) \mid I \in \mathrm{Ideal}^{\perp} \boldsymbol{A} \right\},$$

With the obvious operations and justification by part (e) of the previous proposition, we convert it in the next proposition to an algebra:

$$\mathbf{Con}^{\perp} \boldsymbol{A} = \langle \mathrm{Con}^{\perp} \boldsymbol{A}; \cap, {}^{\perp}, 0 \rangle,$$

where

$$\big(\theta(I)\big)^{\perp} = \theta(I^{\perp}) \quad \text{and} \quad 0 = \theta(A^{\perp}).$$

1.12. PROPOSITION. *In a null-symmetric strict half-shell $\boldsymbol{A}$ without any nilpotents, the algebra $\mathbf{Con}^{\perp} \boldsymbol{A}$ is a complete Boolean subsemilattice of $\mathbf{Con}\,\boldsymbol{A}$, and it is isomorphic to $\mathbf{Ideal}^{\perp} \boldsymbol{A}$.*

PROOF. The first assertion will follow from the second if we make use of the corresponding result for $\mathbf{Ideal}^{\perp} \boldsymbol{A}$ in Proposition 1.8. The isomorphism is the map, $I \mapsto \theta(I)$, which we already know is bijective. By the previous proposition, this map preserves intersection, complementation and the nullity. □

Before proceeding any further, it would be good to take stock of the three suprema, or infinitary joins, of a set N of congruences in $\mathbf{Con}^{\perp} \boldsymbol{A}$ that we have accumulated:

$$\bigsqcup N \supseteq \bigvee N \supseteq \bigcup N.$$

The first is the new join in $\mathbf{Con}^{\perp} \boldsymbol{A}$, which exists because this lattice is complete whenever $\boldsymbol{A}$ is a null-symmetric half-shell without nilpotents. The second is the traditional join of congruences in $\mathbf{Con}\, \boldsymbol{A}$. The third is set union.

1.13. LEMMA. *Assume that* $\boldsymbol{A}$ *is a null-symmetric, strict half-shell and* N *is any ideal of* $\mathbf{Con}^{\perp} \boldsymbol{A}$.[2] *Then,*
(a) $\bigvee N = \bigcup N$,
(b) $\frac{0}{\bigcup N} = \bigcup \frac{0}{N}$ *where* $\frac{0}{N} = \{\frac{0}{\theta} \mid \theta \in N\}$.

PROOF. (a) See Proposition V.1.5.
(b) If $a \in 0/\bigcup N$, then there is a congruence θ in N such that $a\, \theta\, 0$, and hence $a \in \bigcup(0/N)$; and conversely. □

Now it follows from Theorem V.2.1, that any null-symmetric half-shell without nilpotents is a subalgebra of the half-shell of all global sections of a sheaf over $\mathbf{Spec\, Con}^{\perp} \boldsymbol{A}$. This leads to the third and last part of the proof of Theorem 1.3, which is about the lack of zero divisors in the stalks.

1.14. PROPOSITION. *For any null-symmetric strict half-shell* $\boldsymbol{A}$ *without nilpotents, the stalks of its sheaf* $\mathcal{A}$ *have no divisors of zero. Here, the sheaf is:* $\mathcal{A} = \boldsymbol{A} /\!/ \mathbf{Con}^{\perp} \boldsymbol{A}$; *and the base space is:* $\boldsymbol{X} = \mathbf{Spec\, Con}^{\perp} \boldsymbol{A}$.

PROOF. Any stalk of the sheaf in Theorem V.2.1 is of the form $\boldsymbol{A} / \bigvee P$ where P is a prime ideal of $\mathbf{Con}^{\perp} \boldsymbol{A}$. Rather than working solely in the Boolean subsemilattice $\mathbf{Con}^{\perp} \boldsymbol{A}$, we find it convenient to work also with its isomorphic cousin $\mathbf{Ideal}^{\perp} \boldsymbol{A}$. So let $\mathcal{P}$ be the corresponding prime ideal of $\mathbf{Ideal}^{\perp} \boldsymbol{A}$, that is, $\mathcal{P} = 0/P$.

To prove that any stalk $\boldsymbol{A} / \bigvee P$ has no divisors of zero, let us suppose that $(a/\bigvee P) \times (b/\bigvee P) = 0/\bigvee P$, with the intent of deriving that $a/\bigvee P = 0$ or $b/\bigvee P = 0$. Since $\bigvee P = \bigcup P$ by Lemma 1.13, there exists a θ in P such that $ab\, \theta\, 0$. Let I be $0/\theta$ in $\text{Ideal}^{\perp} \boldsymbol{A}$. Therefore, $(ab)^{\perp\perp} \subseteq I^{\perp\perp} = I$. Hence, $(ab)^{\perp\perp} \in \mathcal{P}$, since $\mathcal{P}$, as an ideal, is closed to downward inclusion.

[2]There is a subtlety here that might mislead the reader. By an ideal, do we mean in $\mathbf{Con}^{\perp} \boldsymbol{A}$ as a half-shell or as a Boolean algebra? We mean the latter as in Chap. V.

But $(ab)^{\perp\perp} = a^{\perp\perp} \cap b^{\perp\perp}$ by Lemma 1.9. Since $\boldsymbol{P}$ is prime in $\mathbf{Con}^{\perp} \boldsymbol{A}$, then $\mathcal{P}$ must be prime in $\mathbf{Ideal}^{\perp} \boldsymbol{A}$, and therefore,

$$a^{\perp\perp} \in \mathcal{P} \text{ or } b^{\perp\perp} \in \mathcal{P}.$$

Thus,

$$a^{\perp\perp} \subseteq \bigcup \frac{0}{P} \text{ or } b^{\perp\perp} \subseteq \bigcup \frac{0}{P}.$$

Hence,

$$a \in a^{\perp\perp} \subseteq \frac{0}{\bigvee P} \text{ or } b \in a^{\perp\perp} \subseteq \frac{0}{\bigvee P}.$$

Therefore, $\boldsymbol{A} / \bigvee P$ has no divisors of zero. □

PROOF OF THEOREM 1.3, A RECAPITULATION. We assume that $\boldsymbol{A}$ is a null-symmetric strict half-shell without nilpotents. In the first part of the proof, Proposition 1.6 and Lemma 1.7 set up a calculus of annihilators of subsets of A, leading to Proposition 1.8, which asserts that the collection of all annihilators is a complete Boolean algebra.

In the second part, we switch in Proposition 1.10 to the collection of congruences generated by these annihilators, which by Proposition 1.12 is a complete Boolean subsemilattice $\boldsymbol{B}$ of $\operatorname{Con} \boldsymbol{A}$. By Theorem V.2.1, $\boldsymbol{A}$ is isomorphic to a subalgebra of the algebra of all global sections of a sheaf over a Boolean space $\boldsymbol{X}$, with stalks induced by the spectrum of $\boldsymbol{B}$.

In the third part, Lemma 1.13 relates unions and joins of ideals and congruences. We prove in Proposition 1.14, with the help of Lemma 1.9, that the stalks have no zero divisors. Finally, $\boldsymbol{X}$ is an extremally disconnected Boolean space because it is dual to the complete Boolean algebra $\mathbf{Con}^{\perp} \boldsymbol{A}$. See [**Halm63**, p. 90 ...] for a proof. □

1.15. PROBLEM. A conclusion of Theorem 1.3 is that the base space $\boldsymbol{X}$ is extremally disconnected, but this is never used to draw any conclusions about the space of global sections. Are there any to be drawn? Compare this with the more limited Proposition 2.2 in the next section.

An application of Theorem 1.12 is to commutative semigroups with a nullity but without nilpotents. Any such semigroup is isomorphic to a subsemigroup of the commutative semigroup of all global sections of a sheaf over a Boolean space with stalks having no zero divisors. Meet semilattices with a least element are examples of such half-shells. Now it is already known that any semilattice is a subdirect power of the two-element semilattice [**Grät79**, p. 155, ex. 27]. This stalk has no divisors of zero, and any power of it is a sheaf over a Boolean space, as in Stone's representation theorem for Boolean algebras. However, in general, this traditional representation gives a different sheaf from that developed in this section. For example, any chain has no divisors of zero and so will not decompose by our techniques since there are no nontrivial annihilators.

A half-shell $\boldsymbol{A}$ of Theorem 1.3 can be a proper subalgebra of $\boldsymbol{\Gamma}(\mathcal{A})$. To see, consider the meet semilattice, $\boldsymbol{A} = \boldsymbol{SL}_3$ given by the Hasse diagram:

$\boldsymbol{SL}_3$

Then, $\boldsymbol{\Gamma}(\mathcal{A})$ will have four elements, so it cannot be isomorphic to $\boldsymbol{A}$.

This example also shows how divisors of zero cease to be so when they are spread across the stalks. For $1 \wedge 2 = 0$; and $1^\perp = \{0,2\}$ and $2^\perp = \{0,1\}$, generating the prime ideals in $\mathbf{Ideal}^\perp \boldsymbol{A}$ that give the two points of the base space of the sheaf. The stalks over these points each have two elements, with no divisors of zero.

Another semilattice demonstrates that the supremum of a proper ideal of $\mathbf{Con}^\perp \boldsymbol{A}$ is not necessarily proper. To see this, consider the infinite counterpart of the preceding three-element semilattice:

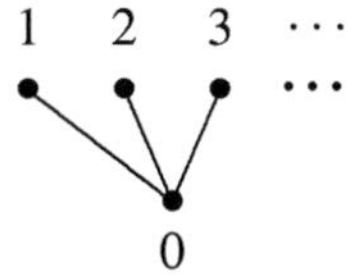

$\boldsymbol{SL}_\omega$

It is easy to see that $B^\perp = (\omega \sim B) \cup \{0\}$ when $B \subseteq \omega$ (where $\omega = \{0,1,2,3,\ldots\}$). Thus, $\mathrm{Ideal}^\perp \boldsymbol{SL}_\omega = \{C \subseteq \omega \mid 0 \in C\}$. The set F of finite members of $\mathrm{Ideal}^\perp \boldsymbol{SL}_\omega$ is an ideal; this certainly has an extension to a prime ideal P of the Boolean algebra $\mathbf{Ideal}^\perp \boldsymbol{SL}_\omega$. But $\bigvee P = \omega$. All this translates readily over to $\mathbf{Con}^\perp \boldsymbol{SL}_\omega$. Also, since there are many different extensions of F to prime ideals, this examples shows that there are distinct prime ideals P and Q such that $\bigvee P = \omega = \bigvee Q$. It is not hard to continue the calculations and see that each principal prime ideal contributes a two-element stalk to the sheaf representation of $\boldsymbol{SL}_\omega$ and each non-principal prime ideal contributes a trivial stalk.

1.16. Problem. Is the set of all non-principal prime ideals of the preceding example an open set in the base space? If so, then this would give us a brace $\langle \boldsymbol{A}, \mathbf{Con}^\perp \boldsymbol{A} \rangle$ that is not reduced, that is, in the language of Sect. V.4, axiom RB is violated: $\mathrm{Int}\,\mathrm{Triv}\langle \boldsymbol{A}, \mathbf{Con}^\perp \boldsymbol{A} \rangle \neq \varnothing$.

We close with some comments on the all-pervasive notion of ideal, of which there are at least three definitions in the literature. The first is the idéal, defined variously for half-shells and shells, and reminiscent of traditional ideals in ring theory. The second is the coset o/θ for any congruence θ of a pointed algebra $\langle \boldsymbol{A}, o \rangle$, which was defined earlier and which works well for factoring, a chief concern of this book. The ideals in this second sense include the idéals of strict half-shells. The third is the term-oriented definition of ideal, originated by Peter Gumm and Aldo Ursini

[**GumUr84**] and also found in the book of Chajda, Eigenthaler and Langer [**ChaEL03**, p. 137], which we will call a term-ideal, to differentiate it from our congruence-oriented definition. It is also motivated by ideals in ring theory. Their definition comes in two parts: an **ideal term** is a term with $m + n$ variables such that $t(\dot{o}, \vec{a}) = o$ where $\dot{o} = \langle o, o, \ldots, o \rangle$ repeated m times and $\vec{a} \in A^n$; and a subset I of A is a **term-ideal** if $t(\vec{i}, \vec{a}) \in I$ whenever $\vec{i} \in I^m$ and $\vec{a} \in A^n$. This is a good definition since it agrees with normal subgroups in group theory, the usual ideals in ring theory, and the idéals for the null-symmetric strict half-shells of this section; however, it does not always agree with our notion of factor ideal, and it does not give the usual ideals in lattice theory unless the lattices are bounded and distributive.

2. Converses and Axiomatics

This section fills out the previous one with three results. The first is a partial converse to Theorem 1.3. It shows that, if the sheaf $\mathcal{A}$ is a Boolean product with no divisors of zero in the stalks, then $\boldsymbol{\Gamma}(\mathcal{A})$ is Baer–Stone. The second proves another partial converse to Sect. 1: a strict half-shell has no divisors of zero if, and only if, it has no nilpotents and is null-symmetric. The third concludes that, if the strict half-shell is null-symmetric and has no nilpotents, then any null product implies any other null product in the same or more variables.

Recall that a **factor ideal** I of a half-shell $\boldsymbol{A}$ comes from a factor congruence θ of $\boldsymbol{A}$ where $I = 0/\theta$. Factor ideals and congruences will be needed whenever we want to show that we have embedded a half-shell into a Baer–Stone half-shell. Annihilators play their role in its definition.

2.1. DEFINITION. Remember that a half-shell $\boldsymbol{A}$ is called a **Baer–Stone** half-shell if the right and left annihilators of every single element of A are factor ideals. It is called **finitely Baer–Stone** if the annihilators on both sides of any finite subset of A are factor ideals. And it is called **completely Baer–Stone** if this is true for any subset of A.

An arbitrary subalgebra of a half-shell need not be a factor ideal. And we should not expect the congruences generated by the annihilators $\theta(I^{\perp})$ and $\theta(I^{\perp\perp})$ to be a pair of complementary factor congruences.

Note that even when θ is a factor ideal of a half-shell it can happen that $\theta \neq \theta(0/\theta)$. An example of this occurs in this four-element meet-semilattice.

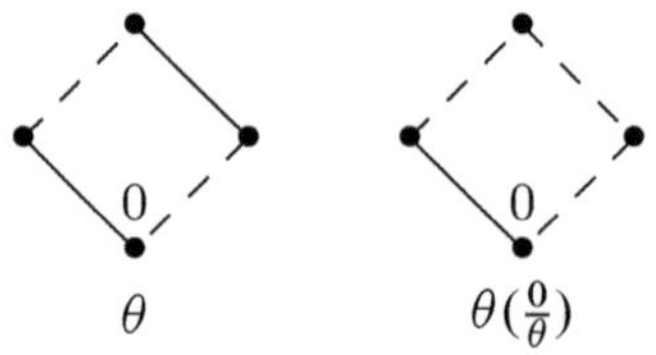

A solid line connecting two elements means they are related. Of course, $\theta \supseteq \theta(0/\theta)$ for any factor congruence θ of a half-shell.

We state a proposition concluding that $\boldsymbol{\Gamma}(\mathcal{A})$ is completely Baer-Stone. For details about extremal disconnectedness in (b), see the end of Sect. III.4.

2.2. PROPOSITION. *Let $\mathcal{A}$ be a Boolean product of half-shells whose stalks have no divisors of zero.*

(a) *Then $\boldsymbol{\Gamma}(\mathcal{A})$ is a finitely Baer-Stone half-shell.*
(b) *If further, the base space $\boldsymbol{X}$ is extremally disconnected, then $\boldsymbol{\Gamma}(\mathcal{A})$ is completely Baer-Stone.*

PROOF. Let σ and τ be global sections in $\boldsymbol{\Gamma}(\mathcal{A})$. Recall that in a Boolean product the equalizers $\sigma:\tau$ are closed, as well as open, and so are the inequalizers clopen. We prove part (b) first, since a shortening of this argument will immediately give a proof of (a). Abbreviate $\boldsymbol{\Gamma}(\mathcal{A})$ to $\boldsymbol{\Gamma}$.

(b) Consider any subset Σ of Γ; we must show that its annihilator $\Sigma^{\perp}$ is a factor ideal. It suffices to find a factor congruence θ of $\boldsymbol{\Gamma}$ such that $0/\theta = \Sigma^{\perp}$. To that end define

$$U = \bigcup_{\sigma\in\Sigma} \sigma;0.$$

This is the set of indices in the base space $\boldsymbol{X}$ where some element σ of Σ has a component different from 0. Because each stalk of $\mathcal{A}$ has no zero divisors, for the product $\sigma\tau$ to be 0 it is necessary and sufficient that in each stalk the component of σ vanish, or that the component of τ vanish. Therefore, each global section of $\Sigma^{\perp}$ must be 0 on all those components whose indices are in U. Thus,

$$\Sigma^{\perp} = \{\sigma \in \Gamma \mid U \subseteq \sigma:0\}.$$

But before doing anything else, we need a topological adjustment. Since each inequalizer $\sigma;0$ is open, so is the union U. Since the base space $\boldsymbol{X}$ is extremally disconnected, the closure $\overline{U}$ of U is clopen. Now define the congruences θ and η by

$$\sigma\ \theta\ \tau \text{ iff } \overline{U} \subseteq \sigma:\tau, \text{ and}$$
$$\sigma\ \eta\ \tau \text{ iff } X \sim \overline{U} \subseteq \sigma:\tau.$$

Let us first establish that θ and η are a complementary pair of factor congruences. That they are congruences follows directly from the defining properties of a complex. To show they are complementary, we prove that

$$\theta \cap \eta = 0_{\mathbf{Con}\,\boldsymbol{\Gamma}}, \quad \text{and} \quad \theta \circ \eta = 1_{\mathbf{Con}\,\boldsymbol{\Gamma}}.$$

The intersection property is derived directly from their definitions. To demonstrate the composition property, assume $\sigma, \tau \in \Gamma$. Define $\rho: X \to \mathcal{A}$ by patching:

$$\rho(x) = \begin{cases} \sigma(x) \text{ if } x \in \overline{U}, \\ \tau(x) \text{ if } x \in X \sim \overline{U}, \end{cases} .$$

Since $\overline{U}$ is clopen, $\rho \in \Gamma$ by patching. Clearly, $\overline{U} \subseteq \sigma : \rho$, and so $\sigma\ \theta\ \rho$. Similarly, $\rho\ \eta\ \tau$. Therefore, $\sigma\ (\theta \circ \eta)\ \tau$.

We claim now that $\Sigma^{\perp} = 0/\theta$. With the definition of θ in mind, we see that, if $\sigma \in 0/\theta$, then $\sigma : 0 \supseteq \overline{U} \supseteq U$, and thus $\sigma \in \Sigma^{\perp}$. Hence, $0/\theta \subseteq \Sigma^{\perp}$. To prove the opposite direction of inclusion, suppose that $\sigma \in \Sigma^{\perp}$. Then $\sigma : 0 \supseteq U$. But remember that $\sigma : 0$ is clopen. Since $\overline{U}$ is the intersection of all closed sets containing U, then $\sigma : 0 \supseteq \overline{U}$. Thus $\sigma \in 0/\theta$.

Therefore, the annihilator $\Sigma^{\perp}$ of any subset Σ of $\Gamma(\mathcal{A})$ is a factor ideal, and hence $\boldsymbol{\Gamma}(\mathcal{A})$ is completely Baer-Stone.

(a) Consider any finite subset Σ of Γ; we must show that $\Sigma^{\perp}$ is a factor ideal. Proceeding as in part (b) above, we observe that it suffices to find a complementary pair of factor congruences θ and η of $\boldsymbol{\Gamma}$ such that $0/\theta = \Sigma^{\perp}$. The crucial difference with (b) is that U is now a finite union of clopen subsets $\sigma ; 0$ of $\boldsymbol{X}$, and hence is clopen itself. Thus, no topological adjustment is needed, that is, $\overline{U} = U$, without the need for extremal disconnectedness. The argument finishes as before. □

Unfortunately, this proposition is marred by its requiring a Boolean product. Apparently, we can enlarge $\boldsymbol{\Gamma}(\mathcal{A})$ into a completely Baer–Stone half-shell in general only at the expense of the compactness of $\boldsymbol{X}$. In other words, $\boldsymbol{\Gamma}(\mathcal{A})$ is always a subalgebra of the full product, $\boldsymbol{P} = \prod_{x \in X} \boldsymbol{A}_x$, which can be considered a sheaf when X is given the discrete topology. As all subsets of X are clopen, $\boldsymbol{P}$ is completely Baer–Stone. However, $\boldsymbol{X}$ may not be compact, and so is not a Boolean space.

But we do have a logical equivalence within subdirect products. Recall that a half-shell, $\boldsymbol{A} = \langle A, \times, 0 \rangle$, is semi-integral if it is a subdirect product of integral half-shells; that is, $\boldsymbol{A} \subseteq_{s.d.} \prod_{i \in I} \boldsymbol{A}_i$ where each $\boldsymbol{A}_i$ has no divisors of zero.

2.3. THEOREM. *A strict two-sided half-shell,* $\boldsymbol{A} = \langle A, \times, 0 \rangle$, *is semi-integral if, and only if,* $\boldsymbol{A}$ *is null-symmetric and without nilpotents.*

PROOF. $\Rightarrow$. Consider a clause defining null-symmetry: for example,

$$\text{if } ab = 0, \text{ then } ba = 0.$$

If $a, b \in A$ and $ab = 0$, then this is true on every integral quotient of the subdirect product, say, indexed by i in I. Thus, $a_i b_i = 0$. By integrality, $a_i = 0$ or $b_i = 0$. Hence $b_i a_i = 0$ for all i in I. Therefore, $ba = 0$. The other clauses are proven similarly.

To show that $\boldsymbol{A}$ has no nilpotents, suppose that $a \in A$ and $a^n = 0$. Then in each stalk, $a_i{}^n = 0$. Multiply by a_i enough times to obtain $a_i{}^{2^k} = 0$ for some positive integer k. Rearrange by null-symmetry to rewrite this as $a_i{}^{2^{k-1}} \times a_i{}^{2^{k-1}} = 0$. Since there are no divisors of zero in the stalk, $a_i{}^{2^{k-1}} = 0$. By descent, eventually $a_i = 0$ in each stalk. Therefore, $a = 0$.
$\Leftarrow$. By Theorem 1.14. □

Putting together previous results, we create a corollary.

2.4. COROLLARY. *Let* $\boldsymbol{A}$ *be a null-symmetric strict half-shell without nilpotents whose sheaf* $\mathcal{A}$*, as constructed from* $\mathbf{Con}^{\perp} \boldsymbol{A}$ *in Sect. 1, is a Boolean product. Then its extension* $\boldsymbol{\Gamma}(\mathcal{A})$*, the half-shell of global sections of* $\mathcal{A}$*, is completely Baer–Stone and has no divisors of zero.*

PROOF. By Theorem 1.3, $\boldsymbol{\Gamma}(\mathcal{A})$ has no divisors of zero. By Proposition 1.12, $\mathbf{Con}^{\perp} \boldsymbol{A}$ is a complete Boolean subsemilattice. Thus, $\mathbf{Spec}(\mathbf{Con}^{\perp} \boldsymbol{A})$, the base space $\boldsymbol{X}$ of $\mathcal{A}$, is extremally disconnected. By Proposition 2.2, $\boldsymbol{\Gamma}(\mathcal{A})$ is completely Baer–Stone. □

Here are suggestions for further research.

2.5. PROBLEM. In Chap. VIII, Baer–Stone shells were proven to be isomorphic to the algebra of all global sections of a sheaf over a Boolean space. In this chapter, while weakening the hypothesis to no nilpotents, only isomorphism to a subalgebra of the global sections could be insured. In the case of commutative unitary rings, it is shown in [**Kist69**] that the former is a special case of the latter. Now obtain for shells, à la Kist, the isomorphism for Baer–Stone shells as a special case of the proper embedding in a subdirect product of stalks without any divisors of zero. Are these representations semiprime in his sense? If so, fit Chap. VIII into this chapter.

Let us expand more on this. Kist's theorem 9.6 for commutative rings in [**Kist63**] exhibits the isomorphism of the Boolean algebra of idempotents with the algebra dual to $\mathcal{M}(\boldsymbol{R})$, the space of all minimal prime ideals. This says that the base space of our sheaf is homeomorphic to the base space of Kist's sheaf, at least in the case of commutative rings. It would be interesting to know if the congruences creating the stalks of our sheaf of shells correspond to minimal prime ideals in some sense.

In summary, it would be notable if one could redevelop this chapter along the lines of Kist's treatment of minimal prime ideals [**Kist63, Kist69**]. For a guide as to how this might be done see the paper of Picavet [**Pica80**].

The remainder of this section demonstrates how axioms I to IV in Sect. 1 imply all similar implications between null products of any number of elements. The next two lemmas prepare us for the proposition to follow by building up terms. By $\mathcal{P}_n$ we mean the set of all terms p in the one binary operation $\times$ and the n variables, $x_1, x_2, \ldots, x_n$, each variable appearing exactly once. For example, $x_1(x_2x_3)$ and $(x_3x_1)x_2$ belong to $\mathcal{P}_3$, but $\big(x_1(x_2x_1)\big)x_3\big)$ and x_1x_2 do not.

2.6. LEMMA. *In any strict half-shell,* $\boldsymbol{A} = \langle A; \times, 0\rangle$, *the following three statements are equivalent:*

(a) $\boldsymbol{A}$ *is null-symmetric;*
(b) *For all* $a_1, \ldots, a_n$ *in* A *and for all* p, q *in* $\mathcal{P}_n$,

$$\textit{if } p(a_1, \ldots, a_n) = 0, \ \textit{then } q(a_1, \ldots, a_n) = 0;$$

(c) *For all* $a_1, \ldots, a_n$ *in* A *and for all* p *in* $\mathcal{P}_m$ *and* q *in* $\mathcal{P}_n$ *with* $m \leq n$,

$$\textit{if } p(a_1, \ldots, a_m) = 0, \ \textit{then } q(a_1, \ldots, a_m, a_{m+1}, \ldots, a_n) = 0.$$

PROOF. We go full circle: (a) $\Rightarrow$ (b) $\Rightarrow$ (c) $\Rightarrow$ (a).

(a) $\Rightarrow$ (b). This is the hardest implication, and will be split into three parts. We convince ourselves that: first, it is true when $n \leq 4$; second, any new association of terms of a null product is possible; and third, that any permutation of the variables is also possible. The last two parts are established by induction when $n \geq 5$. Details are omitted, and $p(a_1, \ldots, a_m)$ is abbreviated as p, etc.

Clearly, repeated application of (II), (IIIa) and (IIIb) shows that (b) is true when $n = 3$. When is $n = 4$, it is also clear that any new association is possible except perhaps with inside cases such as:

$$\text{if } a_1\big((a_2a_3)a_4\big) = 0, \ \text{then } a_1\big(a_2(a_3a_4)\big) = 0.$$

For this, from the hypothesis, and (IIIb) and (IIIa), we get $(a_1a_4)(a_2a_3) = 0$. We obtain the conclusion by applying (IV), (IIIb) and (IIIa), in that order. In general, for permuting variables when $n = 4$, the polynomial p may be first balanced to $(a_1a_2)(a_3a_4)$ by reassociation. Then null-symmetry may be applied enough times to permute the variables in any way whatsoever.

To do induction on n, assume that $n \geq 5$. First, we show that in a null product the parenthesis may be pushed around at will (within the limits, naturally, of well-formed terms). This will be done by proving with induction on n that an equation of the form $p = 0$ in n variables implies an equation of the form $r = 0$ in the same variables but where the operations of r are now all associated to the right. It will follow that $q = 0$, since the implications defining null-symmetry are reversible.

Now the base step when $n \leq 4$ has already been taken care of. Remember that the variables are not repeated in this lemma. Without loss of generality, assume that the variables in p appear in the order $a_1, a_2, \ldots, a_n$; this order will not change in this paragraph.. By associating differently, if necessary, the four most outer subterms, we may assume from what was proven for $n = 4$ that we are starting with

$$(p_1p_2)(p_3p_4) = 0.$$

By thinking of p_1p_2 as a single new variable, we can apply the induction hypothesis to re-associate p_3p_4 to the right. Now interchange the roles of the two outer terms and end up associating the whole product to the right. Likewise associate q to the right and equate with p.

By induction on n again we show with the implications available that any permutation of the indices is possible. To this end it suffices to show that any two adjacent variables may be transposed, say a_i and a_j. Rearranging the association to

$$p_1\big((a_ia_j)p_2\big) = 0,$$

we may apply the already established case of $n \leq 4$ to transpose a_i and a_j.

(b) $\Rightarrow$ (c). Since $p(a_1,\ldots,a_m) = 0$, it follows that $p(a_1,\ldots,a_m)\times(a_{m+1}\times \cdots \times a_n) = 0$. By the previous part, $q(a_1,\ldots,a_n) = 0$.

(c) $\Rightarrow$ (a). Trivial. □

2.7. LEMMA. *In any strict half-shell, $\boldsymbol{A} = \langle A; \times, 0\rangle$, the following statements are equivalent:*

(a) *$\boldsymbol{A}$ is null-symmetric with no nilpotents;*
(b) *For any polynomial p in $x_1,\ldots,x_n$, without repetition of these variables, and for any other polynomial q in the same variables but with repetitions allowed,*

$$p(a_1,\ldots,a_n) = 0 \quad \textit{iff} \quad q(a_1,\ldots,a_n) = 0.$$

PROOF. There are two directions of implication to establish.

$\Rightarrow$. Multiplying p by additional elements on the right preserves its nullity. Any necessary re-associations are taken care of by the previous lemma.

$\Leftarrow$. Superfluous repetitions are gotten rid of by the following trick. Assume that $q = 0$. Multiply q by enough additional duplications of the elements already occurring there so that each element occurs the same number of times. Rearrange and intermingle these occurrences so that this enlarged q takes on the appearance p^k for some positive integer k. By Lemma 2.6, $p^k = 0$. (Bear in mind that, although variables must appear only once in the terms of Lemma 2.6, the elements substituted for them may be equal.) Making use of axiom I that there are no nilpotents, we deduce that $p = 0$. If too many occurrences of the a_i were eliminated, they can be put in again with the previous lemma. □

The next proposition can be viewed in two lights. It can be seen as a consequence of Theorem 1.3 that any null polynomial is implied by any other with the same or more variables. Or it can be considered a direct proof of this fact without using Sect. 1.

2.8. PROPOSITION. *Suppose p is a term in $\times$ and variables $x_1,\ldots,x_m$, and q is a term in $\times$ and variables $x_1,\ldots,x_m,x_{m+1},\ldots,x_n$ with $m \leq n$.*

Assume these variables occur at least once in p and q, perhaps repeatedly. In a null-symmetric half-shell $\boldsymbol{A}$ having no nilpotents, with $a_1, \ldots, a_m$, $a_{m+1}, \ldots, a_n$ in A,

$$\text{if } p(a_1, \ldots, a_m) = 0, \text{ then } q(a_1, \ldots, a_m, a_{m+1}, \ldots, a_n) = 0.$$

PROOF. Lemma 2.7 eliminates repeated elements. Lemma 2.6 adds new elements. Lemma 2.7 adds repeated elements. □

2.9. PROBLEM. Curiously, in order to prove Theorem 1.3 we did not need implications between null products with more than three variables. The proofs were written so that it would be easy to check that (IV) was never used. But its conclusion of no zero divisors implies null-symmetry by Theorem 2.3. Is it possible to prove directly that (IV) follows from (I), (II), (IIIa), and (IIIb)?

3. Adding a Unity or a Loop

We pass to half-shells with a unity or a loop; this simplifies Sect. 1 in several different ways. Note that the addition of a unity makes no difference in the formulation of the result of Sect. 1, which gives stalks that have no divisors of zero. But factor ideals will now be principal and easily described. Recall from Sect. VII.2 that factor objects in a unital half-shell come in corresponding pairs: congruences θ and θ'; elements e and e' where $0\,\theta\,'e\,\theta\,1$ and $0\,\theta\,'e'\,\theta\,1$; and ideals $I = 0/\theta = e'A = [e']$ and $I' = 0/\theta' = eA = [e]$. The first part of this section examines the components of factor elements in subdirect products of integral unital half-shells: we might expect them to be 0 or 1, but this is not always the case.

The second part includes in the half-shell an addition that is a loop, and assumes that the multiplication distribute over the addition. If this algebra is null-symmetric and without nilpotents, then the conclusions of Proposition 2.2 follow, but without assuming the structure of a Boolean product.

3.1. PROPOSITION. *Let $\boldsymbol{A}$ be a semi-integral, two-sided, strict and unital half-shell $\langle A; \times, 0, 1\rangle$, that is, assume that*

$$\boldsymbol{A} \underset{s.d.}{\subseteq} \prod_{x \in X} \boldsymbol{A}_x,$$

where each quotient $\boldsymbol{A}_x$ has no divisors of zero. Let θ and θ' be complimentary factor congruences of $\boldsymbol{A}$ generated by factor elements ε and ε'. Then,

(a) $[\varepsilon] = \varepsilon'^{\perp}$ *and* $[\varepsilon'] = \varepsilon^{\perp}$;

(b) $[\varepsilon] = \{\sigma \in A \mid \sigma;0 \subseteq \varepsilon;0\}$, *and* $[\varepsilon'] = \{\sigma \in A \mid \sigma;0 \subseteq \varepsilon';0\}$;

(c) $\varepsilon(x) = 0_x$ *if* $x \in \varepsilon';0$, *and* $\varepsilon'(x) = 0_x$ *if* $x \in \varepsilon;0$ $\quad (x \in X)$.

PROOF. By Theorem 2.3, $\boldsymbol{A}$ is null-symmetric and without nilpotents. Recall from Lemma VII.2.3 that $\theta = \theta_\varepsilon$ and $\theta' = \theta_{\varepsilon'}$, and from Theorem VII.2.11 that $0/\theta = \varepsilon' A = [\varepsilon']$ and $0/\theta' = \varepsilon A = [\varepsilon]$. We demonstrate only the first half of each part.

(a) $\subseteq$. Since $[\varepsilon]$ is the smallest ideal containing ε, we may prove this inclusion by induction. That is, first we show that ε belongs to the right side, and then show that, if r belongs to it, then so do ar and ra for any a in A. That $\varepsilon\varepsilon' = 0$ implies that $\varepsilon \in \varepsilon'^{\perp}$. To establish the induction step, assume that $r \in \varepsilon'^{\perp}$ and $a \in A$. Then $\varepsilon' r = 0$, and hence $\varepsilon'(ra) = 0$ by null-symmetry. So $ra \in \varepsilon'^{\perp}$ and similarly $ar \in \varepsilon'^{\perp}$.

$\supseteq$. If $r \in \varepsilon'^{\perp}$, then $\varepsilon' r = 0 = \varepsilon' 0$. Therefore, $r\ \theta'\ 0$, and $r \in [\varepsilon]$.

(b) By the preceding part, $[\varepsilon]$ and $[\varepsilon']$ are each other's annihilators.

$\subseteq$. This direction of inclusion is trivial since $\sigma\varepsilon = \sigma$ whenever $\sigma \in [\varepsilon]$, and hence $\sigma{;}0 \subseteq \varepsilon{;}0$.

$\supseteq$. For this direction, we need only prove that any σ on the right belongs to $[\varepsilon]$. By Corollary VII.2.17, $\sigma \in [\varepsilon]$ iff $\varepsilon'\sigma = 0$. But this last equality is clear since $\sigma{;}0 \subseteq \varepsilon{;}0 \subseteq \varepsilon'{:}0$.

(c) This is true since $\varepsilon\varepsilon' = 0$. □

3.2. COROLLARY. *Let $\boldsymbol{A}$ be a half-shell satisfying the hypotheses of Proposition 3.1. For any pair of complementary factor elements ε and ε', the index set X is a disjoint union of three subsets:*

$$X = (\varepsilon{;}0 \cap \varepsilon'{:}0) \cup (\varepsilon{:}0 \cap \varepsilon'{:}0) \cup (\varepsilon{:}0 \cap \varepsilon'{;}0). \tag{3.1}$$

The stalks $\boldsymbol{A}_x$ from the middle meet are trivial, where $\varepsilon(x) = 0 = \varepsilon'(x)$.

PROOF. Joining $\varepsilon{;}0 \cap \varepsilon'{;}0$ to the right side of (3.1) gives the whole space $\boldsymbol{X}$, by distributivity. But $\varepsilon{;}0 \cap \varepsilon'{;}0 = \varnothing$ by semi-integrality. □

3.3. COUNTEREXAMPLE. One might have expected a stronger conclusion in part (c) of the proposition:

$$\varepsilon(x) = \begin{cases} 1 & \text{if } x \in \varepsilon{;}0, \\ 0 & \text{if } x \in \varepsilon'{;}0, \end{cases} \tag{3.2}$$

and similarly for ε'. This would give us part of the original definition of 'reduced' in Pierce's monograph for unital rings, where each factor element takes only the values 0 or 1 on each stalk of a sheaf [**Pier67**, p. 67]. It is easy enough to see that this holds for semi-integral, unital, two-sided shells, for there $\varepsilon + \varepsilon' = 1$, or equivalently when the stalks are directly indecomposable [**Pier67**, lemma 4.2].

But for semi-integral unital half-shells, this need not be, as the following counterexample shows. Let $\boldsymbol{A}$ be the four-element subsemilattice $\{\mathsf{0}, \mathsf{e}, \mathsf{f}, \mathsf{1}\}$

of the finite meet-semilattice that is the square of a three-element chain $\boldsymbol{C}_3$ with carrier $\{0, m, 1\}$:

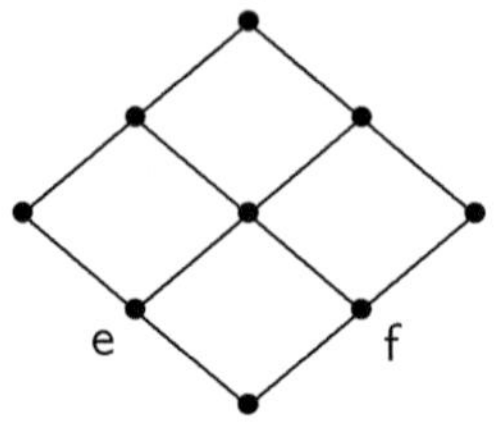

Thus, $0 = \langle 0,0 \rangle$, $\mathsf{e} = \langle m, 0\rangle$, $\mathsf{f} = \langle 0, m\rangle$, $1 = \langle 1,1\rangle$; and so $\boldsymbol{A}$ is a subdirect power of $\boldsymbol{C}_3$. Let $I = \{0, \mathsf{e}\}$ and $J = \{0, \mathsf{f}\}$. Then I and J are complementary factor ideals of $\boldsymbol{A}$ generated by factor elements, e and f. But $\mathsf{e} = \langle m, 0\rangle \neq \langle 1, 0\rangle$.

We mention, before moving on, one realization of (3.2), originated by Irving Sussman in the context of rings [**Suss58**], but it is readily definable more generally. After him we label a unital half-shell to be **associate** if it is a subdirect product of integral unital half-shells $\boldsymbol{A}_x$, and whenever σ is in A, its **associate** σ° is also in A, where, for the xth component,

$$\sigma^\circ(x) = \begin{cases} 1 & \text{if } \sigma(x) \neq 0, \\ 0 & \text{if } \sigma(x) = 0. \end{cases}$$

We obtain the following result.

3.4. PROPOSITION. *If $\boldsymbol{A}$ is an associate, semi-integral, two-sided, strict, unital half-shell, then any factor element ε is its own associate, and thus it has the form of (3.2).*

PROOF. Let ε° be the associate of a factor element ε with complement ε'. Now $\varepsilon = \varepsilon^\circ\varepsilon$ and $\varepsilon^\circ\varepsilon' = 0$ by the previous definition; hence $\varepsilon^\circ \in \varepsilon'^\perp$, and $\varepsilon'^\perp = [\varepsilon]$ by the previous proposition. Thus $\varepsilon^\circ = \varepsilon^\circ\varepsilon = \varepsilon$. □

Sussman proves that any von Neumann regular ring without nilpotents is associate. An open question is whether every semi-integral unital half-shell or shell has a representation that is associate. The counterexample above shows that being a semi-integral unital half-shell may not insure it being associate for this representation. More sharply, even though a particular representation of a semi-integral unital half-shell as a subdirect product may not be associate (see also [**Suss58**, p. 328], for a transparent example), does our sheaf constructed for $\boldsymbol{A}$ always give an associate subdirect product?

Of some historical interest, as examples of associate rings, are the periodic rings of Sussman and Foster [**SusFo60**]. We present the concepts of their paper, if not all the results, more generally.

3.5. DEFINITION. A half-shell is **power-associative** if

$$a(aa) = (aa)a$$

for all elements a in it. It is **periodic** if it is power-associative and for every element a there is a positive integer $n(a)$ greater than one such that

$$a^{n(a)} = a.$$

Forerunners of periodic rings are the p-rings of McCoy and Montgomery [**McCMo37**] and the p^k-rings of [**McCoy38**].

3.6. PROPOSITION. *If a power-associative half-shell is periodic, then it has no nilpotents; it also is von Neumann regular.*

PROOF. Suppose for some a, by way of contradiction, that $a^n = a$ and $a^m = 0$, where $n, m \geq 2$. Choose a positive integer k so that $n^k > m$. Then

$$a = a^n = (a^n)^n = \ldots = a^{n^k} = 0. \square$$

Forsythe and McCoy have shown that any regular ring without nilpotents is a subdirect product of division rings [**ForMc46**]. But Jacobson [**Jaco45**] has proven that any periodic ring is commutative; therefore, even better, it is a subdirect product of fields. Chinburg and Henriksen proved more in [**ChiHe76**]: every periodic ring is the union of a countable ascending chain of rings $\boldsymbol{R}(n)$ in which $a^n = a$ with $n \geq 2$, and n is independent of a in each $\boldsymbol{R}(n)$.

3.7. EXERCISE. Adding a unity is powerful. Show that a two-sided, strict, unital half-shell $\boldsymbol{A}$ is semi-integral if, and only if, these two statements are true:

$$\text{if } a^2 = 0, \text{ then } a = 0 \qquad (a \in A);$$
$$\text{if } (ab)(cd) = 0, \text{ then } (ac)(db) = 0 \qquad (a, b, c, d \in A).$$

To reprove the representation theorem of Sect. 1 for shells without unities requires adding a few of the ring axioms, but not all of them. Distributivity is needed to insure that any annihilator is closed to addition and therefore an ideal. Addition must be like a group operation, or more generally a loop, in order for each annihilator to be the equivalence class of some congruence. Although written additively, the loop operation need not be commutative. Loops were defined and discussed toward the end of Sect. VIII.1.

3.8. THEOREM. *Let the algebra,* $\boldsymbol{A} = \langle A; +, \times, 0 \rangle$, *be of type* $\langle 2,2,0 \rangle$, *and assume that it has three properties:*

(i) $\langle A; +, 0 \rangle$ *is a loop, with right and left differences,* $/$ *and* $\backslash$*;*
(ii) $\langle A; \times, 0 \rangle$ *is a null-symmetric half-shell with no nilpotents; and*
(iii) *multiplication* $\times$ *distributes over addition* $+$ *on both sides.*

Two conclusions follow.

(a) *The algebra $\boldsymbol{A}$ is isomorphic to a subalgebra of the algebra $\boldsymbol{\Gamma}(\mathcal{A})$ of all global sections of a sheaf $\mathcal{A}$ of algebras of the same type but without divisors of zero in any of the stalks. The base space of this sheaf is an extremally disconnected Boolean space; it is dual to the complete Boolean algebra of annihilators of subsets of $\boldsymbol{A}$.*

(b) *Further, if $\sigma{:}0$ is clopen for each global section σ, then $\mathcal{A}$ is a Boolean product and $\boldsymbol{\Gamma}(\mathcal{A})$ is completely Baer–Stone.*

PROOF. Before starting the proof proper, we give a brief outline of it. First, we show that annihilators are idéals, in a sense to be made precise. Second, using the rather special structure of congruences in loops, we characterize in an especially simple way the minimal congruences generated by these ideals. Third, we demonstrate that these congruences form a complete Boolean subsemilattice of **Con** $\boldsymbol{A}$. We finish by invoking the structure theory developed earlier.

(a) Extend the earlier definition of idéal: a subset I of A is an **idéal** if

(i) $0 \in I$;
(ii) if $a \in A$ and $i \in I$, then $ai \in I$ and $ia \in I$; and
(iii) if $i, j \in I$, then $i + j \in I$, $i/j \in I$, and $i\backslash j \in I$.

To prove that each annihilator $K^{\perp}$ is an idéal, only sums and differences are new to check. The sum condition follows readily from distributivity: if $ki = 0$ and $kj = 0$ whenever $k \in K$, then $k(i + j) = ki + kj = 0 + 0 = 0$. The difference condition is proven in the same way, providing we know some distributive laws for the subtractions over multiplication:

$$k(i\backslash j) = (ki)\backslash(kj), \text{ and } k(i/j) = (ki)/(kj).$$

To prove the last, it suffices to show by distributivity for addition that

$$k(i/j) + kj = k\big((i/j) + j\big) = ki.$$

The set, $\mathrm{Ideal}^{\perp} \boldsymbol{A} = \{I \subseteq A \mid I^{\perp\perp} = I\}$, forms a Boolean subsemilattice, $\mathbf{Ideal}^{\perp} \boldsymbol{A} = \langle \mathrm{Ideal}^{\perp} \boldsymbol{A}; \cap, {}^{\perp}, 0\rangle$, of the semilattice **Idéal** $\boldsymbol{A}$ of idéals. This is so since the statement about $\mathrm{Ideal}^{\perp} \boldsymbol{A}$ being a Boolean subsemilattice is really only about the strict half-shell of annihilators, $\langle \mathrm{Ideal}^{\perp} \boldsymbol{A}; \cap, 0\rangle$, studied earlier in Proposition 1.8.

To each such annihilator we associate a congruence of $\boldsymbol{A}$. This should be the smallest congruence $\theta(I)$ such that all elements of the idéal I are related by it. Because of the presence of a loop operation, the congruence must take a form different from that of the Rees congruence in half-shells in the preceding section. For each I in $\mathrm{Ideal}^{\perp} \boldsymbol{A}$, define a relation μ_I — a temporary notation for what will prove to be $\theta(I)$ — by

$$a\, \mu_I\, b \quad \text{iff} \quad a/b \in I.$$

This is analogous to the relationship between ideals and congruences in ring theory. It is useful to have an alternative characterization:

$$a\, \mu_I\, b \quad \text{if} \quad ka = kb \quad \text{for all} \quad k \in I^{\perp},$$

proven by distributivity:

$$\begin{aligned} a\, \mu_I b \quad & \text{iff} \quad a \diagup b \in I \\ & \text{iff} \quad k(a \diagup b) = 0 \qquad (k \in I^{\perp}) \\ & \text{iff} \quad (ka) \diagup (kb) = 0 \qquad (k \in I^{\perp}) \\ & \text{iff} \quad ka = kb \qquad (k \in I^{\perp}). \end{aligned}$$

Using what has gone before, we demonstrate, with some work, that μ_I is also a congruence of $\boldsymbol{A}$. The alternative definition quickly yields that it is an equivalence relation and that addition is preserved. To establish preservation of multiplication, we use the original definition of μ_I. If $a\, \mu_I\, b$, then $(ca) \diagup (cb) = c(a \diagup b) \in I$. Hence, $ca\, \mu_I\, cb$, and similarly for multiplication on the other side. Thus μ_I is a congruence.

Next, on our way to proving that $\mu_I = \theta(I)$, let us show that $0/\mu_I = I$. On the one hand, if $i \in 0/\mu_I$, then $ki = k0 = 0$ whenever $k \in I^{\perp}$; and hence $i \in I^{\perp\perp}$. On the other, if $i \in I$, then $i \diagup 0 = i$ implies $i\, \mu_I\, 0$.

To verify that μ_I is really the smallest congruence relating all the elements of I, suppose that η is a congruence of $\boldsymbol{A}$ such that $0/\eta \supseteq I$. Is $\eta \supseteq \mu_I$? Whenever $a\, \mu_I\, b$, then $a \diagup b \in I$, and hence by assumption $a \diagup b\ \eta\ 0$. We deduce that $a = a \diagup b + b\ \ \eta\ \ 0 + b = b$, proving that $\eta \supseteq \mu_I$.

We have just proven that μ_I is the minimal congruence such that $0/\mu_I = I$. It is also the maximal congruence. In fact, it is the unique congruence with this property. This is so because $+$ is a loop. Thus, we are justified in writing $\mu_I = \theta(I)$, and dispensing with the temporary notation.

In any case, it immediately follows from the definition of μ_I that

$$\begin{gathered} \theta(I) \cap \theta(J) = \theta(I \cap J), \\ \theta(I) = \theta(J) \ \text{ iff } \ I = J. \end{gathered}$$

Thus, the Boolean subsemilattice,

$$\mathbf{Ideal}^{\perp} \boldsymbol{A} = \langle \mathrm{Ideal}^{\perp} \boldsymbol{A}; \cap, {}^{\perp}, 0 \rangle,$$

of **Idéal** $\boldsymbol{A}$ is isomorphic to the Boolean subsemilattice,

$$\mathbf{Con}^{\perp} \boldsymbol{A} = \langle \mathrm{Con}^{\perp} \boldsymbol{A}; \cap, ', 0 \rangle,$$

of $\mathbf{Con}\, \boldsymbol{A}$, where $\mathrm{Con}^{\perp} \boldsymbol{A} = \left\{ \theta(I) \mid I \in \mathrm{Ideal}^{\perp} \boldsymbol{A} \right\}$ and $\left(\theta(I)\right)' = \theta(I^{\perp})$.

Leaning on Theorem V.2.1, Theorem 1.3 gives us a sheaf $\mathcal{A}$ of algebras over an extremaly connected Boolean space in which the stalks have no zero divisors, and $\boldsymbol{A}$ is isomorphic to a subalgebra of $\boldsymbol{\Gamma}(\mathcal{A})$. While the annihilators of the present proof and those of Theorem 1.3 are the same, their congruences are usually different; but this is not a concern since the last few assertions depend only on annihilators.

(b) We can bootstrap $\sigma : 0$ being clopen to any $\sigma : \tau$ being clopen, since $\langle A; +, 0\rangle$ is a loop. This shows that $\mathcal{A}$ is a Boolean product. By Proposition 2.2b, $\boldsymbol{\Gamma}(\mathcal{A})$ is completely Baer–Stone. □

3.9. EXAMPLE. Commutative unital rings without nilpotents are examples of algebras satisfying the hypothesis of the theorem: the stalks have no divisors of zero. Another representation of such rings, but only as a subdirect product, was obtained by [**McCoy38**] using a theorem of [**Krull29**, p. 735], and also by Birkhoff [**Birk44**]; the stalks in these representations are actually fields. Kist used minimal prime ideals to obtain a sheaf [**Kist69**]. However, these different proofs, including the one in this chapter, give different stalks. For example, our proof does not decompose the ring $\mathbb{Z}$ of integers, that is, the base space has only one element; but $\mathbb{Z}$ does decompose as a nontrivial subdirect product of fields $\mathbb{Z}_p$.

3.10. COUNTEREXAMPLE. In the statement of the theorem, at least one of the two hypotheses — that $\times$ distributes over $+$ or that $+$ is a loop — is necessary. A counterexample is the five-element non-modular lattice $\boldsymbol{N}_5$, which is subdirectly irreducible but has divisors of zero.

We make this observation about our motivation. In order to extend the techniques for working with semi-integrality from half-shells to shells, we fastened onto the most obvious assumptions of a ring-like nature. Unfortunately, these are not enough to handle additional operations. What is needed to handle the theory of this section, but is not known, is a condition guaranteeing that annihilators extend uniquely to congruences.

It would be good at this point to see how the theorem of this section applies to unital rings that are not necessarily commutative. (All rings to the end of this section are unital.) A classical result [**Cohn77**, p. 443] about prime ideals states that:

$$\text{a ring } \boldsymbol{R} \text{ is semi-prime if, and only if,}$$
$$I^2 = \{0\} \Rightarrow I = \{0\} \qquad (I \text{ any ideal of } \boldsymbol{R}).$$

Any integral ideal is prime, but not conversely [**Hung74**, theorem III.2.15]. Hence, any semi-integral ring is semi-prime. We can exhibit these classes in the Venn diagram of Fig. 1.

The relationship of these classes with von Neumann regular rings is a bit curious. On the one hand, any regular ring is semiprime [**Goode79**, theorem 1.17]. On the other hand, as already mentioned in Sect. VIII.2, the matrix ring $\boldsymbol{M}_2(\mathbb{Q})$ is simple and regular but with nilpotents and divisors of zero. Hence we have an example of a regular ring that can not be

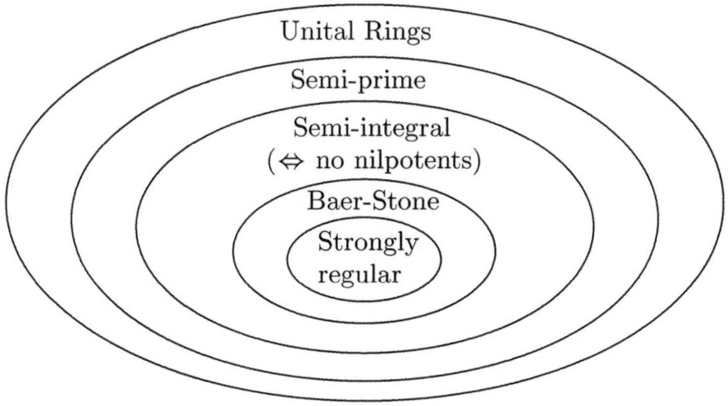

FIGURE 1. Special unital rings

represented as a subdirect product of rings without divisors of zero. Thus, $\boldsymbol{M}_2(\mathbb{Q})$ is an example of a (semi)prime ring that is not (semi)integral.

Thus von Neumann regularity, if it were put into our picture, would cut across the interior lines. But two of the regions so cut off would be vacuous, since a regular ring is strongly regular iff it has no nilpotents. In this way we see again that strongly regular rings are as far as we can push our representation theorem into the class of regular rings.[3]

In commutative rings, many of these classes coalesce, for here integral ideals and prime ideals are one and the same. As noted earlier, [**Kist69**] has proven a similar theorem for commutative rings using the spectrum of minimal prime ideals: but this spectrum falls short of being a Boolean space since it need not be compact. It is tantalizing to reflect on whether this use of minimal prime ideals could be extended to the noncommutative case or even further.

Clusters are another generalization of rings close to what we are studying. As originated by [**Good48**], they are algebras $\langle A; +, -, 0, \times\rangle$ such that $\langle A; +, -, 0\rangle$ is a group, not necessarily commutative, and $\times$ distributes over $+$ on both sides. One may demonstrate (most easily in this order) that:

$$a(b-c) = ab - ac, \qquad (a-b)c = ac - bc,$$
$$a0 = 0, \qquad 0a = a,$$
$$(-a)b = -(ab), \qquad a(-b) = -(ab).$$

[3][**Goode79**, theorems 3.2 and 3.5]. See pp. 35–36 for pertinent historical notes. See also [**Cohn77**, p. 112, exercises (4) and (5)].

Thus, the additive unity is a multiplicative nullity. Also, once one knows the trick in [**Good48**], it is easy to show that, in any cluster with a multiplicative unity, addition is commutative. In other words, a unital cluster is a non-associative ring.

3.11. COROLLARY (to Theorem 3.8). *A cluster* $\boldsymbol{A}$ *is semi-integral if, and only if, it is null-symmetric and has no nilpotents.*

PROOF. $\Rightarrow$. Use the proof of Theorem 2.3.
$\Leftarrow$. Use Theorem 3.8. $\square$

X

VARIETIES GENERATED BY PREPRIMAL ALGEBRAS

In a tour de force, Ivo Rosenberg completed the discovery of preprimal algebras, those one operation away from being primal [**Rose70**]. This chapter tests the theory of previous chapters to see how readily one may find sheaf representations of algebras in the varieties generated by preprimal algebras. As any primal variety is equivalent to the category of Boolean spaces, it is natural to wonder whether this correspondence might extend in some way to these preprimal varieties, and over what kind of spaces. The results are generally positive, with some similarities to primal varieties, but also with some significant differences. We find sheaf representations for most preprimal varieties, and identify the stalks for some of them. This leaves open for the others the determination of their stalks. This program was put forth in [**Knoe03**].

1. Overview

Here the basic notions are defined, questions about them are posed, common theorems are set out, and their sheaf representations are outlined. A comprehensive source for preprimal algebras is the excellent book by Klaus-Dieter Denecke [**Dene82**].

1.1. DEFINITION. An algebra $\boldsymbol{P}$ is **primal** if it is nontrivial, finite, and every finitary function on P is a term-operation, that is, expressible in terms of the basic operations of $\boldsymbol{P}$. An algebra is **preprimal** if it is not

A. Knoebel, *Sheaves of Algebras over Boolean Spaces*, DOI 10.1007/978-0-8176-4642-4_X, primal but it becomes primal with the addition to its basic operations of any finitary function that is not already a term-operation. A primal or preprimal variety is an equational class generated by such an algebra.

As primal varieties have been represented by Theorem VI.3.22, we move on to preprimals. For each preprimal algebra $\boldsymbol{P}$, Rosenberg [**Rose70**] described an m-ary relation ρ on P such that the n-ary term-operations of $\boldsymbol{P}$ are precisely the n-ary functions φ on P that **preserve** ρ:

$$\text{if } \rho(a_1), \rho(a_2), \ldots, \rho(a_n), \quad \text{then } \rho(\varphi(a^1), \varphi(a^2), \ldots, \varphi(a^m)),$$

for any m by n matrix of elements a^i_j in P with rows a^i and columns a_j, in the matrix notation introduced near the end of Sect. II.1. Write $\boldsymbol{P}_\rho$ for $\boldsymbol{P}$. Its particular presentation by a type is unimportant since the properties of interest are preserved by term-equivalence.

These preprimal algebras fall naturally into seven classes. We list them in Table 1, splitting some classes to better summarize information about their varieties, sheaf representations and stalks. They are ordered roughly by the complexity of their defining relations. The first relations are permutations with cycles all the same prime length. The second come from elementary Abelian p-groups. The third are proper subsets. The fourth are proper, nontrivial equivalence relations. The fifth are bounded partial orders. The last two are central and h-adic relations. Precise definitions and explanations of them will be given in subsequent sections.

Algebras in the varieties that preprimal algebras generate are the subject of investigation. A number of questions spring to mind.

(1) Which algebras have bijective representations by sheaves?
(2) What is the nature of the stalks, the base space, and the sheaf space?
(3) When does a representation agree with Birkhoff's subdirect representation, which applies to all algebras?
(4) What theorems about primal varieties extend to preprimal?

Some of these are answered in Table 1.

Assorted tools help to partly fill in this table. Many of these varieties are congruence-distributive, in some others the algebras are shells. Either guarantees Boolean factor congruences, and consequently by Theorem VI.3.15, each algebra has a bijective, reduced, and factor-transparent sheaf representation. Some of the sheaves are rather special: Boolean products come from the classes induced by permutations and subsets; those from elementary Abelian groups are over Cantor spaces.

But having BFC does not tell us what the stalks are. Although the subdirectly irreducible algebras are known for all preprimal varieties except those coming from partial orders [**Knoe85**], these are not necessarily the stalks. In fact, the bulk of work in the next two sections is proving that the stalks are subdirectly irreducible. Left open are the preprimal varieties where this is unknown.

Class	Form	Congs.	Factor congs.	Sheaf rep.	Subd. irreds.	Stalks
Primal	U2 shell	Arith.	FC	Bijec.	$\boldsymbol{P}_\varnothing$	$\boldsymbol{P}_\varnothing$
Permutation	—	Arith.	FC	Bijec.	$\boldsymbol{P}_\pi$	$\boldsymbol{P}_\pi$
Elem.Ab.grp.	Vec. sp.	Com.	Basis	Bijec.	$\boldsymbol{P}_\alpha$, $\boldsymbol{P}_\lambda$	$\boldsymbol{P}_\alpha$,$\boldsymbol{P}_\lambda$
Subset						
$\|S\| = 1$	U2 shell	Arith.	FC	Bijec.	$\boldsymbol{P}_S$	$\boldsymbol{P}_S$, $\boldsymbol{S}$
$\|S\| \geq 2$	U2 shell	Arith.	FC	Bijec.	$\boldsymbol{P}_S$, $\boldsymbol{S}$	$\boldsymbol{P}_S$, $\boldsymbol{S}$
Equivalence	U2 shell	Arith.	FC	Bijec.	$\boldsymbol{P}_\varepsilon$, $\boldsymbol{P}_\varepsilon/\varepsilon$	???
Partial order						
Dist.	U2 shell	Dist.	FC	Bijec.	Bnd.	???
Not dist.	U2 shell	—	FC	Bijec.	Unbnd.	???
Central, $h \geq 2$	U$\frac{1}{2}$shell	Dist.	FC	Bijec.	$\boldsymbol{P}_\sigma$	???
h-adic	—	—	—	Injec.	Unbnd.	???

TABLE 1. Varieties generated by primal and preprimal algebras. 'U2 shell' means unital two-sided shell; 'U$\frac{1}{2}$shell', unitary half-shell. 'Arith'. means arithmetic, that is, the variety generated is both congruence-distributive and -commutable (= -permutable). FC stands for factorable congruences, which implies BFC. 'Basis' is explained in Sect. 3. 'Bijec'. means bijective. 'Unbnd'. means unbounded in size. The various stalks are defined in Sects. 2–5. A dash means no other property similar to those elsewhere in the column is known. Question marks mean unknown.

Proving the entries in Table 1 variety by variety would be a tedious manipulation of term-operations. In some cases we lift ourselves out of complicated preprimals into simpler ones by exploiting theorem 3.1 of Denecke and Lüders [**DenLü95**], who find up to categorical equivalence unique representatives of the preprimal algebras, except those coming from bounded partial orders. McKenzie breaks the categorical equivalence of algebras and their generated varieties into three more manageable equivalences [**McKe96**]. Then we invoke Theorem VI.4.6, which says categorically equivalent varieties with Boolean factor congruences have similarly related sheaves.

We discuss these preprimal varieties in detail only for the classes when ρ is a permutation, a subset, or comes from a group; for these the stalks are determined. For permutations, the algebras in the generated variety are converted to Boolean algebras. For groups we appeal to vector spaces. For subsets, the stalks are known from the theory of quasi-primal algebras. For the remaining classes, the stalks have yet to be determined.

1.2. PROBLEM. Find a general theorem that tells when the stalks and subdirectly irreducibles are one and the same.

1.3. PROBLEM. Some of these preprimal algebras belong to larger classes such as arithmetic, quasiprimal, etc. Which of the sheaf representations of this section extend to these bigger 'nearly primal' classes? Consult Pixley [1996] for a description of these.

2. From Permutations

A pretty preprimal algebra is $\langle\{0,1\};\mu,'\rangle$ where is μ is the majority function on three arguments and $'$ interchanges 0 and 1. Its clone is all 'self-dual' functions f:

$$f(a',b',\ldots) = \big(f(a,b,\ldots)\big)' \qquad (a,b,\ldots \in \{0,1\}).$$

In other words, the clone is all finitary functions that commute with the permutation $'$. Adding either constant as an operation makes the algebra primal, that is, it becomes term-equivalent to a Boolean algebra.

More generally, suppose π is a permutation of a finite set P that has all cycles of the same prime length. Consider the algebra $\boldsymbol{P}_\pi$ whose operations are all those finitary functions commuting with π. Its type may well be taken as these operations arranged in some order, that is, they may type themselves as operation symbols. As a function, π is also a binary relation. Alternatively then, the algebra $\boldsymbol{P}_\pi$ has as operations those finitary functions preserving π. Note that π itself is among these operations, and it is an automorphism of $\boldsymbol{P}_\pi$.

Another term-operation of $\boldsymbol{P}_\pi$ is the quaternary function:

$$t(a,b,c,d) = \begin{cases} c & \text{if } a = b; \\ d & \text{if } a \neq b. \end{cases} \tag{2.1}$$

Thus, $\boldsymbol{P}_\pi$ is a discriminator algebra, and hence simple.

Pixley proved that this discriminator term also implies that Var $\boldsymbol{P}_\pi$ is congruence-distributive and congruence-commutable [**Pixl63**]. It is essentially a primal variety, although not categorically equivalent to one, since constants are missing. At any rate, each algebra of Var $\boldsymbol{P}_\pi$ is bijectively represented by a Hausdorff sheaf, all of whose stalks are isomorphic to $\boldsymbol{P}_\pi$. This is now stated formally, and may be proven using theorem IV.9.4 of [**BurSa81**] about discriminator varieties, or by the proof given below, which exploits the closeness of $\boldsymbol{P}_\pi$ to primality.

2.1. THEOREM. *Let $\boldsymbol{P}_\pi$ be the preprimal algebra on a carrier of kp elements whose clone is all finitary functions that preserve a permutation ρ with a finite number of k disjoint cycles each of prime length p. Each algebra $\boldsymbol{A}$ in* Var $\boldsymbol{P}_\pi$ *is a Boolean power of $\boldsymbol{P}_\pi$.*

PROOF. Since Var $\boldsymbol{P}_\pi$ is congruence-distributive, it has factorable congruences. By Theorem VI.3.15 it is categorically equivalent to all the reduced and factor-transparent sheaves over Boolean spaces with stalks yet to be verified as $\boldsymbol{P}_\pi$. A more indirect approach will identify these stalks with a constant-free reduct of a primal algebra. Initially, we prove the theorem for preprimal algebras $\boldsymbol{P}_\pi$ coming from permutations π that have just one cycle ($k = 1$).

Set $\mathfrak{P}_\pi = \operatorname{Var} \boldsymbol{P}_\pi$. Expand the type τ of $\mathfrak{P}_\pi$ to $\overline{\tau}$ by adding a constant symbol $\mathfrak{c}$. From any algebra $\boldsymbol{A}$ in $\mathfrak{P}_\pi$ create an algebra $\overline{\boldsymbol{A}}^c$ of type $\overline{\tau}$ by interpreting $\mathfrak{c}$ as an arbitrary element $c = \mathfrak{c}^{\boldsymbol{A}}$ of $\boldsymbol{A}$. Let $\overline{\mathfrak{P}_\pi}$ be the collection of these $\overline{\boldsymbol{A}}^c$, containing an expansion of each $\boldsymbol{A}$ in $\mathfrak{P}_\pi$ by each element c in A, so as to create a variety (it is closed to H, S, and P). Since $\boldsymbol{P}_\pi$ is preprimal without constant operations, $\overline{\boldsymbol{P}_\pi}^c$ must be primal.

A crucial claim is that $\operatorname{Var} \overline{\boldsymbol{P}_\pi}^c = \overline{\mathfrak{P}_\pi}$. Since $\operatorname{Var} \overline{\boldsymbol{P}_\pi}^c$ is primal, it has no nontrivial subvarieties. Hence $\overline{\mathfrak{P}_\pi} \supseteq \operatorname{Var} \overline{\boldsymbol{P}_\pi}^c$.

The other direction, $\operatorname{Var} \overline{\boldsymbol{P}_\pi}^c \supseteq \overline{\mathfrak{P}_\pi}$, will be proven by working instead with identities, by proving that $\operatorname{Id} \overline{\boldsymbol{P}_\pi}^c \subseteq \operatorname{Id} \overline{\boldsymbol{A}}^c$ for each $\boldsymbol{A}$ in $\mathfrak{P}_\pi$, courtesy of the Galois connection between models and identities explained in Sect. III.1. To that end, assume that $\boldsymbol{A} \in \mathfrak{P}_\pi$ and $c \in A$, and that

$$\overline{t}(x_1, \ldots, x_n) \approx \overline{w}(x_1, \ldots, x_n) \quad (\operatorname{Id} \overline{\boldsymbol{P}_\pi}^c), \tag{2.2}$$

for some n-ary terms $\overline{t}$ and $\overline{w}$ of type $\overline{\tau}$. Then, $\overline{t}$ is of the form $t(\mathfrak{c}, x_1, \ldots, x_n)$ for some $(n+1)$-ary term of type τ; similarly, for some $\overline{w}$.

Let us first work in the preprimal $\boldsymbol{P}_\pi$ and prove that

$$t^{\boldsymbol{P}_\pi}(a_0, a_1, \ldots, a_n) = w^{\boldsymbol{P}_\pi}(a_0, a_1, \ldots, a_n) \qquad (a_0, a_1, \ldots, a_n \in P_\pi). \tag{2.3}$$

Consider specific $a_0, a_1, \ldots, a_n$. Since π is a cyclic permutation, there is a positive integer e such that $\pi^e(c) = a_0$. By (2.2),

$$t^{\boldsymbol{P}_\pi}(c, \pi^{-e}(a_1), \ldots, \pi^{-e}(a_n)) = w^{\boldsymbol{P}_\pi}(c, \pi^{-e}(a_1), \ldots, \pi^{-e}(a_n)). \tag{2.4}$$

As π is a term of type τ, and as it preserves all term-operations of that type, we may apply it e times to each side of (2.4) and move it inside to obtain (2.3), that is,

$$\begin{aligned} t^{\boldsymbol{P}_\pi}(a_0, a_1, \ldots, a_n) &= \pi^e\big(t^{\boldsymbol{P}_\pi}(c, \pi^{-e}(a_1), \ldots, \pi^{-e}(a_n))\big) \\ &= \pi^e\big(w^{\boldsymbol{P}_\pi}(c, \pi^{-e}(a_1), \ldots, \pi^{-e}(a_n))\big) \\ &= w^{\boldsymbol{P}_\pi}(a_0, a_1, \ldots, a_n). \end{aligned}$$

In other words, $t \equiv w$ ($\operatorname{Id} \boldsymbol{P}_\pi$).

In the equational class generated by $\boldsymbol{P}_\pi$, $\boldsymbol{A}$ also satisfies $t \approx w$. Therefore, by putting $a_0 = c$ in (2.3), we satisfy (2.2) in $\overline{\boldsymbol{A}}^c$. We have proven that $\operatorname{Var} \overline{\boldsymbol{P}_\pi}^c \subseteq \overline{\mathfrak{P}_\pi}$.

So, if $\boldsymbol{A}$ is in $\operatorname{Var} \boldsymbol{P}_\pi$, then $\overline{\boldsymbol{A}}^c$ is in the primal variety $\overline{\mathfrak{P}_\pi}$. Hence, $\overline{\boldsymbol{A}}^c$ is a Boolean power of $\overline{\boldsymbol{P}_\pi}^c$ by Theorem VI.3.22. By removing the constant $\mathfrak{c}$ from the type, we find that $\boldsymbol{A}$ is a Boolean power of $\boldsymbol{P}_\pi$.

Now consider the general case where ρ is a general permutation of k disjoint cycles of cyclic permutations of length p with k being any positive integer, and π is the earlier cyclic permutation of length p. Categorical equivalence, $\mathbf{Var}\,\boldsymbol{P}_\rho \simeq \mathbf{Var}\,\boldsymbol{P}_\pi$, as a matrix power, comes from [**DenLü95**]. This equivalence in turn, by Theorem VI.4.6, gives us Boolean powers in $\mathrm{Var}\,\boldsymbol{P}_\rho$. □

3. From Groups

The third class of preprimal varieties of Rosenberg are created out of finite elementary Abelian p-groups [**Rose70**]. In the varieties generated by them there are two non-isomorphic subdirectly irreducibles, which are also the stalks of a representing sheaf over a Cantor space. Although these varieties do not have Boolean factor congruences, in each algebra there is a maximal Boolean algebra of factor congruences, generated by congruences corresponding to the elements of a basis in the sense of linear algebra. Throughout we use properties of these varieties found in [**Knoe85**, sect. 3].

3.1. DEFINITION. Let $\boldsymbol{P}$ be a finite, non-trivial, elementary Abelian p-group $\langle P; +, -, 0\rangle$. Define on P the quaternary relation α:

$$\langle a,b,c,d\rangle \in \alpha \quad \text{if} \quad a+b=c+d \qquad (a,b,c,d \in P).$$

This determines a preprimal algebra $\boldsymbol{P}_\alpha$ whose clone of operations is all those finitary functions preserving α, the **affine** operations.

In the variety generated by $\boldsymbol{P}_\alpha$ there are two subdirectly irreducibles $\boldsymbol{P}_\alpha$ and $\boldsymbol{P}_\lambda$. The algebra $\boldsymbol{P}_\lambda$ is $\boldsymbol{P}_\alpha$ in which each n-ary term-operation φ is re-interpreted:

$$\varphi^{\boldsymbol{P}_\lambda}(a_1,\ldots,a_n) = \varphi^{\boldsymbol{P}_\alpha}(a_1,\ldots,a_n) - \varphi^{\boldsymbol{P}_\alpha}(0,0,\ldots,0).$$

Alternatively, these are the affine operations preserving 0, or equivalently, those operations preserving the ternary relation λ:

$$\langle a,b,c\rangle \in \lambda \quad \text{if} \quad a+b=c \qquad (a,b,c \in P).$$

Hence these are called **linear**.

These irreducibles turn out to be also the stalks in the sheaf created for any algebra $\boldsymbol{A}$ in $\mathrm{Var}\,\boldsymbol{P}_\alpha$. But to create the sheaf we need a Boolean subsemilattice of $\mathrm{Con}\,\boldsymbol{A}$. There is at least one (although $\boldsymbol{A}$ does not have Boolean factor congruences) since there is always a maximal Boolean algebra of factor congruences. We state this formally.

3.2. THEOREM. *For any algebra $\boldsymbol{A}$ in the variety $\mathrm{Var}\,\boldsymbol{P}_\alpha$ generated by the preprimal algebra $\boldsymbol{P}_\alpha$, there is a bijective sheaf representation of it over a Cantor space on which every stalk is isomorphic to $\boldsymbol{P}_\alpha$ or $\boldsymbol{P}_\lambda$. Any two such representations of $\boldsymbol{A}$ are isomorphic.*

PROOF. We begin by recalling some useful observations about $\boldsymbol{P}_\alpha$ and $\boldsymbol{P}_\lambda$. Any n-ary term-operation φ of $\boldsymbol{P}_\alpha$ may be written:

$$\varphi(a_1, \ldots, a_n) = c_0 + \varepsilon_1(a_1) + \ldots + \varepsilon_n(a_n) \tag{3.1}$$

for unique endomorphisms ε_i of $\boldsymbol{P}$ and a constant, $c_0 = \varphi(0, 0, \ldots, 0)$. When interpreted in $\boldsymbol{P}_\lambda$, $c_0 = 0$. Hence, both $\boldsymbol{P}_\alpha$ and $\boldsymbol{P}_\lambda$ have the original $\boldsymbol{P}$ as a reduct, which is a vector space. So another way of looking at the algebras of Var $\boldsymbol{P}_\alpha$ is that they are vector spaces with additional unary operations, the endomorphisms, which are the same as linear transformations.

Any finite elementary Abelian p-group $\boldsymbol{P}$ is a direct power of the cyclic group $\mathbb{Z}_p$ on p elements:

$$\boldsymbol{P} \cong (\mathbb{Z}_p)^k.$$

There are different cases, according to whether $k = 1$ or $k > 1$. The burden of the proof is when $k = 1$, which we tackle now. The endomorphisms in (3.1) of $\mathbb{Z}_p$ are now multiplications by a constant: $\varepsilon_i(x) = c_i x$. This amounts to repeated addition in $\mathbb{Z}_p$. So for the type of $\boldsymbol{P}_\alpha$ we now need only $+$ and a constant κ that we take to be 1 in $\boldsymbol{P}_\alpha$ and 0 in $\boldsymbol{P}_\lambda$. Since the presence or absence of constants does not affect congruences, the theory simplifies to linear algebra when decomposing an algebra $\boldsymbol{A}$ in Var $\boldsymbol{P}_\alpha$.

Recall that in a vector space, ideals are subspaces, corresponding to congruences and sesquimorphisms (see Exercise II.2.21). Any of these might be used in the proof, but congruences, although unusual in vector spaces, are chosen since previous theorems are phrased in that language.

To find the sheaf representation when $k = 1$, let H be a basis of an algebra $\boldsymbol{A}$ in Var $\boldsymbol{P}_\alpha$. Subsets of H generate subspaces of $\boldsymbol{A}$, or equivalently congruences. All these subsets of H forms a complete and atomic Boolean algebra, corresponding to a maximal Boolean lattice $\boldsymbol{B}$ of factor congruences of $\boldsymbol{P}_\alpha$. They are *factor* congruences since any subspace of a vector space has a complement. The topological space dual to $\boldsymbol{B}$ is a Cantor space, and all Cantor spaces arise in this way (Sect. III.3). Now $\langle \boldsymbol{A}, \boldsymbol{B} \rangle$ is a factorial brace, and Theorem VI.1.8 gives a sheaf $\mathcal{A}$ such that $\Gamma(\mathcal{A}) \cong \boldsymbol{A}$.

Left is the determination that the stalks are $\boldsymbol{P}_\alpha$ and $\boldsymbol{P}_\lambda$. Since $\boldsymbol{B}$ is atomic and complete, its prime ideals are principal, and consequently their suprema are co-atoms θ of **Con** $\boldsymbol{A}$. The quotient $\boldsymbol{A}/\theta$ is the vector space $\mathbb{Z}_p$ whose type is expanded by the constant $\mathfrak{c}$. If $\mathfrak{c}^{\mathbb{Z}_p} = 0$, then $\boldsymbol{A}/\theta$ is $\boldsymbol{P}_\lambda$. If $\mathfrak{c}^{\mathbb{Z}_p} \neq 0$, then $\boldsymbol{A}/\theta$ is $\boldsymbol{P}_\alpha$.

Moving to the general case where $k \geq 1$, we now write α^k for the original α of Definition 3.1, suggesting that α^k is the kth power of α^1. Let $\boldsymbol{B}$ be in Var $\boldsymbol{P}_{\alpha^k}$. We want a sheaf representation for it with stalks $\boldsymbol{P}_{\alpha^k}$ and $\boldsymbol{P}_{\lambda^k}$. Denecke and Lüders [**DenLü95**] tell us that $\boldsymbol{P}_{\alpha^k}$ is a matrix power of $\boldsymbol{P}_{\alpha^1}$ (whose variety was just studied):

$$\boldsymbol{P}_{\alpha^k} \cong \boldsymbol{P}_{\alpha^1}{}^{[k]},$$

implying by Definition III.2.25 and Theorem III.2.28 that their varieties are categorically equivalent [**McKe96**]:

$$\operatorname{Var} \boldsymbol{P}_{\alpha^k} \simeq \operatorname{Var}\left(\boldsymbol{P}_{\alpha^1}{}^{[k]}\right) \simeq \operatorname{Var}(\boldsymbol{P}_{\alpha^1}).$$

Hence, any $\boldsymbol{B}$ in $\operatorname{Var} \boldsymbol{P}_{\alpha^k}$ is isomorphic to some kth matrix power of an algebra $\boldsymbol{A}$ in $\operatorname{Var} \boldsymbol{P}_{\alpha^1}$. Now let $\mathcal{A}$ be its sheaf as constructed as earlier in this proof, that is $\boldsymbol{A} \cong \Gamma(\mathcal{A})$. It follows that their kth matrix powers are isomorphic: $\boldsymbol{A}^{[k]} \cong (\Gamma(\mathcal{A}))^{[k]}$. But $(\Gamma(\mathcal{A}))^{[k]} \cong \Gamma(\mathcal{A}^{[k]})$ by Proposition VI.4.7. Therefore, any algebra of $\operatorname{Var} \boldsymbol{P}_{\alpha^k}$ is represented by a sheaf over a Cantor space on which every stalk is isomorphic to $\boldsymbol{P}_{\alpha^k}$ or $\boldsymbol{P}_{\lambda^k}$. □

4. From Subsets

The preprimal algebra $\boldsymbol{P}_S$ has as term-operations all those finitary functions preserving a proper, nonempty subset S of a finite set P. Foster and Pixley [**FostPi64**] call this a **semiprimal** algebra; they proved that its variety $\operatorname{Var} \boldsymbol{P}_S$ has two subdirectly irreducible algebras up to isomorphism, $\boldsymbol{P}_S$ and $\boldsymbol{S}$ when $|S| \geq 2$, and only $\boldsymbol{P}_S$ when $|S| = 1$. It is easy to find terms showing that $\operatorname{Var} \boldsymbol{P}_S$ is congruence-distributive and congruence-commutable (Proposition III.1.6). Hence, it has factorable congruences (Proposition VI.2.12), which provide bijective sheaf representations (Theorem VI.3.15).

But what are the stalks? The answer comes from realizing that $\boldsymbol{P}_S$ is a discriminator algebra, meaning a quasiprimal. Keimel and Werner [**KeiWe74**] (see also [**BurSa81**, 1981, theorem IV.9.4]) showed that the stalks are among the subalgebras of $\boldsymbol{P}_S$, which are $\boldsymbol{P}_S$ and $\boldsymbol{S}$. Although $\boldsymbol{S}$ may have only one element, it may still be a stalk. We summarize this in a theorem, and leave an independent proof of it as a problem.

4.1. THEOREM. *Each algebra $\boldsymbol{A}$ in* $\operatorname{Var} \boldsymbol{P}_S$ *is a Boolean product of $\boldsymbol{P}_S$ and $\boldsymbol{S}$.*

Incidentally, Denecke and Lüders [**DenLü95**] have shown that there are two categorical representatives for these semiprimal algebras with one subalgebra: $\{0,1,2\}_{\{0,1\}}$ if $|S| \geq 2$, and $\{0,1\}_{\{0\}}$ if $|S| = 1$. For the latter, we have the variety of Boolean rings $\langle A; +, \times, 0\rangle$, lacking an explicit unity (see [**Halm63**, sect. 19, exer. 1]). For the many different ways of representing Boolean rings by sheaves, look toward the end of Sect. XII.5.

4.2. PROBLEM. Find an independent proof of Theorem 4.1 based on a Boolean algebra $\boldsymbol{B}$ residing in each algebra $\boldsymbol{A}$ of $\operatorname{Var} \boldsymbol{P}_S$. Here is a hint when $\boldsymbol{P}_S = \{0,1,2\}_{\{0,1\}}$. Define $\boldsymbol{B}$ as the relativized algebra, $\mu\boldsymbol{A}$, where μ comes from a term operation in $\boldsymbol{P}_S$ defined as

x	0	1	2
$\mu(x)$	0	1	1

Then lift the topological representation of $\boldsymbol{B}$ to one for $\boldsymbol{A}$.

5. Remaining Preprimal Varieties

We mention the remaining four classes of preprimal algebras, which we know less about. Equivalences, partial orders, and central relations generate varieties with Boolean factor congruences. But we do not know the stalks of their canonical sheaves. The last class of h-adic relations is more mysterious; for them we know only the subdirect sheaf representations, which are injective but not necessarily surjective.

EQUIVALENCES. For any nontrivial, proper equivalence relation ε on a finite set P, the preprimal algebra $\boldsymbol{P}_\varepsilon$ is the algebra on P whose clone of operations is the set of all finitary functions preserving ε. Of necessity, it must have at least three elements. It is among the linear hemiprimal algebras of Foster [**Fost70**]. Since Var $\boldsymbol{P}_\varepsilon$ is congruence-distributive (see [**Knoe85**]), it has factorable congruences by Proposition VI.2.12, BFC by Theorem VI.3.9, and canonical sheaf representations over Boolean spaces by Theorem VI.3.15; consequently the sheaves are reduced and factor-transparent. The remaining open question is whether its subdirectly irreducible algebras are the stalks. Denecke and Lüders [**DenLü95**] have shown that the varieties Var $\boldsymbol{P}_\varepsilon$ generated by the $\boldsymbol{P}_\varepsilon$ are all categorically equivalent to the three-element case, which should simplify the analysis.

PARTIAL ORDERS. These relations are easy enough to describe, but the varieties their algebras generate are remarkable. Throughout let $\leq$ be a partial order on a finite set P, with distinct lower and upper bounds of 0 and 1. Examples are finite lattice orders. The simplest nontrivial lattice order is that on two elements. Then the variety Var $\boldsymbol{P}_\leq$ generated is term-equivalent to bounded distributive lattices, which has the two-element lattice as the sole subdirectly irreducible. But the sheaves coming from a canonical BFC decomposition may have other stalks. For example, look at the six-element product in Fig. 1a, which has two- and three-element chains as BFC stalks, but just the two-element chain as a subdirect stalk.

Examples of some non-lattice orders $\leq$ are illustrated in Fig. 1b–d.

The algebra $\boldsymbol{P}_\leq$, whose clone is all operations preserving $\leq$, has a unital two-sided shell reduct $\langle P; +, \times, 0, 1\rangle$: here $+$ satisfies $0 + a = a = a + 0$, and is 0 otherwise; $\times$ satisfies $0 \times a = 0 = a \times 0$ and $1 \times a = a = a \times 1$, and is 1 otherwise. So everything already proven about unital shells applies to any algebra $\boldsymbol{A}$ of Var $\boldsymbol{P}_\leq$. By Theorem VII.3.2, therefore, $\boldsymbol{A}$ has factorable congruences, and by Theorems VI.3.9 and VI.3.15 it is represented bijectively by a reduced and factor-transparent sheaf $\mathcal{A}$ over a Boolean space, that is, $\boldsymbol{A}$ is isomorphic to the unital, two-sided shell of all global sections of $\mathcal{A}$.

In general the stalks are not known, but in some varieties Var $\boldsymbol{P}_\leq$ they may be numerous, as discussed below. Despite this, Var $\boldsymbol{P}_\leq$ is equationally maximal, that is, there is no equational class between it and the trivial variety [**Knoe76, Dene78**].

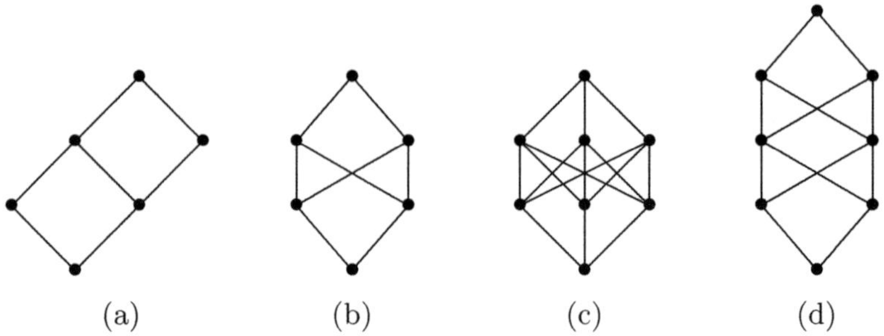

FIGURE 1. Bounded partial orders

These partial orders split into two classes according to whether the variety $\boldsymbol{P}_{\leq}$ is congruence-distributive or not, and this has consequences for the subdirectly irreducible algebras of Var $\boldsymbol{P}_{\leq}$, and consequently also for the sizes of the stalks, which have the subdirectly irreducibles as homomorphic images. If it is distributive, as in Fig. 1b,c, then there is one subdirectly irreducible. If not, as in (d), then McKenzie proved that Var $\boldsymbol{P}_{\leq}$ has subdirectly algebras of arbitrarily large cardinality [**McKe90**].

5.1. PROBLEM. What is the nature of the stalks of the sheaves coming from bounded partial orders?

CENTRAL RELATIONS. Their definition has four clauses. We abbreviate $\langle a_1, a_2, \ldots, a_h \rangle$ as $\vec{a}$.

5.2. DEFINITION. An h-ary relation σ on a finite set A is **central** if

(i) it is **totally symmetric**, that is, for all $\vec{a}$ in A^h and for all permutations π of $\{1, 2, \ldots, h\}$, if $\vec{a} \in \sigma$, then $\langle a_{\pi(1)}, a_{\pi(2)}, \ldots, a_{\pi(h)} \rangle \in \sigma$;
(ii) it is **totally reflexive**, that is, for all $\vec{a}$ in A^h with at least two of the a_i equal, we have that $\vec{a} \in \sigma$;
(iii) there is an a_1 such that for all $a_2, \ldots, a_h$ in A we have $\vec{a} \in \sigma$; and
(iv) $\sigma \neq A^h$.

The **center** of σ is the set of all those a_1 that satisfy (iii).

These clauses work together to force A to have more than h elements. The preprimal algebra $\boldsymbol{P}_\sigma$ has as term-operations all those preserving a central relation σ. Denecke and Lüders have shown that their categorical equivalence depends only on h, one categorical representative for each h, except when $h = 1$ where there are two [**DenLü95**]. We discuss here the case when $h \geq 2$; the case when $h = 1$ was discussed in Sect. 4. When $h \geq 2$, each representative comes from a particular central relation σ_h, now defined.

5.3. DEFINITION. The relation σ_h is the universal h-ary relation on the set $\{0, 1, \ldots, h\}$ of $h + 1$ elements, less all permutations of $\{1, 2, \ldots, h\}$. In other words, σ_h is the only h-ary central relation on $\{0, 1, \ldots, h\}$ with center $\{0\}$.

Each algebra $\boldsymbol{P}_{\sigma_h}$ has an operation $\times$ that is a unital half-shell:

$$a \times b = \begin{cases} b, & \text{if } a = 1; \\ 0, & \text{otherwise.} \end{cases}$$

Here 0 is the nullity, and 1 is the one-sided unity. Denecke [1978] established the congruence-distributivity of Var $\boldsymbol{P}_{\sigma_h}$. Either property tells us that Var $\boldsymbol{P}_{\sigma_h}$ has Boolean factor congruences (Theorem VII.2.2 and Proposition VI.2.12). By Theorem VI.3.15 we have a canonical representation of any algebra in Var $\boldsymbol{P}_{\sigma_h}$ by a reduced and factor-transparent sheaf over a Boolean space.

There is one subdirectly irreducible algebra $\boldsymbol{P}_{\sigma_h}$ in each of these categories [**Dene78, Knoe76**]. However, the stalks of the sheaves may not be this algebra; so the stalks have yet to be determined. Theorem VI.4.6 takes us from the sheaf representations in the special categories Var $\boldsymbol{P}_{\sigma_h}$ to those in any Var $\boldsymbol{P}_{\sigma}$, where σ is an h-ary central relation with $h \geq 2$.

5.4. EXERCISE. (a) Show that any preprimal algebra coming from an h-ary central relation with $h = 2$ has a two-sided unital shell as a term-reduct.

(b) Show that no preprimal algebra coming from an h-ary central relation with $h \geq 3$ can have a two-sided unital shell as a term-reduct.

---

h-ADIC RELATIONS. This last class of preprimal algebras of Rosenberg [**Rose70**] is the most fascinating in a degenerate way. It does not include any classical algebras since there are no binary term-operations that take all values and depend on both arguments. There are many subdirectly irreducible algebras, finite in number up to isomorphism, but growing as h increases. As algebras in h-adic varieties do not necessarily have Boolean factor congruences, as they did in previous preprimal varieties, there are not obvious bijective sheaf representations of them. For that reason this discussion is extremely sketchy, and depends heavily on the analysis in [**Knoe85**]. The definition of h-adic relations comes in three parts; we assume that $3 \leq h < \infty$ and $1 \leq k < \infty$.

5.5. DEFINITION. (i) A **primary h-adic relation** is the h-ary relation ζ on a set A of h elements such that $\langle a_1, a_2, \ldots, a_h \rangle \in \zeta$ iff at least two of the a_i are equal.

(ii) An **elementary h-adic relation** is a kth power, $\eta = \zeta^k$, of a primary h-adic relation ζ. It is on the set, $B = A^k$.

(iii) An h-**adic relation** is a preimage θ of an elementary h-adic relation η. In symbols, $\theta = \{\langle c_1, \ldots, c_h\rangle \mid \langle \varphi(c_1), \ldots, \varphi(c_h)\rangle \in \eta\}$ for some surjective function $\varphi: C \to B$.

As before, we designate P_ζ, P_η, and P_θ as the preprimal algebras on a set P of the right size whose operations preserve these relations.

Denecke and Lüders [**DenLü95**, theorem 3.1] establish one categorical representative for each h: namely, Var $\boldsymbol{P}_\zeta$ for ζ the primary h-adic relation on the set $\{1, 2, \ldots, h\}$. This reduction should aid in the study of these illusive varieties. The clone of $\boldsymbol{P}_\zeta$ is the set of all unary functions, all projections, and all other finitary functions with ranges smaller than the carrier.

These h-adic varieties do not have Boolean factor congruences. To see this, consider the simplest case of ζ being the 3-adic primary relation on the set $\{0, 1, 2\}$. Among the subdirectly irreducible algebras in the variety V_ζ is $\boldsymbol{A}_{S_3}$, as designated in [**Knoe85**], which has just two elements, say 0 and ∞, and whose clone is term equivalent to the constant function ∞. So the variety generated by $\boldsymbol{A}_{S_3}$ is that of pointed sets with a designated element ∞. In Var $\boldsymbol{A}_{S_3}$ look at the three element algebra $\boldsymbol{A}$ on $\{0, 1, \infty\}$. This does not have BFC.

All we can say generally is that these algebras have injective subdirect sheaf representations (Theorem IV.2.7), leaving open whether they are ever surjective.

5.6. PROBLEM. Do the algebras in h-adic varieties have bijective sheaf representations that are natural and useful.

5.7. PROBLEM. Find the stalks in the sheaf representations for the remaining preprimal varieties.

XI
Return to General Algebras

Previous chapters have presented clear applications with firm conclusions. When we left Chap. VI earlier, we turned away from general algebras so that we could obtain specific results in shells that would directly generalize theorems about rings, semigroups and other specific algebras. We return now to arbitrary algebras, speculating about what else might be proven, and formulating a few theorems that do not yet lead to any useful applications.

The first section shows how the sheaf representation of Chap. VI may be iterated until all the stalks become directly indecomposable. This extends the paper of Burgess and Stephenson [**BurgSt78**] about noncommutative rings.

The starting point for the second section is the observation that $\mathbf{Con}\,\boldsymbol{A}$ is a shell; we use this to bootstrap the main theorem of Chap. VII to any algebra $\boldsymbol{A}$ of any type. This is possible since those congruences of an algebra $\boldsymbol{A}$ that are factor *elements* of $\mathbf{Con}\,\boldsymbol{A}$ form a Boolean subsemilattice of it. A sheaf is created over the Boolean space $\bigvee\mathbf{Spec}\,\mathbf{Elem}'\,\mathbf{Con}\,\boldsymbol{A}$, the space of suprema of prime ideals of these factor elements. Then $\boldsymbol{A}$ is isomorphic to the algebra of all its global sections.

1. Iteration

This short section iterates the construction of sheaves over Boolean spaces as announced in [**Knoe91b**]. The idea is to create a sheaf for each stalk of an already constructed sheaf, and iterate the process until it can be

A. Knoebel, *Sheaves of Algebras over Boolean Spaces*, DOI 10.1007/978-0-8176-4642-4_XI, continued no further. The rationale for this is the surprise that, even for the Boolean spectrum of a unitary ring, the stalks may be directly decomposable. Our presentation follows closely that of Burgess and Stephenson [**BurgSt78**], who are the originators of this construction in the context of noncommutative rings.

As noted by Pierce [**Pier67**, Lemma 4.2], commutative unitary rings do not exhibit this phenomenon, that is, the stalks are directly indecomposable. The proof of this depends, at a crucial point, on the fact that factor elements in such a ring are merely idempotents. The Baer–Stone shells of Chap. VIII are another class of algebras whose stalks are directly indecomposable, by virtue of having no divisors of zero.

We develop this iteration for those algebras with Boolean factor congruences whose homomorphic images are also algebras with BFC. So it works for algebras in varieties with BFC. These include unital shells and half-shells. When extended as far as possible, until all the stalks are directly indecomposable, we will have our original algebra expressed as a subdirect product of directly indecomposable algebras. Of course, if this is all that we wanted, we could use instead Birkhoff's theorem to get such a decomposition for any algebra (Theorem II.2.40, [**Birk44**]). But even better for this new construction, the stalks will now be optimal, in a certain sense to be explained shortly.

As shorthand, we will say for two comparable congruences, $\theta \subset \eta$, of $\boldsymbol{A}$ that A/η is a **Pierce stalk** of A/θ whenever $\eta/\theta \in \bigvee \operatorname{Spec} \mathbf{Con}'(\boldsymbol{A}/\theta)$,[1] that is, A/θ can be decomposed further in the sense of Chap. VI and $(A/\theta)/(\eta/\theta)$ is one of the stalks of this new sheaf.

This is in anticipation of recursively decomposing an algebra; we will stay on one level by invoking the cancellation Theorem II.1.3: $(A/\theta)/(\eta/\theta) \cong A/\eta$, whenever $\theta \subseteq \eta$.

1.1. DEFINITION. Let γ be an ordinal and let Θ be a family of proper congruences of an algebra $\boldsymbol{A}$, indexed by ordinals less than γ:

$$\Theta = \{\theta_\beta \mid 0 \le \beta < \gamma\}.$$

Then Θ will be called a **Pierce chain** whenever

(i) $\theta_0 = 0_{\mathbf{Con}\,\boldsymbol{A}}$;
(ii) if $\alpha < \beta < \gamma$, then $\theta_\alpha \subset \theta_\beta$;
(iii) if $\beta < \gamma$ and β is not a limit ordinal, then A/θ_β is a Pierce stalk of $A/\theta_{\beta-1}$; and
(iv) if $\beta < \gamma$ and β is a limit ordinal, then $\theta_\beta = \bigcup_{\alpha<\beta} \theta_\alpha$.

---

[1] To decipher the dense formula $\bigvee \operatorname{Spec} \mathbf{Con}'(\boldsymbol{A}/\theta)$, recall that an element of $\operatorname{Spec} \mathbf{Con}'(\boldsymbol{A}/\theta)$ is a prime ideal of the Boolean algebra $\operatorname{Con}'(\boldsymbol{A}/\theta)$, whose elements have the form ξ/θ for certain ξ in $\operatorname{Con} \boldsymbol{A}$. The join $\bigvee$, by the convention near the end of Sect. II.1, acts on each prime ideal separately, producing a congruence η/θ of $\boldsymbol{A}/\theta$.

A congruence θ of $\boldsymbol{A}$ is a **Pierce congruence** if it appears in a Pierce chain; the set of Pierce congruences of $\boldsymbol{A}$ is denoted $\mathcal{P}(\boldsymbol{A})$. Denote by $\mathcal{M}(\boldsymbol{A})$ the set of Pierce congruences θ for which $\boldsymbol{A}/\theta$ is directly indecomposable, and call θ **directly indecomposable**.

Because of the strictly monotonic nature of a Pierce chain Θ, we hope that eventually a congruence in it will be directly indecomposable. But this might not be; conceivably the chain could be bounded by a limit ordinal γ and never terminate; consequently we would have that $\bigcup_{\beta<\gamma}\theta_\beta = 1_{\mathbf{Con}\,\boldsymbol{A}}$. We work around this possibility in the next proof.

1.2. THEOREM. *Let* $\boldsymbol{A}$ *be any algebra with* BFC, *all of whose homomorphic images have* BFC. *Then* $\boldsymbol{A}$ *is a subdirect product of directly indecomposable quotients* $\boldsymbol{A}/\theta$ *where* θ *runs over all congruences in* $\mathcal{M}(\boldsymbol{A})$.

PROOF. To show that we have a subdirect product,

$$\bigcap \mathcal{M}(\boldsymbol{A}) = 0_{\mathbf{Con}\,\boldsymbol{A}},$$

it suffices to prove that any pair of distinct elements a and b of A are not related by some congruence of $\mathcal{M}(\boldsymbol{A})$. We use transfinite induction.

To set things up, let the congruences θ^i of $\mathcal{M}(\boldsymbol{A})$ be indexed by i in I. For each θ^i find a Pierce chain Θ^i with θ^i residing in it:

$$\Theta^i = \left\{\theta^i_\beta \mid \beta < \gamma^i\right\},$$

that is, $\theta^i = \theta^i_\beta$ for some β. We construct a maximal Pierce chain Θ,

$$\theta_0^{i_0} \subset \theta_1^{i_1} \subset \theta_2^{i_2} \subset \cdots \subset \theta_\beta^{i_\beta} \subset \cdots \qquad (\beta < \gamma),$$

out of the θ^i_β such that it is not the case that $a\ \theta^{i_\beta}_\beta\ b$ for all β less than γ. Since $\theta^{i_0}_0 = 0_{\mathbf{Con}\,\boldsymbol{A}}$, then not $a\ \theta^{i_0}_0\ b$. Let us assume that the chain has been constructed up to the ordinal β, that is, not $a\ \theta^{i_\alpha}_\alpha\ b$ for all α less than some β. There are two kinds of inductive steps, nonlimit and limit.

On the one hand, if β is a nonlimit ordinal and $\theta^{i_{\beta-1}}_{\beta-1}$ is directly decomposable, then by the nature of a Pierce decomposition

$$\theta^{i_{\beta-1}}_{\beta-1} = \bigcap_{j\in J} \theta^j_\beta$$

where J is the set of those j such that $\theta^{i_{\beta-1}}_{\beta-1}$ and θ^j_β are in the same Pierce chain. In other words, consider all successors to $\theta^{i_{\beta-1}}_{\beta-1}$. Since $a\ \theta^{i_{\beta-1}}_{\beta-1}\ b$ does not hold by hypothesis, there must be a j_β in J such that $a\ \theta^{j_\beta}_\beta\ b$ does not hold. So add it to the chain.

When, on the other hand, β is an infinite limit ordinal, define $\theta^{i_\beta}_\beta$ by:

$$\theta^{i_\beta}_\beta = \bigcup_{\alpha<\beta} \theta^{i_\alpha}_\alpha.$$

Since not $a\,\theta_\alpha^{i_\alpha}\,b$ $(\alpha < \beta)$ by hypothesis, then not $a\,\theta_\beta^{i_\beta}\,b$. Therefore, $\theta_\beta^{i_\beta}$ is proper and it should be added to the Pierce chain being constructed.

This last argument also demonstrates that the length γ of its completion can never be a limit ordinal. In this way, we obtain the required Pierce chain Θ^{i_γ}, whose last congruence $\theta_{\gamma-1}^{i_{\gamma-1}}$ is indecomposable but does not relate a and b. □

1.3. PROBLEM. In this proof we waltzed around the possibility that $\bigcup_{\beta<\gamma} \theta_\beta = 1_{\mathbf{Con}\,\boldsymbol{A}}$ by showing that the length γ of Θ^{i_γ} is never an infinite limit ordinal. Can it be otherwise for other Pierce chains? In other words, does each Pierce chain terminate in an indecomposable congruence?

While the indexing of Pierce chains in the previous proof makes it tempting to think of these congruences θ_α^i as being arranged rectangularly, as in a matrix, what we really have here is a tree, with diverging branches, perhaps infinitely long. The leaves are all distinct and indecomposable; and the θ_0^i are all equal to $0_{\mathbf{Con}\,\boldsymbol{A}}$, the root. In between θ_0^i and the leaves, the nodes branch. Figure 1 suggests this. More can be said about the congruences of $\mathcal{M}(\boldsymbol{A})$; but first a lemma, which helps to shape up the figure.

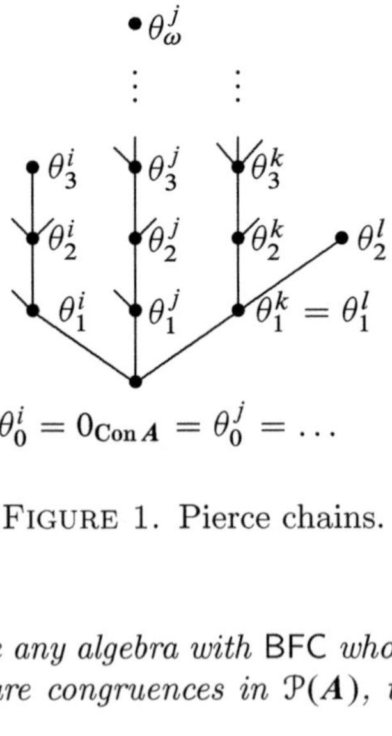

FIGURE 1. Pierce chains.

1.4. LEMMA. *Let* $\boldsymbol{A}$ *be any algebra with* BFC *whose homomorphic images have* BFC. *If* θ *and* η *are congruences in* $\mathcal{P}(\boldsymbol{A})$, *then one of three things happens:*

(a) θ *is in the Pierce chain defining* η, *and the Pierce chains of* θ *and* η *are identical up to* θ;

(b) η *is in the Pierce chain defining* θ, *and the Pierce chains of* θ *and* η *are identical up to* η; *or*

(c) $\theta \circ \eta = 1$.

PROOF. Find Pierce chains Θ and H to which θ and η belong:

$$\theta \in \Theta \quad \text{and} \quad \Theta = \langle \theta_\alpha \mid 0 \leq \alpha < \gamma \rangle,$$
$$\eta \in H \quad \text{and} \quad H = \langle \eta_\alpha \mid 0 \leq \alpha < \delta \rangle.$$

If $\theta = \eta_\alpha$ for some $\alpha < \delta$, or if $\eta = \theta_\alpha$ for some $\alpha < \gamma$, and the chains are identical up to that point, then we are finished. Otherwise, let β be the first ordinal where the chains disagree: $\theta_\beta \neq \eta_\beta$. So $\theta \supseteq \theta_\beta$ and $\eta \supseteq \eta_\beta$. Since a Pierce congruence at a limit ordinal is a union of earlier congruences, β must be a nonlimit ordinal. Therefore, $\theta_{\beta-1} = \eta_{\beta-1}$. This means that θ_β and η_β are unequal congruences yielding stalks in the sheaf for $\boldsymbol{A}/\theta_{\beta-1}$.

It follows that

$$\frac{\theta_\beta}{\theta_{\beta-1}} \circ \frac{\eta_\beta}{\theta_{\beta-1}} = 1_{\mathbf{Con}'(\boldsymbol{A}/\theta_{\beta-1})}. \tag{1.1}$$

This is because there are unequal prime ideals M and P of $\mathbf{Con}'(\boldsymbol{A}/\theta_{\beta-1})$ such that $\theta_\beta/\theta_{\beta-1} = \bigvee M$ and $\eta_\beta/\theta_{\beta-1} = \bigvee P$. Thus, there is a factor congruence ζ of $\boldsymbol{A}/\theta_{\beta-1}$ that is in $M \sim P$, implying that $\zeta' \in P$. Hence, $(\theta_\beta/\theta_{\beta-1}) \circ (\eta_\beta/\theta_{\beta-1}) \supseteq \zeta \circ \zeta' = 1$.

Equation (1.1) implies that $\theta_\beta \circ \eta_\beta = 1_{\mathbf{Con}\,\boldsymbol{A}}$. Hence $\theta \circ \eta \supseteq \theta_\beta \circ \eta_\beta = 1$. □

A consequence of this trichotomy is this: if a proper congruence θ of $\boldsymbol{A}$ is a member of a Pierce chain, then that Pierce chain is uniquely determined up to that point by θ. For, if θ and η each appear in their own Pierce chain but $\theta = \eta$, then by the lemma, $\theta = \theta \circ \eta = 1$, which is improper. Thus, Pierce chains never coalesce once they have separated.

We extend Lemma 1.4 to $\mathcal{M}(\boldsymbol{A})$, where the conclusion is simpler.

1.5. PROPOSITION. *Let $\boldsymbol{A}$ be any algebra with* BFC *whose homomorphic images have* BFC.

(a) *If two different congruences θ and η are in $\mathcal{M}(\boldsymbol{A})$, then $\theta \circ \eta = 1$.*
(b) *A congruence θ of $\boldsymbol{A}$ belongs to $\mathcal{M}(\boldsymbol{A})$ if, and only if, it is a maximal element of $\mathcal{P}(\boldsymbol{A})$.*

PROOF. (a) We see that, if θ and η satisfy (a) or (b) of the lemma, then they must be in the same Pierce chain, whence one has to be directly decomposable and not in $\mathcal{M}(\boldsymbol{A})$. So (c) of the preceding lemma holds again.

(b) By Definition 1.1. □

1.6. PROBLEM. Burgess and Stephenson [**BurgSt78**, proposition 1.8] have a third condition equivalent to the two just given in Proposition 1.5b:

θ is a minimal product-indecomposable congruence.

Does such an equivalence hold more generally than in ring theory?

1.7. PROBLEM. Can the iterated construction of sheaves in Theorem 1.2 be done more flexibly to make it applicable to all algebras, not necessarily to only those having BFC? Could this be done by using Proposition VI.2.3 to find a maximal Boolean algebra of factor congruences at each step?

The original point of this construction of this section was to find a subdirect product of an algebra such that some common properties would transfer back and forth between it and its quotients in this subdirect product. For example, Burgess and Stephenson show that being a von Neumann regular ring passes between a ring and the stalks in its sheaf, where they develop a general method for proving such results. However, in the spirit of this exploratory chapter, we know of no applications beyond those already to be found in their paper.

They have many more applications to ring theory, such as constructing a ring with an infinite Pierce chain, and they show that the obvious way of trying to turn $\mathcal{M}(\boldsymbol{A})$ into a sheaf does not work.

1.8. Problem. Find applications of this iterative process to algebraic systems more general than rings.

2. Self Help

The best results so far have been for specific classes of algebras, such as shells, in which we could uniformly identify Boolean subsemilattices in their lattices of congruences. How much can this be adapted to general algebras without special operations or constants? Remarkably, there is a universal construction working for all algebras. Since it depends on the shells of Chap. VII, it was not presented earlier. It turns on a simple observation: any bounded lattice is a shell; therefore, the lattice of congruences of any algebra $\boldsymbol{A}$ is itself a shell. So, why not use the shell's decomposition by factor objects to create a sheaf for $\boldsymbol{A}$?

This is the direction of our investigation. We show how to use $\mathbf{Con}\,\boldsymbol{A}$, as a shell, to embed any algebra $\boldsymbol{A}$ into the algebra of all continuous sections of a sheaf over a Boolean space. What the significance of this construction is, and whether it might agree with any of our previous ones, or others in the literature, are open questions.

The situation is not as straightforward as we make it sound. For the factor congruences of $\boldsymbol{A}$ may not be the factor elements in the lattice of congruences. In symbols, we do not know in general that

$$\mathrm{Elem}'\,\mathbf{Con}\,\boldsymbol{A} = \mathrm{Con}'\,\boldsymbol{A}.$$

To appreciate what inequality might mean, we look later in more detail at each direction of inclusion.

The congruences in $\mathbf{Elem}'\,\mathbf{Con}\,\boldsymbol{A}$ are characterized by the equations for factor elements taken directly out of Theorem VII.3.4 and rephrased for the two-sided unital shell, $\mathbf{Con}\,\boldsymbol{A}$. They are those congruences θ for which there is a congruence $\overline{\theta}$ such that for all congruences ζ, η in $\mathbf{Con}\,\boldsymbol{A}$:

(i) $$\theta \cap \overline{\theta} = 0,$$

(ii) $$\theta \vee \overline{\theta} = 1,$$

(iii.a) $$\theta \cap (\zeta \vee \eta) = (\theta \cap \zeta) \vee (\theta \cap \eta),$$

(iii.b) $$\overline{\theta} \cap (\zeta \vee \eta) = (\overline{\theta} \cap \zeta) \vee (\overline{\theta} \cap \eta).$$

Any two congruences, θ and $\overline{\theta}$, that satisfy these equations will be called a pair of **complementary Boolean** congruences. Notice that we have omitted all those familiar identities holding in any lattice, but that had to be stated in Chap. VII for factor elements in an arbitrary shell. A consequence of (ii) and (iii.a), which we will use repeatedly, is

(iv) $$(\zeta \cap \theta) \vee (\zeta \cap \overline{\theta}) = \zeta.$$

By way of contrast, recall that complementary factor congruences need satisfy only

(i) $$\theta \cap \theta' = 0,$$

(v) $$\theta \circ \theta' = 1,$$

Recall two things: (1) that that $\theta \vee \eta = 1$ does not guarantee that $\theta \circ \eta = 1$ and (2) that a congruence may not satisfy the distributive laws.

We have to use a notation $\overline{\theta}$ for the new complements in $\mathbf{Elem}'\,\mathbf{Con}\,\boldsymbol{A}$ that is different from the old complements θ' in $\mathrm{Con}'\,\boldsymbol{A}$, when they exist. But the next lemma shows that this is sometimes unnecessary. Remember that $\overline{\theta}$ is already unique since this complement is in a shell.

2.1. LEMMA. *If θ is both a Boolean congruence and a factor congruence of an algebra $\boldsymbol{A}$, then the complement θ' of θ is unique and $\overline{\theta} = \theta'$.*

PROOF. Let θ' be a complement of θ. From (iv) we see that

$$\theta' = (\theta' \cap \theta) \vee (\theta' \cap \overline{\theta}) = 0 \vee (\theta' \cap \overline{\theta}) = \theta' \cap \overline{\theta}.$$

And from (iii.b) we get the same meet for $\overline{\theta}$:

$$\overline{\theta} = \overline{\theta} \cap 1 = \overline{\theta} \cap (\theta \vee \theta') = (\overline{\theta} \cap \theta) \vee (\overline{\theta} \cap \theta') = 0 \vee (\overline{\theta} \cap \theta') = \overline{\theta} \cap \theta'. \quad \square$$

In the following proposition more is said about $\mathbf{Elem}'\,\mathbf{Con}\,\boldsymbol{A}$ than could be said about $\mathbf{Con}'\,\boldsymbol{A}$. It has the theorem after it as a consequence.

2.2. PROPOSITION. *For any algebra $\boldsymbol{A}$, $\mathbf{Elem}'\,\mathbf{Con}\,\boldsymbol{A}$ is a Boolean subsemilattice of $\mathbf{Con}\,\boldsymbol{A}$.*

PROOF. We need only show (Definition V.1.1) that

$$\eta \cap \overline{\theta} = 0 \Leftrightarrow \eta \subseteq \theta \qquad (\theta, \eta \in \mathrm{Elem}'\,\mathbf{Con}\,\boldsymbol{A}).$$

The reverse implication is easy, for if $\eta \subseteq \theta$, then $\eta \cap \overline{\theta} \subseteq \theta \cap \overline{\theta} = 0$.

To prove the forward direction of implication, assume that $\eta \cap \overline{\theta} = 0$ and $a\,\eta\,b$. Since $\eta = (\eta \cap \theta) \vee (\eta \cap \overline{\theta})$, there are $x_1, x_2, \ldots, x_n$ such that

$$a\,(\eta \cap \theta)\,x_1\,(\eta \cap \overline{\theta})\,x_2 \ldots x_{n-1}\,(\eta \cap \theta)\,x_n\,(\eta \cap \overline{\theta})\,b.$$

But $\eta \cap \overline{\theta} = 0$. So $x_1 = x_2$, $x_3 = x_4$, etc. Thus, $a\,(\eta \cap \theta)\,b$. Hence $a\,\theta\,b$. Therefore, $\eta \subseteq \theta$. □

For Theorems 2.3 and 2.5, recall the abbreviation of (V.2.1):

$$\boldsymbol{A}/\!\!/\boldsymbol{B} = \biguplus_{P \in \operatorname{Spec} \boldsymbol{B}} (\boldsymbol{A}/\bigvee P).$$

Here, $\boldsymbol{B}$ is a Boolean subsemilattice of $\mathbf{Con}\,\boldsymbol{A}$.

2.3. THEOREM. *For any algebra $\boldsymbol{A}$, let $\boldsymbol{B}$ be* $\mathbf{Elem}'\,\mathbf{Con}\,\boldsymbol{A}$, *and let* $\mathbf{Spec}\,\boldsymbol{B}$ *be the Boolean space of all prime ideals P of $\boldsymbol{B}$. Then $\boldsymbol{A}$ is isomorphic to a subalgebra of the algebra $\boldsymbol{\Gamma}(\mathcal{A})$ of all global sections of the sheaf $\mathcal{A}$ over* $\mathbf{Spec}\,\boldsymbol{B}$ *where $\mathcal{A} = \boldsymbol{A}/\!\!/\boldsymbol{B}$.*

PROOF. Use Proposition 2.2 and Theorem V.2.1. □

This theorem was announced in [**Knoe92a**]

2.4. PROBLEM. Find consequences of this sheaf representation for an arbitrary algebra $\boldsymbol{A}$ via its Boolean congruences. Consider this as a completion of $\boldsymbol{A}$. Is this extension unique? Does it have any density properties? Do traditional representations follow from this, as well as some new ones?

We now investigate when an inclusion in one direction or the other holds between $\mathrm{Con}'\,\boldsymbol{A}$ and $\mathrm{Elem}'\,\mathbf{Con}\,\boldsymbol{A}$.

2.5. THEOREM. *Suppose every factor congruence in an algebra $\boldsymbol{A}$ is a Boolean congruence:*

$$\mathrm{Con}'\,\boldsymbol{A} \subseteq \mathrm{Elem}'\,\mathbf{Con}\,\boldsymbol{A}.$$

Then $\mathbf{Con}'\,\boldsymbol{A}$ *is a Boolean algebra; that is, $\boldsymbol{A}$ has* BFC. *Therefore, $\boldsymbol{A} \cong \boldsymbol{\Gamma}\big(\boldsymbol{A}/\!\!/(\mathbf{Con}'\,\boldsymbol{A})\big)$.*

PROOF. Clearly, $\mathrm{Con}'\,\boldsymbol{A}$ is distributive by (iii.a) and complemented by Lemma 2.1. Consequently, it is a Boolean lattice. We finish with the representation Theorem VI.3.15. □

Distributivity of $\mathbf{Con}\,\boldsymbol{A}$ affords an example of this.

2.6. PROPOSITION. *Let $\boldsymbol{A}$ be a congruence-distributive algebra.*

(a) *Then a congruence θ of $\boldsymbol{A}$ is Boolean if, and only if, there is another congruence $\overline{\theta}$ of $\boldsymbol{A}$ such that*

$$\theta \cap \overline{\theta} = 0, \text{ and } \theta \vee \overline{\theta} = 1.$$

(b) *Hence, every factor congruence is a Boolean congruence with the same complement.*

PROOF. (a) This reduction of the original definition of Boolean is clear from distributivity, (iii.a) and (iii.b).

(b) Always $\theta \circ \theta' \subseteq \theta \vee \theta'$; so $\theta \circ \theta' = 1$ implies $\theta \vee \theta' = 1$. □

As usual, $\boldsymbol{SL}_3$ gives the lie to the converse of (b). Without distributivity, but still assuming (b), we have already reached in Theorem 2.5 a conclusion about the sheaf representation. But for commuting congruences (=permutable congruences) the inclusion goes the other way.

2.7. PROPOSITION. *In an algebra* $\boldsymbol{A}$ *with commuting congruences, every Boolean congruence is a factor congruence with the same complement:*

$$\text{(2.1)} \qquad \mathrm{Elem}' \, \mathbf{Con}\, \boldsymbol{A} \subseteq \mathrm{Con}' \, \boldsymbol{A}.$$

PROOF. Commutativity means that $\theta \circ \eta = \eta \circ \theta$ and hence $\theta \circ \eta = \theta \vee \eta$. Therefore, join may be replaced by composition. □

2.8. COROLLARY. *If an algebra* $\boldsymbol{A}$ *is arithmetic, that is to say,* **Con** A *has both distributive and commuting congruences, then the lattices of Boolean congruences and factor congruences are the same:* $\mathbf{Elem}' \, \mathbf{Con}\, \boldsymbol{A} = \mathbf{Con}' \, \boldsymbol{A}$.

2.9. PROBLEM. Let's take some of the conclusions as hypotheses, and see what follows. We have already done this in Theorem 2.5 when $\mathrm{Con}' \, \boldsymbol{A} \subseteq \mathrm{Elem}' \, \mathbf{Con}\, \boldsymbol{A}$.

(a) When $\mathbf{Con}\, \boldsymbol{A}$ is commutative, we proved in Proposition 2.7 that every Boolean congruence is factorial. Assuming (2.1), we should try to reprove theorems now established only for congruence-commutable varieties.

(b) As another example, consider the equality: $\mathrm{Con}' \, \boldsymbol{A} = \mathrm{Elem}' \, \mathbf{Con}\, \boldsymbol{A}$. Assuming this, we might try to show that the equational classes generated by such an algebra behave like arithmetical equational classes.

XII

Further Examples Pointing to Future Research

This chapter might well be called an epilogue, but even better it should be called a prologue. It looks forward in time. It lists many applications of sheaf theory, and uses them to point to potential research. Specific problems are rarely pinpointed; rather it paints five broad landscapes, into which the interested reader may venture.

The first is classical algebra: useful sheaf representations in classical ring and lattice theory extended to shells and beyond. The second is the invention of algebras capturing various logics. The third is model theory: preservation of properties, decidability of first-order theories, and model completeness. The fourth is three topics that are loosely interrelated: generalized metrics, the weakening of Boolean algebras, and the mixing of duality theory with sheaf theory. The fifth is a continuation of the last and concerns the diverse sheaves that exist for a each algebra: it is well illustrated by Boolean rings.

Proofs are omitted, but given references.

1. From Classical Algebra

This monograph approaches an algebra, from the top down, to decompose it as a sheaf. The more traditional approach is to start with a small desirable set of common algebras and see which algebras can be built, from the bottom up, with these as stalks. There is a gap. When do these two paths,

A. Knoebel, *Sheaves of Algebras over Boolean Spaces*,
DOI 10.1007/978-0-8176-4642-4_XII,

coming from different directions, merge and give the same theorem? Listed here are possibilities in three areas: rings, lattices, and categories.

Rings. Now that we have generalized some concepts and theorems of ring theory, what about others? What other results about regular rings can be extended to shells? Goodearl is a gold mine of theorems that might be broadened [**Goode79**]. For example, a regular ring is characterized as Abelian if it has no nilpotents; these are the so called strongly regular rings. Perhaps this is a way to enlarge on our brief discussion of these rings in Chap. VIII. In another direction, how much of Keimel's [1971] article [**Keim71**], Kist's [1969] paper [**Kist69**] and Peercy's [1970] thesis [**Peer70**] may be extended to shells and half-shells? In what sense, if any, are these notions of Chap. VII minimal with respect to sheaf representations?

How much of the theory of Chap. VII holds for shells without unities? In Boolean rings, the existence of a unity is closely related to compactness. As pointed out by Halmos in [**Halm63**, p. 83, ex. 1], adding a unity to a Boolean ring without one is a one-point compactification. In general, could we get along with locally compact spaces?

Traditionally, modules have been the object of study in sheaves of algebras. We have neglected part II of Pierce's monograph [**Pier67**] since we don't know to define a module over a shell. Would anything worthwhile turn up by studying semilattices as modules over a fixed bounded lattice?

Lattices. Many papers explore sheaves of these, for example, Crown, Harding and Janowitz [**CroHJ96**]. With a canonical sheaf representation, they start by exploiting the set of factor elements, called the center in lattice theory, and continue by giving conditions equivalent to it being a Boolean product. Papers using sheaf theory to represent expanses of lattices are those of Cignoli [**Cign78**], George Georgescu [**Geor88**], and Swamy and Manikyamba [**SwaMa80**].

1.1. PROBLEM. Are there sheaf representations for the well-studied residuated lattices in Blount and Tsinakis [**BloTs03**], and in Jipsen and Tsinakis [**JipTs02**]?

Categories. In earlier chapters, we have reformulated our representation theorems categorically, first as adjoint situations, and then as equivalences between categories of algebras and sheaves. But there are other ways category theory may enter into sheaf theory.

Looking over these sheaves reveals an essential feature of their base spaces: they are complete lattices of open sets. These have been abstracted into what are variously called locales, frames, or complete Heyting algebras. This is not category theory, but in the next step it enters in the works of Paweł Idziak [**Idzi89**], Michael Fourman and Dana Scott [**FouSc79**], and Peter Johnstone [**John82**]. All of these new style 'sheaves' over a fixed locale form a category. The last step is to abstract the crucial properties of such categories: these are called toposes. See for example the expository articles of Saunders MacLane [**MacL75**] and Johnstone [**John83**].

The representation of an equational class generated by a primal algebra was effected historically through the Boolean powers of Foster [**Fost53**], but it can also be done much more differently by means of a categorical argument. This second use of category theory would be to develop our results anew by applying the techniques of Banaschewski and Nelson [**BanNe80**], Hu [**Hu69**], and Lambek and Rattray [**LamRa79**] to the category of shells. We have already adapted some of many examples of categorical equivalences that appear in Pierce's monograph [**Pier67**]. See Diers's book [**Diers86**] for a general categorical approach.

Using sheaf theory, Gerhard Gierz characterizes the Morita equivalence of varieties generated by single quasi-primal algebras in terms of their inverse semigroups of inner automorphisms [**Gierz96**]. (Two algebras are **Morita equivalent** if the varieties they generate are categorically equivalent, with the equivalence taking one algebra into the other.)

1.2. PROBLEM. Use sheaf representations to find other Morita equivalences.

We close with two miscellanea.

1.3. PROBLEM. The extension of an algebra $\boldsymbol{A}$ obtained by embedding it into the algebra $\Gamma(\mathcal{A})$ of all global sections of a sheaf $\mathcal{A}$ can be thought of as a completion, called the global completion by Krauss and Clark: $\boldsymbol{A} \subseteq \Gamma(\mathcal{A})$ [**KraCl79**]. Further, rational completions of rings are characterized as complete Baer extensions by Peercy [**Peer70**]. Express this functorially and extend it beyond rings, if possible. Are these completions unique?

Using Boolean spaces, William Hanf [**Hanf57**] finds two denumerable Boolean algebras $\boldsymbol{A}$ and $\boldsymbol{B}$ such that $\boldsymbol{A} \cong \boldsymbol{A} \times \boldsymbol{B} \times \boldsymbol{B}$ but $\boldsymbol{A} \not\cong \boldsymbol{A} \times \boldsymbol{B}$. For other examples and references, see the publications of Burris [**Burr75**, theorem 6.1(i)], of Jiří Adámek, Václav Koubek, and Věra Trnková [**AdáKT75**], and of McKenzie, McNulty, and Taylor [**McMcT87**, sect. 5.7].

1.4. PROBLEM. Can these results on pathological Boolean algebras be adapted by sheaf methods to other algebras, such as those having Boolean factor congruences?

2. Algebras from Logic

Cultivating logics into algebras yields a rich harvest of representation theorems. The oldest is Stone's theorem for Boolean algebras, invoked repeatedly in this book. The logics in this section come from introducing additional truth values, often with modal connotations. More often than not, these new algebras are bounded distributive lattices satisfying extra axioms, possibly with additional operations. As lattices have BFC, the representation theorem of Sect. VI.3 applies to these algebras.

First discussed are representations for P-algebras and their relatives; the 'P' stands for a generalization of Post algebras. Then MV-algebras come next; 'MV' stands for multivalued logic.

Typically, the stalks are of a straightforward nature: simple, chains, no divisors of zero, etc. Sometimes the sheaf is Hausdorff, that is, we have a Boolean product. There are three things to do: fit the diverse language of the literature to ours, state the representations, and characterize the stalks.

All the riches in this field could easily fill their own monograph. We only sample. The corresponding logics are not described, and proofs are omitted.

George Georgescu delineates those bounded distributive lattices whose canonical sheaf has stalks with no divisors of zero [**Geor88**, proposition 2.3]. For a subset B of a lattice $\boldsymbol{L}$, call the lattice **B-normal** if, whenever $a \wedge b = 0$ $(a, b \in L)$, there exist u and v in B such that $u \vee v = 1$ and $a \wedge u = 0 = b \wedge v$. To say there are **no divisors of zero** means that $a \wedge b = 0$ implies $a = 0$ of $b = 0$. For u to be in $\mathrm{Elem}' \boldsymbol{L}$ is the same as for it to have a complement.

2.1. PROPOSITION. *Let $\boldsymbol{L}$ be a bounded distributive lattice with $B = \mathrm{Elem}' \boldsymbol{L}$. Then $\boldsymbol{L}$ is B-normal if, and only if, $\boldsymbol{L}$ is isomorphic to the lattice of global sections of a sheaf over a Boolean space whose stalks have no divisors of zero.*

Earlier Cignoli looked at a smaller class of algebras where the stalks are bounded chains [**Cign78**, theorem 3.4]; this is restated in [**Geor93**, proposition 2.8]. For a subset B of a lattice $\boldsymbol{L}$, call the lattice **completely B-normal** if for all a and b in L there exist u and v in B such that $u \vee v = 1$, $a \wedge u \leq b$, and $b \wedge v \leq a$.

2.2. PROPOSITION. *Let $\boldsymbol{L}$ be a bounded distributive lattice with $B = \mathrm{Elem}' \boldsymbol{L}$. Then $\boldsymbol{L}$ is completely B-normal if, and only if, $\boldsymbol{L}$ is isomorphic to the lattice of global sections of a sheaf over a Boolean space whose stalks are chains.*

Cignoli further restricted this to P-algebras, achieving a Boolean product [**Cign78**, theorem 3.6]. A **P-algebra** is a bounded distributive lattice in which:

(i) For all a and b there is a largest c in $\mathrm{Elem}' \boldsymbol{L}$ such that $a \wedge c \leq b$; we write c as $a \Rightarrow b$.

(ii) $(a \Rightarrow b) \vee (b \Rightarrow a) = 1 \qquad (a, b \in L)$.

2.3. PROPOSITION. *A bounded distributive lattice is a P-algebra if, and only if, it is isomorphic to the lattice of global sections of a Hausdorff sheaf of bounded chains over a Boolean space.*

Swamy and Manikyamba give many results of this type [**SwaMa80**]. Here is their theorem 2.1 (pseudocomplementation is defined after Definition V.1.1).

2.4. PROPOSITION. *A bounded distributive lattice is pseudocomplemented if, and only if, each stalk of its canonical sheaf is pseudocomplemented, and pseudocomplementation is continuous in the sheaf.*

MV-algebras are to multivalued logic as Boolean algebras are to two-valued propositional logic. They are defined as algebras $\langle A; \oplus, \neg, 0\rangle$ of type $\langle 2, 1, 0\rangle$ satisfying these equations:

$$\begin{aligned} 0 \oplus a &= a, & \neg 0 \oplus a &= \neg 0, \\ a \oplus b &= b \oplus a, & \neg\neg a &= a, \\ a \oplus (b \oplus c) &= (a \oplus b) \oplus c, & \neg(\neg a \oplus b) \oplus b &= \neg(\neg b \oplus a) \oplus a. \end{aligned}$$

MV-algebras are term-equivalent to Wasjberg algebras.

There are many representation theorems for MV-algebras and subclasses of them. Typically, the stalks have relatively simple descriptions. For example, any MV-algebra is isomorphic to the algebra of all global sections of a sheaf of directly indecomposable MV-algebras over a Boolean space. A smaller class are the quasi local MV-algebras, each of which is bijectively representable by a sheaf over a Boolean space with stalks that are local MV-algebras. An even smaller class are the dual Stone MV-algebras, whose stalks are now chains. An excellent survey of these representations and many more, together with definitions of the preceding MV-algebras, is given in Di Nola, Esposito, and Gerla [**DiNEG07**]. Some of these results appear earlier in Cignoli and Torrens Torrell [**CigTo96**], and Filipoiu and Georgescu [**FilGe95**].

As any MV-algebra has a bounded lattice as a reduct, it is a two-sided unital shell; so the representation theorem of Sect. VII.3 applies, as well as all the supporting apparatus of factor congruences, bands, sesquimorphisms, ideals, and elements. However, it is not clear to what extent the representation theorems in the literature agree with those in this book.

2.5. PROBLEM. Compare the applicable theorems in this book with those cited in this section.

BL-algebras (BL stands for 'basic logic') are another variety of algebras from logic that have a rich assortment of representations by sheaves (Di Nola and Leuştean [**DiNLe03**]). These are algebras $\langle A; \vee, \wedge, \odot, \rightarrow, 0, 1\rangle$ of type $\langle 2, 2, 1, 1, 0, 0\rangle$ satisfying these conditions:

(i) $\langle A; \vee, \wedge, 0, 1\rangle$ is a bounded lattice;
(ii) $\langle A; \odot, 1\rangle$ is a commutative monoid;
(iii) $c \leq a \rightarrow b$ iff $a \odot c \leq b$;
(iv) $a \wedge b = a \odot (a \rightarrow b)$;
(v) $(a \rightarrow b) \vee (a \rightarrow b) = 1$.

By (i) they are shells. But the representation theorems of Di Nola and Leuştean appear to be different from what would be obtained by considering BL-algebras as shells and applying the theorems of this book.

2.6. PROBLEM. The classes of BL-algebras and MV-algebras have been broadened; see Jipsen and Montagna [**JipMo06**], and Galatos and Tsinakis [**GalTs05**]. Do some of the sheaf representations mentioned for BL- and MV-algebras extend to these larger classes?

Gramaglia and Vaggione list many more algebras from logic and elsewhere, analyze them in detail, and identify the stalks in their sheaf representations [**GramVa96, GramVa97**].

3. From Model Theory

Sheaves, and modifications of them, have played a significant role in establishing decidability, elementary equivalence, and model completeness. A full exposition of the concepts, techniques, and proofs would need a small volume. So we merely sample a few theorems of this kind, and encourage the reader to go further.

We assume here a given type of algebras and varieties of them. This gives us a first-order language of formulas built from equations, logical connectives, and quantifiers. The **theory** of a set of algebras is the set all first-order sentences satisfied by the set. Definitions not given here will be found in the literature cited and [**Hodg97**].

PRESERVATION AND TRANSFER. We have, in previous chapters, stated results of the nature that the truth of certain sentences is preserved in the passage from an algebra to its stalks, and back. For example, we already have Proposition III.1.2 and Corollary IV.2.2 about the preservation of identities in subdirect products and complexes. Further there is Proposition V.2.7 about $\forall\exists$ equations preserved in sheaves over Boolean spaces [**Burr75**], and Proposition V.3.13 about Horn sentences in Boolean products. Left open is the question of what are the most general sentences to be preserved for particular sheaves. Hugo Volger solves this for the class of all sheaves over all spaces; his syntactical description of the preserved sentences is too involved to state here [**Volg79**].

When not all sentences are preserved in passing from an algebra to the stalks of its sheaf, it still may by possible to transform them so decidability and other properties are passed along. The paper [**FefVa59**] of Solomon Feferman and Robert Vaught does this. It is extended by [**Comer74**] and [**Volg76**].

DECIDABILITY. The decidability, or lack of it, of the first-order theories of varieties is still open for research. A theory is **decidable** if there is an algorithm that takes as an input any sentence and produces a YES or NO according to whether it is a consequence of the theory. Early positive successes without sheaves were Abelian groups [**Tars49**] and Boolean algebras [**Szmi55**].

By extending the transfer theorem of [**FefVa59**] to sheaves, Comer proves that the theories of the following sets of algebras are decidable [**Comer74**]:

(a) all Boolean extensions of a finite algebra,
(b) those generated by a primal algebra,
(c) post algebras of fixed order n,
(d) unital rings that satisfy $x^p = x$ and $px = 0$ for a fixed prime p,
(e) unital rings that satisfy $x^m = x$ (fixed $m > 1$).

At the same time Christopher Ash [**Ash74**] also proved with sheaves that the theory of Post algebras of order n is decidable. Vincent Astier [**Asti08**] established the decidability of regular, real-closed, commutative unital rings. And Werner does this for residually small discriminator varieties [**Wern78**].

McKenzie and Valeriote tell us when a locally finite variety is decidable [**McVa89**]. On the other hand, the theories of unital rings, semilattices and distributive lattices are not decidable [**BurSa81**, Sect. V.5]. Paweł Idziak has many references [**Idzi99**].

ELEMENTARY EQUIVALENCE AND EMBEDDING.

As an example of the use of sheaf theory we state a theorem of Vincent Astier [**Asti08**].

3.1. DEFINITION. Two algebras, $\boldsymbol{A}_1$ and $\boldsymbol{A}_2$, are if their theories are the same ($\boldsymbol{A}_1 \equiv \boldsymbol{A}_2$). An **elementary embedding** of one algebra $\boldsymbol{A}_1$ into another $\boldsymbol{A}_2$ is an injective homomorphism with the same theories ($\boldsymbol{A}_1 \prec \boldsymbol{A}_2$).

A real-closed ring is a ring with an ordering that generalizes real fields. Its definition is too technical to state briefly; it is in [**Cars89**, definition 5.6]. The elementary equivalences and embeddings are with respect to the ring operations alone, not the ordering.

3.2. THEOREM. *Let $\boldsymbol{R}_1$ and $\boldsymbol{R}_2$ be regular, real-closed, commutative unital rings, and $\boldsymbol{B}_1$ and $\boldsymbol{B}_2$ their respective Boolean rings of idempotents.*

(a) $\boldsymbol{R}_1 \equiv \boldsymbol{R}_2$ *iff* $\boldsymbol{B}_1 \equiv \boldsymbol{B}_2$.
(b) $\boldsymbol{R}_1 \prec \boldsymbol{R}_2$ *iff* $\boldsymbol{R}_1 \subseteq \boldsymbol{R}_2$ *and* $\boldsymbol{B}_1 \prec \boldsymbol{B}_2$.

MODEL COMPLETENESS.

A first-order theory is **model-complete** if any embedding of one model of it into another is elementary. A theorem of Angus Macintyre [**Maci77**, theorem 34] illustrates this extensively developed field.

3.3. THEOREM. *Let T' be a model-complete theory of fields. Let T be the theory of rings of global sections of sheaves with stalks models of T' and over a Boolean space without isolated points. Then T is model-complete.*

Further results are found in Andrew Carson's book [**Cars89**]; some of his completions use sections over open regular sets rather than global sections over the whole space. See also [**BurWe79**], [**Cars73**], [**Maci73**] and [**Weis75**].

3.4. PROBLEM. Try to generalize the model-completeness of commutative von Neumann regular rings to Baer–Stone shells.

4. Beyond Sheaves over Boolean Spaces

Expanding the notion of sheaf ought to enrich the crop of representation theorems. We discuss three ways, leaving the exploration of these paths to the reader. They are generalized metrics, the lifting of a known representation of a simpler algebra to a more complicated one, and merging sheaf theory with duality theory.

GENERALIZED METRICS. The axioms for a complex, or more correctly their duals, look like those for a metric space, except that, in place of the real numbers as a measure of distance, we have open sets to gauge nearness in a topological space. Can this analogy be exploited rigorously? Replace the topological space by an algebraic structure, inspired by the real numbers and emphasizing the aspects of a metric space, that nevertheless, unlike the field of reals, but like Boolean algebras, would be a subdirect product of simpler algebras.

When the representation is a Boolean product, we need only clopen sets to obtain a sheaf representation of the given algebra. When the sheaf is over a Boolean space, which is a weak Boolean product, we need the bigger lattice of open sets, which leads to locales, already introduced in Sect. 1. More broadly we wonder, if the lattice were to be replaced by a semilattice, would another kind of sheaf structure arise? Ralph Kopperman's article [**Kopp88**], on generalized metrics, may be relevant here, although his value semigroup and our semilattice of open sets are not directly comparable. Even more generally, consult the booklet [**TriAl78**] of Enrique Trillas and Claudi Alsina.

BOOT STRAP. The is similar to the previously modified metrics, but now we propose modifying the choice of congruences. The core we have usually looked for in any algebra $\boldsymbol{A}$ is a Boolean subsemilattice of $\mathbf{Con}\,\boldsymbol{A}$, a Boolean algebra, because its representation as a Boolean space, that is, a sheaf with two-element stalks, can be lifted to a representation of the original algebra. But there are many classes of algebras with almost equally nice representations. Might these be used to lift their sheaves to some less amenable algebras?

In other words, we should try to broaden the results of this monograph by algebraically extending the sheaf construction over the dual of a Boolean algebra to constructions over wider equational classes. For example, choose the congruences to be in non-Boolean lattices, such as distributive lattices, or even fragments like semilattices, and use these to induce sheaf structures. For instance, Werner [**Wern79**] considers algebras with a distributive bounded sublattice $\boldsymbol{L}$ of permuting congruences; Cornish [**Corn77**],

Swamy [**Swam74**] and Wolf [**Wolf74**] have earlier results along this line, and Michael Johnson and Shu Hao Sun [**JohMSu92**] a later one. For such an algebra the Gel'fand morphism γ is bijective, that is, it is an isomorphism to the algebra of all global sections of a sheaf whose base space is the dual of $\boldsymbol{L}$. See also Problem V.3.4.

DUALITY THEORY. David Clark and Brian Davey's book [**ClaDa98**] on natural dualities is almost disjoint from this book on sheaf theory. In duality theory, typically all stalks are isomorphic to a kernel that can appear, with modification, on the other side of a categorical equivalence; whereas this book has been concerned principally with sheaves with many non-isomorphic stalks. The classical example is Priestley duality for distributive lattices. The set of prime ideals of a distributive lattice may be topologized, as done for Boolean algebras, but now non-isomorphic lattices may yield the same space. Hilary Priestley [**Prie70**] remedied this by introducing an order on the spaces that served to differentiate lattices. In general, one starts with a finite kernel that has two guises: a purely algebraic one, and a discrete topological one with relations added as necessary to achieve a categorical equivalence. Although this does not work for all classes of algebras, when it does it represents many common algebras in a useful quasi-topological manner.

By way of contrast, this book treats bounded distributive lattices as unital shells, which have many directly indecomposables, such as chains. The next problem asks if there is a middle ground.

4.1. PROBLEM. Create new dualities by mixing these two approaches. Consider a variety with sheaf representations of its algebras whose base spaces alone are insufficient to differentiate them. Add relations and operations to the base spaces to remedy this. One already sees this in the example of Boolean rings, to be developed below, where one singles out a special point to be preserved by the continuous maps determining the global sections.

4.2. PROBLEM. It might be worthwhile to rework some of the examples in [**ClaDa98**] in the spirit of the preceding problem by comparing their structured Boolean spaces with the alternative of locally compact spaces having no relations or operations on them, as in Sect. 5.

See also the writings of Romanowska and Smith [**RomSm97**] for another kind of duality, and Vaggione [**Vagg92, Vagg95**] for studies beyond Krauss's and Clark's [**KraCl79**].

The last two problems leads naturally into the next section.

5. Many Choices

Often the same algebra may be represented by several different sheaves. We list such variation for three classes of algebras, and then, by way of a detailed example, give four representations of another: Boolean rings.

BOUNDED DISTRIBUTIVE LATTICES may be represented by spaces created via BFC or prime ideals. For example, there are the distributive pseudo-complemented lattices of Cornish [**Corn77a**].

UNITARY RINGS again use BFC or prime ideals. Bergman [**Berg73**, sect. 5] gives three representations studied by Hochster [**Hoch69**]. See also the articles by Christopher Mulvey [**Mulv79**] and Harold Simmons [**Simm84**].

BIREGULAR RINGS have even more ways to be represented by sheaves. Some require a unity. Theorem VIII.2.3 represents biregular unitary rings by sheaves over Boolean spaces with simple stalks. In other representations, the base space is weaker than a Boolean space. In one, it is a *locally* compact, totally disconnected, Hausdorff space. Theorems 7.10 and 7.11 of Krauss and Clark [**KraCl79**] give some of these. Burris and Sankappanavar [**BurSa81**, p. 163] have another.

BOOLEAN RINGS serve as a rich source of topological representations, several of them outside the scope of this book. These are commutative rings $\langle R; +, \times, 0\rangle$, not necessarily unital, that satisfy $r+r=0$ and $r \times r = r$; the operation $+$ corresponds to symmetric difference in Boolean algebras. As such they are almost term-equivalent to Boolean algebras. For background see Alexander Abian's introduction to Boolean rings [**Abian76**].

Each Boolean ring may be doubled to make a Boolean algebra, and each nontrivial Boolean algebra may be halved to make a Boolean ring. This subtle difference is enough to skew their sheaf representations.

There are many ways of representing a Boolean ring, which illustrate a diversity of dualities:

(a) as a Boolean product with stalks that are the subalgebras of a semiprimal algebra,
(b) as a sub-Boolean power of the two-element Boolean ring, coming from the sheaf for the associated Boolean algebra,
(c) as a sheaf over a locally compact, Hausdorff and totally disconnected space, created by dropping one point from the Boolean space in (b),
(d) as the dual of a pointed Boolean space, by way of Clark–Davey duality.

We discuss these in turn, using ideals instead of congruences. To that end let $\mathbf{2}$ be the two-element Boolean ring $\langle\{0,1\}; +, \times, 0\rangle$; and let $\mathbf{2}_1$ be the two-element unital Boolean ring $\langle\{0,1\}; +, \times, 0, 1\rangle$. Then the latter is term-equivalent to the two-element Boolean algebra, and the properties we need transfer from one to the other. Each Boolean ring is a subdirect power of $\mathbf{2}$ as shown in [**Abian76**]; hence, the variety of Boolean rings is $\operatorname{Var}\mathbf{2}$.

(a) Now $\mathbf{2}$ has two subalgebras, $\{0\}$ and $\mathbf{2}$, and all operations preserving these are term-operations. Thus, $\mathbf{2}$ is semiprimal and covered by Sect. X.4. Consequently, Boolean rings are Boolean products of $\{0\}$ and $\mathbf{2}$.

(b) This representation and the next require the ability to pass between Boolean rings and unital Boolean rings. Roughly, one can paste on top of any Boolean ring an 'anti-isomorphic' copy, where the original ring is turned up-side-down, with its 0 becoming the 1 of the enlarged ring. This is a special case in ring theory of adding a unity to any ring that may not have one [**Hung74**, theorem III.1.10]. To see this, let $\boldsymbol{R}$ be a Boolean ring $\langle R; +, \times, 0\rangle$ that is to be enlarged to a unital Boolean ring, $\boldsymbol{R}_1 = \boldsymbol{R} \rtimes \mathbf{2}_1$, with a carrier $R \times \{0, 1\}$ and with new operations defined for r_i in R and b_i in $\{0, 1\}$:

$$\begin{aligned}\langle r_1, b_1\rangle +^{\boldsymbol{R}_1} \langle r_2, b_2\rangle &= \langle r_1 + r_2, b_1 + b_2\rangle,\\ \langle r_1, b_1\rangle \times^{\boldsymbol{R}_1} \langle r_2, b_2\rangle &= \langle r_1 r_2 + r_1 b_2 + r_2 b_1, b_1 b_2\rangle,\\ 0^{\boldsymbol{R}_1} &= \langle 0, 0\rangle,\\ 1^{\boldsymbol{R}_1} &= \langle 0, 1\rangle.\end{aligned}$$

By the Stone representation theorem for Boolean algebras, $\boldsymbol{R}_1$ is a Boolean power of $\mathbf{2}_1$ over the Boolean space $\boldsymbol{X}_1$ of all maximal ideals of $\boldsymbol{R}_1$. Therefore, $\boldsymbol{R}$ is a (non-unital) subring of this power in which patching holds; that is, $\boldsymbol{R}$ is a so-called sub-Boolean power of $\mathbf{2}$. Details are left to the imagination. Note that $\boldsymbol{R}$ is a maximal ideal of $\boldsymbol{R}_1$, and any ideal of $\boldsymbol{R}_1$ is a Boolean ring. If an ideal I of $\boldsymbol{R}_1$ is proper, then the quotient $\boldsymbol{R}_1/I$ is nontrivial; if the ideal is maximal, then its quotient is isomorphic to $\mathbf{2}_1$. In short, for any Boolean ring $\boldsymbol{R}$ there is the short exact sequence:

$$\{0\} \to \boldsymbol{R} \to \boldsymbol{R}_1 \to \mathbf{2} \to \{0\}.$$

(c) A flaw in (b) is that, while $\boldsymbol{R}$ is enlarged to $\boldsymbol{R}_1$, the base space $\boldsymbol{X}_1$ for $\boldsymbol{R}_1$ does not shrink when returning to a representation for $\boldsymbol{R}$. This is remedied by removing a particular point from $\boldsymbol{X}_1$ at the expense of its compactness. That point is the maximal ideal R; call the new base space $\boldsymbol{X}$, a subspace of $\boldsymbol{X}_1$, and the new sheaf space $\mathcal{A}$, shrunk from the old sheaf space $\mathcal{A}_1$ for $\boldsymbol{R}_1$ by removing its stalk over R. One may check that now $\boldsymbol{R}$ is isomorphic to $\boldsymbol{\Gamma}(\mathcal{A})$, and $\boldsymbol{X}$ is locally compact, Hausdorff and totally disconnected (see [**Halm63**, p. 83, exercise 1]).

Here is a transparent example showing that the new base space $\boldsymbol{X}$ need not be compact. Let S be an infinite set, and R the set of all finite subsets of S. Then $\langle R; \cup, \cap, \varnothing\rangle$ is a Boolean ring $\boldsymbol{R}$, missing the unity S. Adding the cofinite subsets of S creates a unital Boolean ring isomorphic to $\boldsymbol{R}_1$. To demonstrate noncompactness of $\boldsymbol{X}$, it suffices to find a collection of open subsets of $\boldsymbol{X}$ covering it such that no finite subcollection covers it. To that end, let the covering $\mathcal{C}$ be all open subsets of the form, $U_F = \{M \in X \mid F \notin M\}$ for all F in R. For any finite subcollection,

$$U_{F_1} \cup U_{F_2} \cup \cdots \cup U_{F_n} = U_{F_1 \cup F_2 \cup \cdots \cup F_n}.$$

By Zorn's lemma, there is a maximal ideal M containing the union of the F_i. Hence, $\boldsymbol{X}$ is not covered by any such finite subcollection.

(d) The last representation is a duality found in [**ClaDa98**, Table 10.2], but we do not need all their machinery to explain it. To describe it, we reconsider how the representation of (c) was created. The troublesome maximal ideal R of $\boldsymbol{R}_1$ was eliminated from the space $\boldsymbol{X}_1$, truncating it to $\boldsymbol{X}$. Equivalently, this may achieved by fixing the point R in $\boldsymbol{X}_1$, effectively cutting it from the clopen sets that one does not want in $\boldsymbol{X}$, as explained next.

From the topological side of the duality, designate and fix a single point in each Boolean space, to be preserved by all continuous functions. To see that these pointed topological spaces $\boldsymbol{X}_0$ faithfully represent Boolean rings $\boldsymbol{R}$, look at their open sets that avoid the fixed point. These correspond one-to-one to continuous maps from $\boldsymbol{X}_0$ to $\mathbf{2}_0$, the two-element topological space $\{0, 1\}$ with the discrete topology and the fixed point 0. Just as clopen sets may be combined with symmetric differences and intersections, so may homomorphisms be combined, creating a Boolean ring.

5.1. PROBLEM. As a starting point for research, find diverse representations within the table of varieties generated by two-element algebras, listed in [**ClaDa98**, sect. 10.7]. A few have been discussed in Chap. X.

5.2. PROBLEM. Which of the many sheaf representations in this section might be generalized to shells and beyond?

To close this book, let us go back to Leibniz's dream of a universal logic, which would automatically answer all objective questions. In its spirit, Boole's algebra serves to parse general algebras by sheaves over Boolean spaces. Their ideas have evolved into useful tools for solving substantial problems.

List of Symbols

Chapter II — Algebra

$\boldsymbol{A}$, A, algebra and carrier, 19
ω, ω_i, operation of algebra, 19
$\boldsymbol{n}$, type of $\boldsymbol{A}$, 20
π_i^n, projection in $\boldsymbol{A}$, 20
$\vec{a} = \langle a_1, a_2, \ldots, a_n \rangle$, sequence, 20
$\mathbb{Z} = \langle \mathbb{Z}; +, \times, 0, 1 \rangle$, unital ring of integers, 20
$\mathbb{Z}_n = \langle \mathbb{Z}_n; +, \times, 0, 1 \rangle$, unital ring of integers mod n, 20
$\boldsymbol{SL}_3$, a three-element semilattice, 21
$\bigwedge T, \bigvee T$, infimum, supremum, 21
$\boldsymbol{L} = \langle L; \vee, \wedge \rangle$, lattice, 21
$\boldsymbol{L} = \langle L; \vee, \wedge, 0, 1 \rangle$, bounded lattice, 22
$\boldsymbol{C}_2$, $\boldsymbol{C}_3$, $\boldsymbol{M}_2$, $\boldsymbol{M}_3$, $\boldsymbol{N}_5$, lattices, 22
a', complement of a, 22
$\boldsymbol{A} \subseteq \boldsymbol{B}$, subalgebra, 23
$\varphi: \boldsymbol{A} \to \boldsymbol{B}$, homomorphism of algebras, 23
$\boldsymbol{A} \cong \boldsymbol{B}$, isomorphism of algebras., 23
$a \equiv b \ (\theta)$, congruence modulo θ, 24
$\boldsymbol{A}/\theta$, A/θ, quotient algebra, class, 24
ω/θ, quotient operation, 24
$\ker \psi$, kernel of homomorphism, 25
$\mathbf{Con}\, \boldsymbol{A}$, congruence lattice, 25

A. Knoebel, *Sheaves of Algebras over Boolean Spaces*,
DOI 10.1007/978-0-8176-4642-4,

Chapter III — Tools

Chapter IV — Complexes and their Sheaves

Chapter V — Boolean Subsemilattices

Chapter VI — Sheaves from Factor Congruences

Chapter VII — Shells

Chapter VIII — Baer-Stone Shells

Chapter IX — Strict Shells

Chapter X — Varieties Generated by Preprimal Algebras

Chapter XI — Return to General Algebras

Chapter XII — Further Examples Pointing to Future Research

References

[Abian76] ABIAN, Alexander. *Boolean Rings.* Branden Press, Boston, MA, 1976.

[AdáKT75] ADÁMEK, Jiří, KOUBEK, Václav and TRNKOVÁ, Věra. Sums of spaces represent every group. *Pacific J. Math.* **61** (1975), pp. 1–6.

[AreKa48] ARENS, Richard F. and KAPLANSKY, Irving. Topological representation of algebras. *Trans. Amer. Math. Soc.* **63** (1948), pp. 457–481.

[Ash74] ASH, Christopher J. Reduced powers and Boolean extensions. *J. London Math. Soc.* (2) **9** (1974/75), pp. 429–432.

[Asti08] ASTIER, Vincent. Elementary equivalence of some rings of definable functions. *Arch. Math. Logic* **47** (2008), pp. 327–340.

[Aull85] AULL, Charles E., ed. *Rings of Continuous Functions.* Proceedings of the special session held in Cincinnati, Ohio, 1982. Lecture Notes in Pure and Applied Math., **95**. Dekker, New York, 1985.

[Baer52] BAER, Reinhold. *Linear Algebra and Projective Geometry.* Academic, New York, 1952.

[BalHo70] BALBES, Raymond and HORN, Alfred. Stone lattices. *Duke Math. J.* **37** (1970), pp. 537–545.

[BanNe80] BANASCHEWSKI, Bernhard & NELSON, Evelyn. Boolean Powers as Algebras of Continuous Functions. *Dissertationes Math. (Rozprawy Mat.)* **179** (1980), 51 pp.

A. Knoebel, *Sheaves of Algebras over Boolean Spaces*,
DOI 10.1007/978-0-8176-4642-4,

[Berg72] BERGMAN, George M. Boolean rings of projection maps. *J. London Math. Soc.* (2), **4** (1972), pp. 593–598.

[Berg73] BERGMAN, George M. The Zariski topology and its generalizations. *Proceedings of the International Symposium on Topology and its Applications (Budva, 1972)*, pp. 32-39. Savez Drutava Mat. Fiz. i Astronom., Belgrade, 1973.

[Berg91] BERGMAN, George M. Actions of Boolean rings on sets. *Algebra Universalis* **28** (1991), pp. 153–187.

[Berg98] BERGMAN, George M. *An Invitation to General Algebra and Universal Constructions.* Henry Helson, Berkeley, 1998.

[BigBu90] BIGELOW, David and BURRIS, Stanley. Boolean algebras of factor congruences. *Acta Sci. Math. (Szeged)* **54** (1990), pp. 11–20.

[Birk35] BIRKHOFF, Garrett. On the structure of abstract algebras. *Proc. Cambridge Philos. Soc.* **31** (1935), pp. 117–138.

[Birk44] BIRKHOFF, Garrett. Subdirect unions in universal algebras. *Bull. Amer. Math. Soc.* **50** (1944), pp. 764–768.

[Birk58] BIRKHOFF, Garrett. Von Neumann and lattice theory. *Bull. Amer. Math. Soc.* **64** (1958), pp. 50–56.

[Birk67] BIRKHOFF, Garrett. *Lattice Theory,* 3rd ed. Amer. Math. Soc. Colloquium Pub. **25**, 1967.

[BloÉM90] BLOOM, Stephen, ÉSIK, Zoltán and MANES, Ernest. A Cayley Theorem for Boolean Algebras. *Amer. Math. Monthly* **97** (1990), pp. 831–833.

[BloTs03] BLOUNT, Kevin and TSINAKIS, Constantine. The structure of residuated lattices. *Internat. J. Algebra Comput.* **13** (2003), pp. 437–461.

[Bool47] BOOLE, George. *The Mathematical Analysis of Logic, Being an Essay Towards a Calculus of Deductive Reasoning.* Macmillan, Barclay and Macmillan, 1847. Reprinted by Philosophical Library, New York, NY, 1948.

[Bool54] BOOLE, George. *An Investigation of the Laws of Thought, on Which Are Founded the Mathematical Theories of Logic and Probabilities.* MacMillan, 1854. Reprinted by Dover, New York, 1957.

[Borc94] BORCEUX, Francis. *Handbook of Categorical Algebra.* **3**. *Categories of Sheaves.* Encyclopedia of Mathematics and its Applications, **52**. Cambridge University Press, Cambridge, UK, 1994.

[BorVa91] BORCEUX, Francis and VAN DEN BOSSCHE, Gilberte. A generic sheaf representation for rings. *Category theory (Como, 1990)*, pp. 30–42, Lecture Notes in Math. **1488**. Springer, Berlin, 1991.

[Borel64] BOREL, Armand. *Cohomologie des espaces localement compacts d'après J. Leray.* Lecture Notes in Mathematics **2**. Springer, Berlin, 1964.

[Bred97] BREDON, Glen E. *Sheaf Theory,* 2nd ed. Springer, Berlin, 1997.

[Bruck58] BRUCK, E. Hubert. *A Survey of Binary Systems.* Springer, Berlin, 1958.

[BulWe77] BULMAN-FLEMING, Sydney and WERNER, Heinrich. Equational compactness in quasiprimal varieties. *Algebra Universalis* **7** (1977), pp. 33–46.

[BurgSt78] BURGESS, Walter D. and STEPHENSON, William. An analogue of the Pierce sheaf for non-commutative rings. *Comm. Algebra* **6** (1978), pp. 863–886.

[Burr75] BURRIS, Stanley. Boolean powers. *Algebra Universalis* **5** (1975), pp. 341–360.

[Burr86] BURRIS, Stanley. Remarks on the Fraser-Horn property. *Algebra Universalis* **23** (1986), pp. 19–21.

[Burr00] BURRIS, Stanley. The Algebra of Boole. ms (2000), 9 pp.

[BurMc81] BURRIS, Stanley and MCKENZIE, Ralph. Decidability and Boolean Representations. *Memoirs Amer. Math. Soc.* **32** (1981), no. 246.

[BurSa81] BURRIS, Stanley and SANKAPPANAVAR, Hanamantagouda. *A Course in Universal Algebra.* Springer, Berlin, 1981. Internet: `http://www.math.uwaterloo.ca/~snburris/htdocs/ualg.html`. 1981.

[BurWe79] BURRIS, Stanley and WERNER, Heinrich. Sheaf constructions and their elementary properties. *Trans. Amer. Math. Soc.* **248** (1979), pp. 269–309.

[Cars73] CARSON, Andrew B. The model completion of the theory of commutative regular rings. *J. Algebra* **27** (1973), pp. 136–146.

[Cars89] CARSON, Andrew B. *Model completions, ring representations and the topology of the Pierce sheaf.* Pitman Research Notes in Mathematics Series, 209. Longman Scientific & Technical, Harlow, 1989.

[Cart49] CARTAN, Henri. Théorie des faisceaux. *Séminaire Cartan* (Topologie Algébrique), Ire année (1948–1949), exposés **12–17** (1949).

[Cayl54] CAYLEY, Arthur. On the theory of groups, as depending on the symbolic equation $\theta^n = 1$. *Philosophical Magazine* **7** (1854), pp. 40–47.

[Cha94] CHAJDA, Ivan. Congruences permutable with factor and decomposing congruences. *General Algebra and Ordered Sets (Horní Lipová, 1994)*, pp. 10–18, Palacký Univ. Olomouc, Olomouc, 199?.

[ChaEL03] CHAJDA, Ivan, EIGENTHALER, Günther and LÄNGER, Helmut. *Congruence Classes in Universal Algebra.* Research and Exposition in Mathematics, **26**. Heldermann Verlag, Lemgo, Germany, 2003.

[Chang58] CHANG, Chen-Chung. Algebraic analysis of many valued logics. *Trans. Amer. Math. Soc.* **88** (1958), pp. 467–490.

[ChangJT64] CHANG, Chen-Chung, JÓNSSON, Bjarni and TARSKI, Alfred. Refinement properties for relational structures. *Fund. Math.* **55** (1964), pp. 249–281.

[Chen77] CHEN, Chuan-Chong. On the unique factorization of algebras. *Nanta Math.* **10** (1977), pp. 149–157.

[ChiHe76] CHINBURG, Ted and HENRIKSEN, Melvin. Multiplicatively periodic rings. *Amer. Math. Monthly* **83** (1976), pp. 547–549.

[Cign78] CIGNOLI, Roberto. The lattice of global sections of chains over Boolean spaces. *Algebra Universalis* **8** (1978), pp. 357–373.

[CigTo96] CIGNOLI, Roberto and TORRENS TORRELL, Antoni. Boolean products of MV-algebras: hypernormal MV-algebras. *J. Math. Anal. Appl.* **199** (1996), pp. 637–653.

[ClaDa98] CLARK, David M. and DAVEY, Brian A. *Natural Dualities for the Working Algebraist.* Cambridge University Press, Cambridge, UK, 1998.

[Cohn77] COHN, Paul M. *Algebra*, vol. 2. Wiley, New York, 1977.

[Cohn81] COHN, Paul M. *Universal Algebra*, 2nd ed. D. Reidel, Boston, MA, 1981.

[Comer71] COMER, Stephen D. Representations by algebras of sections over Boolean spaces. *Pacific J. Math.* **38** (1971), pp. 29–38.

[Comer72] COMER, Stephen D. A sheaf-theoretic duality theory for cylindric algebras. *Trans. Amer. Math. Soc.* **169** (1972), pp. 75–87.

[Comer74] COMER, Stephen D. Elementary properties of structures of sections. *Bol. Soc. Mat. Mexicana* (2) **19** (1974), pp. 78–85.

[Cont88] CONTESSA, Maria. A note on Baer rings. *J. Algebra* **118** (1988), pp. 20–32.

[Corn77] CORNISH, William H. The Chinese remainder theorem and sheaf representations. *Fund. Math.* **46** (1977), pp. 177–187.

[Corn77a] CORNISH, William H. O-ideals, congruences, and sheaf representations of distributive lattices. *Rev. Roumaine Math. Pures Appl.* **22** (1977), pp. 1059–1067.

[CroHJ96] CROWN, Gary D., HARDING, John and JANOWITZ, Melvin F. Boolean products of lattices. *Order* **13** (1996), pp. 175–205.

[DauHo66] DAUNS, John and HOFMANN, Karl Heinrich. The representation of biregular rings by sheaves. *Math. Z.* **91** (1966), pp. 103–123.

[Davey73] DAVEY, Brian A. Sheaf spaces and sheaves of universal algebras. *Math. Z.* **134** (1973), pp. 275–290.

[DavPr02] DAVEY, Brian A. and PRIESTLEY, H. A. *Introduction to lattices and order*, 2nd ed. Cambridge University Press, Cambridge, UK, 2002.

[Dede97] DEDEKIND, Richard. Über Zerlegungen von Zahlen durch ihre grössten gemeinsamen Teiler. *Festschrift Techn. Hochschule Braunschweig bei Gelegenheit der 69. Versammlung Deutscher Naturforscher und Arzte*, pp. 1–40. 1897. Also in *Gesammelte mathematische Werke* **2**, pp. 103–148. Vieweg, Braunschweig, 1931.

[DeMo47] DE MORGAN, Augustus. *Formal Logic.* Taylor and Walton, London, 1847.

[Dene78] DENECKE, Klaus-Dieter. *Algebraische Fragen Nichtklassicher Aussagenkalküle.* Doctoral dissertation, Pädagogischen Hochschule 'Karl Liebknecht', Potsdam, 1978.

[Dene82] DENECKE, Klaus-Dieter. *Preprimal Algebras.* Mathematical Research, **11**. Akademie-Verlag, Berlin, 1982.

[DenLü95] DENECKE, Klaus-Dieter and LÜDERS, Otfried. Category equivalences of clones. *Algebra Universalis.* **34** (1995), pp. 608–618.

[Diers86] DIERS, Yves. *Categories of Boolean Sheaves of Simple Algebras.* Springer, Berlin, 1986.

[DiNEG07] DI NOLA, Antonio, ESPOSITO, I. and GERLA, Brunella. Local algebras in the representation of MV-algebras. *Algebra Universalis* **56** (2007), pp. 133–164.

[DiNLe03] DI NOLA, Antonio and LEUŞTEAN, Laurenţiu. Compact representations of BL-algebras. *Arch. Math. Logic* **4** (2003), pp. 737–761.

[EilMa45] EILENBERG, Samuel and MACLANE, Saunders. General theory of natural equivalences. *Transactions Amer. Math. Soc.* **58** (1945), pp. 231–294.

[Fear82] FEARNLEY-SANDER, Desmond. Hermann Grassmann and the prehistory of universal algebra. *Amer. Math. Monthly* **89** (1982), pp. 161–166.

[FefVa59] FEFERMAN, Solomon and VAUGHT, Robert L. The first order properties of products of algebraic systems. *Fund. Math.* **47** (1959), pp. 57–103.

[FilGe95] FILIPOIU, Alexandru and GEORGESCU, George. Compact and Pierce representations of MV-algebras. *Rev. Roumaine Math. Pures Appl.* **40** (1995), pp. 599–618.

[ForMc46] FORSYTHE, Alexandra and McCOY, Neal H. On the commutativity of certain rings. *Bull. Amer. Math. Soc.* **52** (1946), pp. 523–526.

[Fost51] FOSTER, Alfred L. p-rings and their Boolean-vector representation. *Acta Math.* **84** (1951), pp. 231–261.

[Fost53] FOSTER, Alfred L. Generalized "Boolean" theory of universal algebras. Part I. Subdirect sums and normal representation theorem. *Math. Z.* **58** (1953), pp. 306–336.

[Fost53a] FOSTER, Alfred L. Generalized "Boolean" theory of universal algebras. Part II. Identities and subdirect sums in functionally complete algebras. *Math. Z.* **59** (1953), pp. 191–199.

[Fost70] FOSTER, Alfred L. Congruence relations and functional completeness in universal algebras; structure theory of hemiprimals, I. *Math. Z.* **113** (1970), pp. 293–208.

[FostPi64] FOSTER, Alfred L. and PIXLEY, Alden. Semi-categorical algebras. I. Semi-primal algebras. *Math. Z.* **83** (1964), pp. 147–169.

[FouSc79] FOURMAN, Michael P. and SCOTT, Dana. S. Sheaves and logic. *Applications of sheaves*, Lecture Notes in Math., **753**, pp. 302–401, Springer, Berlin, 1979.

[FraHo70] FRASER, Grant A. and HORN, Alfred. Congruence relations in direct products. *Proc. Amer. Math. Soc.* **26** (1970), pp. 390–394.

[FreMc87] FREESE, Ralph and McKENZIE, Ralph. *Commutator Theory for Congruence Modular Varieties.* Cambridge University Press, Cambridge, UK, 1987.

[Frink62] FRINK, Orrin. Pseudocomplements in semi-lattices. *Duke Math. J.* **29** (1962), pp. 505–514.

[Gabr62] GABRIEL, Pierre. Des categories abéliennes. *Bull. Soc. Math. de France* **90** (1962), pp. 323–448.

[GalTs05] GALATOS, Nikolaos and TSINAKIS, Constantine. Generalized MV-algebras. *J. Algebra* **283** (2005), pp. 254–291.

[GehJó04] GEHRKE, Mai and JÓNSSON, Bjarni. Bounded distributive lattice expansions. *Math. Scand.* **94** (2004), pp. 13–45.

[Gel'f41] GEL'FAND, Israel M. Normierte Ringe. *Mat. Sb.* **9** (51) (1941), pp. 3–24.

[Gel'Ši41] GEL'FAND, Israel M. and ŠILOV, George E. Über verschiedene Methoden der Einführung der Topologie in die Menge der maximalen Ideale eines normierten Ringes. *Rec. Math. [Mat. Sbornik] N. S.* **9** (**51**) (1941), pp. 25–39.

[Geor88] GEORGESCU, George. Some sheaf constructions for distributive lattices. *Bull. Math. Soc. Sci. Math. R. S. Roumanie (N.S.)* **32**(**80**) (1988), pp. 299–303.

[Geor93] GEORGESCU, George. Pierce representations of distributive lattices. *Kobe J. Math.* **10** (1993), pp. 1–11.

[Gibb81] GIBBS, J. Williard. *Elements of Vector Analysis.* (privately printed) New Haven, 1881.

[Gierz96] GIERZ, Gerhard. Morita equivalence of quasi-primal algebras and sheaves. *Algebra Universalis* **35** (1996), pp. 570–576.

[Gill85] GILLMAN, Leonard. Rings of continuous functions are rings: a survey. *Rings of continuous functions (Cincinnati, Ohio, 1982)*, pp. 143–147, Lecture Notes in Pure and Appl. Math., **95**, Dekker, New York, 1985.

[Głaz85] GŁAZEK, Kazimierz. *A Short Guide through the Literature on Semirings.* Math. Institute, Univ. Wrocław, preprint no. 39, 1985.

[Gode58] GODEMENT, Roger. *Topologie Algébrique et Theorie des Faisceaux*, Publ. Inst. Math. Univ. Strasbourg, no. 13, Hermann, 1958 (3rd ed. 1973).

[Good48] GOOD, Richard A. On the theory of clusters. *Trans. Amer. Math. Soc.* **63** (1948), pp. 482–513.

[Goode79] GOODEARL, Kenneth R. *Von Neumann Regular Rings.* Pitman, Boston, MA, 1979.

[GouGr67] GOULD, Matthew I. and GRÄTZER, George. Boolean Extensions and Normal Subdirect Powers of Finite Universal Algebras. *Math. Z.* **99** (1967), pp. 16–25.

[GramVa96] GRAMAGLIA, Hector and VAGGIONE, Diego J. Birkhoff-like sheaf representation for varieties of lattice expansions. *Studia Logica* **56** (1996), pp. 111–131.

[GramVa97] GRAMAGLIA, Hector and VAGGIONE, Diego J. (Finitely) subdirectly irreducibles and Birkhoff-like sheaf representation for certain varieties of lattice ordered structures. *Algebra Universalis* **38** (1997), pp. 56–91.

[Gras44] GRASSMANN, Herman *Die lineale Ausdehnungslehre, ein neuer Zweig der Mathematik.* Otto Wigand, Leipzig, 1844.

[Grät69] GRÄTZER, George Stone algebras form an equational class. Remarks on lattice theory. III. *J. Austral. Math. Soc.* **9** (1969), pp. 308–309.

[Grät98] GRÄTZER, George. *General Lattice Theory*, 2nd ed. Birkhäuser, Basel, 1998.

[Grät79] GRÄTZER, George *Universal Algebra*, 2nd ed. Springer, Berlin, 1979.

[GrätSc57] GRÄTZER, George and SCHMIDT, Elégius T. On a problem of M. H. Stone. *Acta Math. Acad. Sci. Hungar.* **8** (1957), pp. 455–460.

[GrätSc63] GRÄTZER, George and SCHMIDT, Elégius T. Characterizations of congruence lattices of abstract algebras. *Acta Sci. Math. (Szeged)* **24** (1963), pp. 34–59.

[Gray79] GRAY, John W. Fragments of the history of sheaf theory. *Applications of Sheaves*, eds. M. P. Fourman, C. J. Mulvey, D. S. Scott, pp. 1–79. Lecture Notes in Mathematics **753**, Springer-Verlag, Berlin, 1979.

[GroDi60] GROTHENDIECK, Alexander and DIEUDONNÉ, Jean A. *Eléments de Géométrie Algébrique, tome I: le language des schémas.* Inst. Hautes Études Sci. Publ. Math., no. 4, Paris, 1960. (2nd ed., 1971, Springer, Berlin).

[GumUr84] GUMM, Peter and URSINI, Aldo. Ideals in universal algebras. *Algebra Universalis* **19** (1984), pp. 45–54.

[Halm63] HALMOS, Paul R. *Lectures on Boolean Algebras.* Van Nostrand, Princeton, 1963.

[Hami44] HAMILTON, William R. On a new species of imaginary quantities connected with the theory of quaternions. *Proceedings of the Royal Irish Academy* **2** (1844), pp. 424–434.

[Hanf57] HANF, William. On some fundamental problems concerning isomorphism of Boolean algebras. *Math. Scand.* **5** (1957), pp. 205–217.

[Heav93] HEAVISIDE, Oliver. Electromagnetic Theory **1**. Electrician Printing, London, 1893. Also reprinted by Dover, New York, 1950. pp. 132–305.

[HenkMT71] HENKIN, Leon, MONK, J. Donald and TARSKI, Alfred. *Cylindric Algebras*, vol. 1. North-Holland, Amsterdam, 1971.

[HenrJe65] HENRIKSEN, Melvin and JERISON, Meyer. The space of minimal prime ideals of a commutative ring. *Trans. Amer. Math. Soc.* **115** (1965), pp. 110–130.

[HerSt79] HERRLICH, Horst and STRECKER, George E. *Category Theory: an Introduction.* Heldermann, Berlin, 1979.

[Hilb96] HILBERT, David. Über die Theorie der algebraischen Invarianten. *Proceedings Math. Congress Chicago*, pp. 116–124. MacMillan, New York, 1896.

[HobMc88] HOBBY, David and McKENZIE, Ralph. *The Structure of Finite Algebras.* American Mathematical Society, Providence, RI, 1988.

[Hoch69] HOCHSTER, Melvin. Prime ideal structure in commutative rings. *Trans. Amer. Math. Soc.* **142** (1969), pp. 43–60.

[Hodg97] HODGES, Wilfrid. *A shorter model theory.* Cambridge University Press, Cambridge, 1997.

[Hofm72] HOFMANN, Karl Heinrich. Representation of algebras by continuous sections. *Bull. Amer. Math. Soc.* **78** (1972), pp. 291–373. (And see next item.)

[Hofm74] HOFMANN, Karl Heinrich. *Some bibliographical remarks on "Representations of algebras by continuous sections" (Bull. Amer. Math. Soc. 78 (1972), 291–373).* Mem. Amer. Math. Soc., No. 148, pp. 177–182. Amer. Math. Soc., Providence, RI, 1974.

[Houz98] HOUZEL, Christian. Histoire de la théorie des faisceaux. *Matériaux pour l'histoire des mathématiques au XXe siècle (Nice, 1996).* Sémin. Congr., **3**, Soc. Math. France, Paris, pp. 101–119, 1998.

[Hu69] HU, Tah-Kai. Stone duality for primal algebra theory. *Math. Z.* **110** (1969), pp. 180–198.

[Hung74] HUNGERFORD, Thomas W. *Algebra.* Holt, Rinehart and Winston, New York, 1974.

[Hunt33] HUNTINGTON, Edward V. New sets of independent postulates for the algebra of logic, with special reference to Whitehead and Russell's Principia Mathematics. *Trans. Amer. Math. Soc.* **35** (1933), pp. 274–304, 557–558, 971.

[Idzi89] IDZIAK, Paweł M. Sheaves in universal algebra and model theory. *Rep. Math. Logic* no. 23 (1989), Part I, pp. 39–65; no. 24 (1990), Part II, pp. 61–86.

[Idzi99] IDZIAK, Paweł M. Decidability and structure. *Logic, Algebra, and Computer Science* (Warsaw, 1996), pp. 125–135. Banach Center Publ., **46** (1999), Polish Acad. Sci.

[Iska96] ISKANDER, Awad A. Factorable congruences and strict refinement. *Acta Math. Univ. Comenianae* **65** (1996), pp. 101–109.

[Jaco45] JACOBSON, Nathan. A topology for the set of primitive ideals in an arbitrary ring. *Proc. Nat. Acad Sci.* **31** (1945), pp. 333–338.

[Jaco80] JACOBSON, Nathan. *Basic Algebra* W. H. Freeman, San Francisco, 1980.

[Jeri85] JERISON, Meyer. Prime ideals in function rings. *Rings of Continuous Functions (Cincinnati, Ohio, 1982),* pp. 203–206. Lecture Notes in Pure and Appl. Math., **95**, Dekker, New York, 1985.

[JipMo06] JIPSEN, Peter and MONTAGNA, Franco. On the structure of generalized BL-algebras. *Algebra Universalis* **55** (2006), pp. 227–238.

[JipTs02] JIPSEN, Peter and TSINAKIS, Constantine. A survey of residuated lattices. *Ordered algebraic structures* (Gainesville, Florida, 2001), pp. 19–56. Developments in Mathematics, **7**, Kluwer, Dordrecht, 2002.

[JohMSu92] JOHNSON, Michael and SUN, Shu Hao. Remarks on representations of universal algebras by sheaves of quotient algebras. *Category theory 1991*, pp. 299–307, CMS Conf. Proc., **13**, Amer. Math. Soc., Providence, RI, 1992.

[John82] JOHNSTONE, Peter T. *Stone Spaces.* Cambridge University Press, Cambridge, UK, 1982.

[John83] JOHNSTONE, Peter T. The point of pointless topology. *Bull. Amer. Math. Soc. (N. S.)* **8** (1983), pp. 41–53.

[JónTa47] JÓNSSON, Bjarni and TARSKI, Alfred. *Direct Decompositions of Finite Algebraic Systems.* Notre Dame Mathematical Lectures No. 5, University of Notre Dame, Notre Dame, Ind., 1947.

[Kapl68] KAPLANSKY, Irving. *Rings of Operators.* Benjamin, New York, 1968.

[Katz98] KATZ, Victor J. *A History of Mathematics: An Introduction.* Addison-Wesley, Reading, MA, 1998.

[Keim70] KEIMEL, Klaus. Darstellung von Halbgruppen und universellen Algebren durch Schnitte in Garben; bireguläre Halbgruppen. *Math. Nach.* **45** (1970), pp. 81–96.

[Keim71] KEIMEL, Klaus. The representation of lattice-ordered groups and rings by sections in sheaves. *Lectures on the applications of sheaves to ring theory. (Tulane Univ. Ring and Operator Theory Year, 1970-1971, Vol. III)*, pp. 1-98. Lecture Notes in Math., **248**, Springer, Berlin, 1971.

[Keim71a] KEIMEL, Klaus. Baer extensions of rings and Stone extensions of semigroups. *Semigroup Forum* **2** (1971a), pp. 55–63.

[Keim74] KEIMEL, Klaus. Représentation des algèbres universelles par des faisceaux. *Séminaire P. Dubreil (27e année: 1973/74)*, Algèbre, Fasc. 2, Exp. No. 12. Secrétariat Mathématique, Paris, 1974.

[KeiWe74] KEIMEL, Klaus and WERNER, Heinrich. Stone duality for varieties generated by quasi-primal algebras. *Recent advances in the representation theory of rings and* C^**-algebras by continuous sections (Sem., Tulane Univ., New Orleans, La., 1973)*, pp. 59-85. Mem. Amer. Math. Soc., No. 148, Amer. Math. Soc., Providence, RI, 1974.

[Kell55] KELLEY, John L. *General Topology.* Van Nostrand, Princeton, NJ, 1955.

[Kenn76] KENNISON, John F. Integral domain type representations in sheaves and other topoi. *Math. Z.* **151** (1976), pp. 35–36.

[Kenn81] KENNISON, John F. Triples and compact sheaf representation. *J. Pure Appl. Algebra* **20** (1981), pp. 13–38.

[Kist63] KIST, Joseph. Minimal prime ideals in commutative semigroups. *Proc. London Math. Soc.* (3) **13** (1963), pp. 31–50.

[Kist69] KIST, Joseph. Compact spaces of minimal prime ideals. *Math. Z.* **111** (1969), pp. 151–158.

[Kist69a] KIST, Joseph. Representing Rings by Sections: Complexes. Unpublished manuscript, 1969.

[Knoe72] KNOEBEL, Arthur. Sheaves of algebras over Boolean spaces. *Notices Amer. Math. Soc.* **19** (1972), p. A-690.

[Knoe73] KNOEBEL, Arthur. Products of independent algebras with finitely generated identities. *Algebra Universalis* **3** (1973), pp. 147–151.

[Knoe76] KNOEBEL, Arthur. Representation theorems for varieties generated by single functionally precomplete algebras. *Notices Amer. Math. Soc.* **23** (1976), p. A-514.

[Knoe82] KNOEBEL, Arthur. A new proof for the product decomposition of Newman algebras. *Algebra Universalis* **14** (1982), pp. 135–139.

[Knoe83] KNOEBEL, Arthur. Free algebras—are they categorically definable? *Quaestiones Math.* **6** (1983), no. 4, pp. 333–341.

[Knoe85] KNOEBEL, Arthur. *The Equational Classes Generated by Single Functionally Precomplete Algebras.* Memoir Amer. Math. Soc. **57**, no. 332, 1985.

[Knoe91a] KNOEBEL, Arthur. Sheaves of shells with loop addition. *Abstracts Amer. Math. Soc.* **12** (1991), p. 194.

[Knoe91b] KNOEBEL, Arthur. The iteration of sheaves over Boolean spaces. *Abstracts Amer. Math. Soc.* **12** (1991), p. 282.

[Knoe92a] KNOEBEL, Arthur. A sheaf constructed over a Boolean space for each general algebra. *Abstracts Amer. Math. Soc.* **13** (1992), p. 288.

[Knoe92b] KNOEBEL, Arthur. The categories of algebras, complexes and sheaves. *Abstracts Amer. Math. Soc.* **13** (1992), pp. 503–504.

[Knoe92c] KNOEBEL, Arthur. The equivalence of some categories of algebras and sheaves. *Abstracts Amer. Math. Soc.* **13** (1992), p. 528.

[Knoe00] KNOEBEL, Arthur. The shells of Vaggione: some comments and questions. *Algebra Universalis* **44** (2000), pp. 27–38.

[Knoe02] KNOEBEL, Arthur. Separator algebras. *Atlas Mathematical Conference Abstracts*, at.yorku.ca/amca. International Conference on Modern Algebra, Nashville, Tennessee, 2002.

[Knoe03] KNOEBEL, Arthur. Sheaf representations of preprimal algebras. *Abstracts Amer. Math. Soc.* **24** (2003), p. 238.

[Knoe07a] KNOEBEL, Arthur. Sesquimorphisms. *Abstracts Amer. Math. Soc.* (2007), #07T-08-20.

[Knoe07b] KNOEBEL, Arthur. Shells. *Abstracts Amer. Math. Soc.* (2007), #1032-08-178.

[Kopp88] KOPPERMAN, Ralph. All topologies come from generalized metrics. *Amer. Math. Monthly* **95** (1988), pp. 89–97.

[Köthe30] KÖTHE, Gottfried. Abstrakte Theorie nichtkommutativer Ringe mit einer Anwendung auf die Darstellungstheorie kontinuierlicher Gruppen. *Math. Ann.* **103** (1930), pp. 545–572.

[KraCl79] KRAUSS, Peter H. and CLARK, David M. *Global subdirect products.* Memoirs Amer. Math. Soc. **17**, #210 (1979).

[Krull29] KRULL, Wolfgang. Idealtheorie in Ringen ohne Endlichkeitsbedingung. *Math. Ann.* **101** (1929), pp. 729–744.

[Krull35] KRULL, Wolfgang. *Idealtheorie.* Ergeb. Math. Grenzgeb. **4**. Springer, Berlin, 1935.

[Krus69] KRUSE, Arthur. *Localization and Iteration of Axiomatic Set Theory.* Wayne State University Press, Detroit, 1969.

[Kuro63] KUROSH, A. G. *Lectures on General Algebra.* Chelsea, New York, 1963.

[LamRa78] LAMBEK, Joachim and RATTRAY, Basil A. Functional completeness and Stone duality. *Studies in Foundations and Combinatorics* pp. 1-9, Adv. in Math. Suppl. Stud., **1**, Academic, New York, 1978.

[LamRa79] LAMBEK, Joachim and RATTRAY, Basil A. A general Stone-Gelfand duality. *Trans. Amer. Math. Soc.* **248** (1979), pp. 1–35.

[Ledb77] LEDBETTER, Carl Scotius, Jr. *Sheaf Representations and First Order Conditions.* Ph. D. Dissertation, Clark University, Worcester, MA, University of Michigan Press, 1977.

[Leib66] LEIBNIZ, Gottfried Wilhelm von. *Dissertatio de Arte Combinatoria.* Lipsiae, Apud J. S. Fickium et J. P. Seuboldum 1966.

[Leray50] LERAY, Jean. L'anneau spectral et l'anneau filtre d'homologie d'un espace localement compact et d'une application continue. *J. Math. Pures Appl .* **29** (1950), pp. 1–80, 81–139.

[MacH85] MacHALE, Desmond. *George Boole: His Life and Work.* Boole Press, Dublin, 1985.

[Maci73] MACINTYRE, Angus. Model-completeness for sheaves of structures. *Fund. Math.* **81** (1973/74), pp. 73–89.

[Maci77] MACINTYRE, Angus. Model Completeness. Chapter A.4, *Handbook of Mathematical Logic*, Jon Barwise, ed. North-Holland, Amsterdam, 1977.

[MacL65] MacLANE, Saunders. Categorical algebra. *Bulletin Amer. Math. Soc.* **71** (1965), pp. 40–106.

[MacL71] MacLANE, Saunders. *Categories for the Working Mathematician.* Springer, Berlin, 1971.

[MacL75] MacLANE, Saunders. Sets, topoi, and internal logic in categories. *Logic Colloquium '73 (Bristol, 1973)*, pp. 119–134. Studies in Logic and the Foundations of Mathematics, **80**, North-Holland, Amsterdam, 1975.

[McCoy38] McCOY, Neal H. Subrings of infinite direct sums. *Duke Math. J.* **4** (1938), pp. 486–494.

[McCoy64] McCOY, Neal H. *The Theory of Rings.* MacMillan, New York, 1964. Reprinted by Chelsea, NewYork, 1973.

[McCMo37] McCOY, Neal H. and MONTGOMERY, D. A representation of generalized Boolean rings. *Duke Math. J.* **3** (1937), pp. 455–459.

[McKe90] McKENZIE, Ralph. Monotone clones, residual smallness and congruence distributivity. *Bull. Austral. Math. Soc.* **41** (1990), pp. 283–300.

[McKe96] McKENZIE, Ralph. An algebraic version of categorical equivalence for varieties and more general algebraic categories. Logic and algebra (Pontignano, 1994), pp. 211–243, *Lecture Notes in Pure and Appl. Math.*, **180**, Dekker, New York, 1996.

[McMcT87] McKENZIE, Ralph N., McNULTY, George F. and TAYLOR, Walter F. *Algebras, Lattices, Varieties* vol. 1. Wadsworth & Brooks/Cole, Monterey, CA, 1987.

[McVa89] McKENZIE, Ralph and VALERIOTE, Matthew. *The structure of decidable locally finite varieties.* Progress in Mathematics, **79**, Birkhuser, Boston, MA, 1989.

[Mend64] MENDELSON, Elliott. *Introduction to Mathematical Logic.* Van Nostrand, Princeton, NJ, 1964.

[Mido65] MIDONICK, Henrietta O. *The Treasury of Mathematics.* Philosophical Library, New York, 1965.

[Mulv79] MULVEY, Christopher J. Representations of rings and modules. *Applications of sheaves*, pp. 542–585. Lecture Notes in Math., **753**. Springer, Berlin, 1979.

[Myci06] MYCIELSKI, Jan. A system of axioms of set theory for the rationalists. *Notices Amer. Math. Soc.* **53** (2006), pp. 206–233.

[Noet21] NOETHER, Emmy. Idealtheorie in Ringbereichen. *Math. Ann.* **83** (1921), pp. 24–66.

[Noet26] NOETHER, Emmy. Abstrakter Aufbau der Idealtheorie in algebraischen Zahl- und Funktionenkörpern. *Math. Annalen* **96** (1926), pp. 26–61.

[Nový73] NOVÝ, Luboš. *Origins of Modern Algebra.* Nordhoof, Leyden, 1973.

[Peac30] PEACOCK, George. *A Treatise on Algebra.* J. & J. J. Deighton, Cambridge, 1830.

[Peer70] PEERCY, David Eugene. *The Complete Baer Extension of a Commutative Ring.* Ph. D. Dissertation, New Mexico State University, Las Cruces, New Mexico, 1970.

[Peir70] PEIRCE, Benjamin. *Linear Associative Algebras.* Washington, D.C., 1870. Also in *Amer. J. Math.* **4** (1881), pp. 97–229.

[Petr77] PETRICH, Mario. *Lectures in Semigroups.* Wiley, New York, 1977.

[Pica80] PICAVET, Gabriel. Ultrafiltres sur un espace spectral – anneaux de Baer – anneaux a spectre minimal compact. *Math. Scand.* **46** (1980), pp. 23–53.

[Pier67] PIERCE, Richard. S. *Modules over commutative regular rings.* Memoir American Mathematical Society, No. 70, Amer. Math. Soc., Providence, RI, 1967.

[Pier68] PIERCE, Richard. S. *Introduction to the Theory of Abstract Algebras.* Holt, Rinehart and Winston, New York, 1968.

[Pilz83] PILZ, Günter. *Near-Rings — The Theory and its Applications*, revised edition. North-Holland, Amsterdam, 1983.

[Pinus93] PINUS, Alexander G. *Boolean Constructions in Universal Algebra.* Kluwer, Dordrecht, 1993.

[Pixl63] PIXLEY, Alden F. Distributivity and permutability of congruence relations in equational classes of algebras. *Proc. Amer. Math. Soc.* **14** (1963), pp. 105–109.

[Pixl96] PIXLEY. Alden F. Boolean Universal Algebra. *Logic and Algebra (Pontignano, 1994)*, pp. 245–266. Lecture Notes in Pure and Applied Mathematics, **180**. Dekker, New York, 1996.

[Post21] POST, Emil L. Introduction to a general theory of elementary propositions. *Amer. J. Math.* **43** (1921), pp. 163–185.

[Prest] PREST, Mike. *Model theory and modules.* London Mathematical Society Lecture Note Series, **130**. Cambridge University Press, Cambridge.

[Prie70] PRIESTLEY, Hillary A. Representation of distributive lattices by means of ordered Stone spaces. *Bull. London Math. Soc.* **2** (1970), pp. 186–190.

[Prie74] PRIESTLEY, Hillary A. Stone lattices: a topological approach. *Fund. Math.* **84** (1974), pp. 127–143.

[Quac79] QUACKENBUSH, Robert W. Primality: the influence of Boolean algebras in universal algebra. Appendix 5 of *Universal Algebra*, 2nd ed. G. Grätzer. Springer, Berlin, 1979.

[Rasi74] RASIOWA, Helena. *An Algebraic Approach to Non-Classical Logics.* North-Holland, Amsterdam, 1974.

[Ribe69] RIBENBOIM, Paulo. Boolean powers. *Fund. Math.* **65** (1969), pp. 243–268.

[RomSm97] ROMANOWSKA, Anna B. and SMITH, Jonathan D. H. Duality for semilattice representations. *J. Pure Appl. Algebra* **115** (1997), pp. 289–308.

[Romo94] ROMO SANTOS, Concepción. Los Orígenes de la teoría de haces (The origins of the theory of sheaves). *Mathematical contributions*, pp. 145–160, Editorial Complutense, Madrid, 1994.

[Rose70] ROSENBERG, Ivo. *Über die funktionale Vollständigkeit in den mehrwertigen Logiken. Struktur der Funktionen von mehreren Veränderlichen auf endlichen Mengen.* Rozpravy Československé Akademie Věd, Řada Matematických a Přírodních Věd **80** (1970), no. 4.

[Rosen42] ROSENBLOOM, Paul C. Post Algebras. I. Postulates and general theory. *Amer. J. Math.* **64** (1942), pp. 167–188.

[Sánc08] SÁNCHEZ TERRAF, Pedro. *Boolean Factor Congruences and Property (*)*. http://arxiv.org/PS_cache/arxiv/pdf/0809/0809.3815v1.pdf, 22 Sep. 2008.

[SánVa09] SÁNCHEZ TERRAF, Pedro, VAGGIONE, Diego J. Varieties with definable factor congruences. *Trans. Amer. Math. Soc.* **361** (2009), pp. 5061-5088.

[SeeSS70] SEEBACH, J. Arthur, Jr., SEEBACH, Linda A. and STEEN, Lynn A. What is a Sheaf? *The American Mathematical Monthly* **77** (1970), pp. 681–703.

[Serre55] SERRE, Jean-Pierre. Faisceaux algébriques cohérents. *Ann. Math.* (2) **61** (1955), pp. 197–278.

[Simm84] SIMMONS, Harold. Three sheaf constructions for noncommutative rings. *Houston J. Math.* **10** (1984), pp. 433–443.

[Smith76] SMITH, Jonathan D. H. *Mal'cev Varieties*. Springer, Berlin, 1976.

[Stone36] STONE, Marshall H. The theory of representations for Boolean algebras. *Trans. Amer. Math. Soc.* **40** (1936), pp. 37–111.

[Stone37] STONE, Marshall H. Applications of the theory of Boolean rings to general topology. *Trans Amer. Math. Soc.* **41**(1937), pp. 375–481.

[Stone37a] STONE, Marshall H. Topological representations of distributive lattices and Browerian logics. *Časopis pešt. Mat.* **67** (1937), pp. 1–25.

[Stone70] STONE, Marshall H. Remarks by Professor Stone. *Functional Analysis and Related Fields. Proceedings of a Conference in honor of Professor Marshall Stone, held at the University of Chicago, May 1968.* Edited by Felix E. Browder. Springer, Berlin, 1970.

[Suss58] SUSSMAN, Irving. A generalization of Boolean rings. *Math. Ann.* **136** (1958), pp. 326–338.

[SusFo60] SUSSMAN, Irving and FOSTER, Alfred L. On rings in which $a^{n(a)} = a$. *Math. Ann.* **140**(1960), pp. 324–333.

[Swam74] SWAMY, U. Maddana. Representation of universal algebras by sheaves. *Proc. Amer. Math. Soc.* **45** (1974), pp. 55–58.

[Swam80] SWAMY, U. Maddana. Baer-Stone Semigroups. *Semigroup Forum* **19** (1980), pp. 385–386.

[SwaMa80] SWAMY, U. Maddana and MANIKYAMBA, P. Representation of certain classes of distributive lattices by sections of sheaves. *Internat. J. Math. Math. Sci.* **3** (1980), pp. 461–476.

[SwaMu81] SWAMY, U. Maddana and MURTI, G. Suryanarayana. Boolean centre of a universal algebra. *Alg. Universalis* **13** (1981), pp. 202–205.

[SwaMu81a] SWAMY, U. Maddana and MURTI, G. Suryanarayana. Boolean centre of a semigroup. *Pure and Appl. Math. Sci.* **13** (1981), pp. 5–6.

[Szeto77] SZETO, George. On sheaf representations of a biregular near-ring. *Canad. Math. Bull.* **20** (1977), pp. 495–500.

[Szeto77a] SZETO, George. On a sheaf representation of a class of near-rings. *J. Austral. Math. Soc.* Ser. A **23** (1977), pp. 78–83.

[Szmi55] SZMIELEW, Wanda. Elementary properties of Abelian groups. *Fund. Math.* **41** (1955), pp. 203-271.

[Tars49] TARSKI, Alfred. Arithmetical classes and types of Boolean algebras. *Bull. Amer. Math. Soc.* **55** (1949), p. 64.

[Tenn75] TENNISON, Barry R. *Sheaf Theory.* Cambridge University Press, Cambridge, UK, 1975.

[TriAl78] TRILLAS, Enrique; ALSINA, Claudi. *Introducción a los Espacios Métricos Generalizados.* Serie Universitaria, **49**. Fundación Juan March, Madrid, 1978.

[Vagg92] VAGGIONE, Diego J. Sheaf representation and Chinese remainder theorems. *Algebra Universalis* **29** (1992), pp. 232–272.

[Vagg95] VAGGIONE, Diego J. Locally Boolean spectra. *Algebra Universalis* **33** (1995), pp. 319–354.

[Vagg96] VAGGIONE, Diego J. Varieties of shells. *Algebra Universalis* **36** (1996), pp. 483–487.

[Vagg96a] VAGGIONE, Diego J. Varieties in which the Pierce stalks are directly indecomposable. *J. Algebra* **184** (1996), pp. 424–434.

[Vagg10] VAGGIONE, Diego J. E-mails to the author, 2010.

[VagSá04] VAGGIONE, Diego J., SÁNCHEZ TERRAF, Pedro. Compact factor congruences imply Boolean factor congruences. *Algebra Universalis* **51** (2004), pp. 207–213.

[vdWa85] van der WAERDEN, B. L. *A History of Algebra. From al-Khwārizmī to Emmy Noether.* Springer, Berlin, 1985.

[Volg76] VOLGER, Hugo. The Feferman-Vaught theorem revisited. *Colloq. Math.* **36** (1976), pp. 111.

[Volg79] VOLGER, Hugo. Preservation theorems for limits of structures and global sections of sheaves of structures. *Math. Z.* **166** (1979), pp. 27–53.

[vonN36] von NEUMANN, John. On regular rings. *Proc. Nat. Acad. Sci. U.S.A.* **22** (1936), pp. 707–713.

[Wade45] WADE, L. I. Post algebras and rings. *Duke Math. J.* **12** (1945), pp. 389–395.

[Weis75] WEISPFENNING, Volker. Model-completeness and elimination of quantifiers for subdirect products of structures. *J. Algebra* **36** (1975), pp. 252–277.

[Weis79] WEISPFENNING, Volker. Lattice products. *Logic Colloquium '78 (Mons, 1978)*, pp. 423-426. Stud. Logic Foundations Math., **97**. North-Holland, Amsterdam, 1979.

[Wern74] WERNER, Heinrich. Congruences on products of algebras and functionally complete algebras. *Algebra Universalis* **4** (1974), pp. 99–105.

[Wern78] WERNER, Heinrich. *Discriminator-algebras. Algebraic representation and model theoretic properties.* Studien zur Algebra und ihre Anwendungen, **6**. Akademie-Verlag, Berlin, 1978.

[Wern79] WERNER, Heinrich. A generalization of Comer's sheaf-representation theorem. *Contributions to general algebra (Proc. Klagenfurt Conf., Klagenfurt, 1978)*, pp. 395–397. Heyn, Klagenfurt, 1979.

[Will90] WILLARD, Ross. Varieties Having Boolean Factor Congruences. *J. Algebra* **132** (1990), pp. 130–153.

[Will94] WILLARD, Ross. An overview of modern universal algebra. *Logic Colloquium 2004*, pp. 197-220, Lect. Notes Log., **29**, Assoc. Symbol. Logic, Chicago, IL, 2008.

[Whit98] WHITEHEAD, Alfred North. *A Treatise on Universal Algebra with Applications.* Cambridge at the University Press, 1898. Reprinted by Hafner, New York, 1960.

[Wolf74] WOLF, Albrecht. Sheaf representation of arithmetical algebras. *Memoirs Amer. Math. Soc.* **148** (1974), pp. 87–93.

Index

A. Knoebel, *Sheaves of Algebras over Boolean Spaces*,
DOI 10.1007/978-0-8176-4642-4,

FSC
www.fsc.org

MIX
Papier aus verantwortungsvollen Quellen
Paper from responsible sources
FSC® C105338

If you have any concerns about our products,
you can contact us on
ProductSafety@springernature.com

In case Publisher is established outside the EU,
the EU authorized representative is:
Springer Nature Customer Service Center GmbH
Europaplatz 3, 69115 Heidelberg, Germany

Printed by Libri Plureos GmbH
in Hamburg, Germany